NATURAL ORGANIC MATTER IN WATER

NATURAL ORGANIC MATTER IN WATER

CHARACTERIZATION, TREATMENT METHODS, AND CLIMATE CHANGE IMPACT

SECOND EDITION

MIKA SILLANPÄÄ
LUT University, Finland

YURI PARK
LUT University, Finland

ELSEVIER

Butterworth-Heinemann
An imprint of Elsevier

Butterworth-Heinemann is an imprint of Elsevier
The Boulevard, Langford Lane, Kidlington, Oxford OX5 1GB, United Kingdom
50 Hampshire Street, 5th Floor, Cambridge, MA 02139, United States

ISBN: 978-0-12-824274-2

For information on all Butterworth-Heinemann publications
visit our website at https://www.elsevier.com/books-and-journals

Publisher: Candice Janco
Acquisitions Editor: Louisa Munro
Editorial Project Manager: Joshua Mearns
Production Project Manager: Paul Prasad Chandramohan
Cover Designer: Christian Bilbow

Typeset by TNQ Technologies

Contents

CHAPTER 1

General introduction

Mika Sillanpää[1,2,3,4] and Yuri Park[5]
[1]Department of Chemical Engineering, School of Mining, Metallurgy and Chemical Engineering, University of Johannesburg, Johannesburg, South Africa; [2]Chemistry Department, College of Science, King Saud University, Riyadh, Saudi Arabia; [3]Department of Applied Physics, Faculty of Science and Technology, Universiti Kebangsaan Malaysia Bangi, Selangor, Malaysia; [4]International Research Centre of Nanotechnology for Himalayan Sustainability (IRCNHS), Shoolini University, Solan, Himachal Pradesh, India; [5]Institute of Environmental Technology, Department of Environmental Engineering, Seoul National University of Science and Technology, Seoul, South Korea

Abstract

Natural organic matter (NOM) is a complex matrix of organic materials and a key component in aquatic environments. As a result of the interactions between the hydrologic cycle and the biosphere and geosphere, the water sources of drinking water generally contain NOM. The amount, character, and properties of NOM vary considerably according to the origins of the waters and depend on the biogeochemical cycles of their surrounding environments. Also, the interrelation between NOM and climate change has attracted a great deal of attention in recent research. NOM has a significant impact on many aspects of water treatment, including the performance of unit processes, necessity for and application of water treatment chemicals, and the biological stability of the water. As a result, NOM affects potable water quality as a carrier of metals and hydrophobic organic chemicals and by contributing to undesirable color, taste, and odor problems. Moreover, NOM has been found to be the major contributor to disinfection by-product (DBP) formation. Changes in NOM quantity and quality have a significant influence on the selection, design, and operation of water treatment processes. These changes also cause operational difficulties in water utilities. High seasonal variability and the trend toward elevated levels of NOM concentration pose challenges to water treatment facilities in terms of operational optimization and proper process control. To improve and optimize these processes, it is vital to characterize and quantify NOM at various stages during the purification and treatment process. It is also essential to be able to understand and predict the reactivity of NOM or its fractions during different phases of the treatment. Once the composition and quantity of NOM in the water source has been examined, suitable methods for efficient NOM removal can be applied. No single process alone can be used to treat NOM due to its high variability.

Keywords: Characterization; Climate change; Coagulation; Drinking water; Natural organic matter (NOM); Water analysis; Water treatment.

Natural Organic Matter in Water
ISBN 978-0-12-824274-2
https://doi.org/10.1016/B978-0-12-824274-2.00003-X

Abbreviations

AOP:	Advanced oxidation process
DBP:	Disinfection by-product
FA:	Fulvic acids
FTICR-MS:	Fourier transform ion cyclotron resonance mass spectrometry
GAC:	Granulated activated carbon
HA:	Humic acids
HAAs:	Haloacetic acids
HMM:	High molecular mass
HMW:	High molecular weight
LMM:	Low molecular mass
LMW:	Low molecular weight
MIEX:	Magnetic ion exchange resin
NF:	Nanofiltration
NMR:	Nuclear magnetic resonance
NOM:	Natural organic matter
Py-GC-MS:	Pyrolysis gas chromatography-mass spectrometry
SEC:	Size exclusion chromatography
SUVA:	Specific UV absorbance
THM:	Trihalomethane
TOC:	Total organic carbon
UV−Vis:	Ultraviolet and visible

Natural organic matter (NOM) is ubiquitous in waters, sediments, and soils. Aquatic NOM is derived both from the breakdown of terrestrial plants and as the by-product of bacteria, algae, and aquatic plants. NOM is defined as a complex matrix of organic materials present in all natural waters.

A wide range of terminology is used to describe NOM in the environment. These terms are listed and comprehensively discussed in previous reviews by Filella (2009) and Uyguner-Demirel and Bekbolet (2011). NOM is a key component in aquatic environments and is present in all fresh waters, particularly surface waters. As a result of the interactions between the hydrologic cycle and the biosphere and geosphere, the water sources of drinking water generally contain NOM. It consists of a range of compounds with a wide variety of chemical charges, from highly charged to uncharged. These also vary widely according to chemical composition and molecular size, molecular weight (the molecular masses of humic substances range from several hundreds to tens of thousands), and structure. NOM compounds are complex mixtures possessing unique combinations of various functional groups, including esteric, phenolic, quinine, carboxylic, hydroxyl, amino, and nitroso, which are usually negatively charged at neutral pH (Gjessing, 1976). Humic substances, which are the major constituents of NOM in waters, are amorphous, dark colored, and acidic in nature. Structurally, they consist of substituted aromatic rings linked by aliphatic chains (Uyguner-Demirel and Bekbolet 2011). NOM also influences the acidity, mobility, and toxicity of metals and organic pollutants and weathering (Winterdahl, 2013).

Water systems often have multiple sources of NOM, thus different organic carbon fractions (Rigobello et al., 2011). The amount, character, and properties of NOM vary considerably according to the origins of the waters and depend on the biogeochemical cycles of their surrounding environments (Fabris et al., 2008). The factors that determine the composition of NOM are location dependent and include the source of organic matter, the water chemistry, temperature, pH, and biological processes (Leenheer and Croué, 2003). Thus, the character of NOM varies with source and season (Sharp et al., 2006a,b; Fabris et al., 2008; Rigobello et al., 2011; Nkambule et al., 2012). Moreover, the range of organic components in NOM can even vary seasonally at the same location (Sharp et al., 2006a,b; Smith and Kamal, 2009), for example, due to rainfall, snowmelt runoff, floods, or droughts. Floods and droughts are the main indications of the impact of climate change on water availability and quality. It has been suggested that these changes may explain the increase in the total amount of NOM (Delpla et al., 2009; Evans et al., 2005), leading to browning of natural waters. This widespread phenomenon has been observed in several parts of the world during the past 20 years (Eikebrokk et al., 2004; Evans et al., 2005; Korth et al., 2004; De Wit et al., 2007; Monteith et al., 2007; Worral and Burt, 2007, 2009; Couture et al., 2012; Jarvis et al., 2012; Winterdahl, 2013; Gough et al., 2014; de Wit et al., 2016; O'Driscoll et al., 2018; McDonough et al., 2020).

NOM is an integral part of the carbon cycle, which may have indirect climatic effects (Winterdahl, 2013; McDonough et al., 2020). The interrelation between NOM and climate change has attracted a great deal of attention in recent research, since this consequence affects water quality as well as the aquatic ecosystems, which is associated with health risks and the production of clean and safe water (Maurice et al., 2002; Hejzlar et al., 2003; Monteith et al., 2007; Granskog et al., 2007; Park et al., 2007; Clements et al., 2008; Soh et al., 2008; Porcal et al., 2009; Sulzberger and Durisch-Kaiser, 2009; Bunting et al., 2010; Chen et al., 2012; Pautler et al., 2012; Brezonik and Arnold, 2012; Diem et al., 2013; Ritson et al., 2014). Not only the quantity of NOM but also its quality have been observed to change when other important characteristics of NOM, such as specific UV absorbance (SUVA), are taken into account (Eikebrokk et al.,

2004). Winterdahl et al. (2013, 2014) have emphasized that intra annual NOM variability is often larger than year-to-year changes by several orders of magnitude. Unraveling the controls on intra-annual NOM dynamics is thus essential to understanding long-term changes, since climate change could alter NOM dynamics in ways not reflected in interannual trends. Several potential factors, including an increase in air and surface water temperature, rainfall intensity, precipitation, and atmospheric CO_2 content and/or a decrease in acid deposition, have been proposed to explain the increased amount of NOM, and there are many scientific research undergone that may in explaining the correlation among the phenomenon by the drivers affected by climate change but still there is yet no scientific consensus on this issue (Delpla et al., 2009; Garmo et al., 2014; Finstad et al., 2016; Sawicka et al., 2017; Houle et al., 2020; Ferretto et al., 2021).

Tang et al. (2014) demonstrated that NOM has complex effects on the environmental behavior and removal of heavy metal ions by nanomaterials. The impacts were reported to be controversial, depending both on the type of nanomaterials and metal ions and on geochemical conditions. They also demonstrated that the presence of NOM can modify the mechanisms controlling metal ion removal and transportation by nanomaterials in heterogeneous aquatic environments (Tang et al., 2014). Another emerging field of research is the effect of NOM on nanoparticle aggregation (Louie et al., 2013). Collin et al. (2014) recently concluded that NOM decreased the toxicity and bioaccumulation of nanoparticles. The impact of photooxidation on the optical, electrochemical, and photochemical properties of NOM has recently been investigated systematically (Sharpless et al., 2014). This sheds light on the relationships between NOM aromaticity, redox state, and photoreactivity.

NOM found in natural waters consists of both hydrophobic and hydrophilic components, of which the largest fraction is generally hydrophobic acids, making up approximately 50% of the total organic carbon (TOC) in water (Thurman, 1985). These hydrophobic acids can be described as humic substances comprising (1) humic acids (HAs), which are soluble in alkali, but insoluble in acid, (2) fulvic acids (FAs), which are soluble in both alkali and acid, and (3) humins, which are insoluble in both alkali and acid. FAs constitute a major fraction of these humic substances, while structurally comparable, vary in molecular size and functional group content (Table 1.1). Hydrophobic NOM is rich in aromatic carbon, phenolic structures, and conjugated double bonds, whereas hydrophilic

Table 1.1 Common properties of humic acid (HA) and fulvic acid (FA) (Snoeyink and Jenkins, 1980; Xing, 2010; Chamoli, 2013).

Property	Humic acid (HA)	Fulvic acid (FA)
Elemental composition (% by weight)		
Carbon	50–60	40–50
Hydrogen	4–6	4–6
Oxygen	30–35	44–50
Nitrogen	2–4	<1–3
Sulfur	1–2	0–2
Solubility in strong acid	Not soluble	Soluble
Apparent molecular weight range (atomic mass units)	Few 100 to several million	180–10,000
Functional group distribution (% of oxygen is indicated in functional groups)		
Carboxyl (–COOH)	14–45	58–65
Phenol (–Ph)	10–38	9–19
Alcohol (–R–OH)	13–15	11–16
Carbonyl (–CO)	4–23	4–11
Methoxyl (–O–CH_3)	1–5	1–2

Aquatic HAs are larger than aquatic FAs, but FAs have more carboxylic functional groups and oxygen, hence less carbon on a mass basis than HAs (Chamoli, 2013). FAs also are more soluble in water, because they have more polar groups per unit mass than HAs.

NOM contains more aliphatic carbon and nitrogenous compounds, such as carbohydrates, sugars, and amino acids. The physical and chemical fractionation of aquatic NOM at specific pH can be used to classify organic solutes into these broadly defined hydrophobic and hydrophilic fractions (Chow et al., 2004; Leenher, 2004; Sharp et al., 2006a,b; Kent et al., 2014). While these fractions are more operationally than structurally defined, organic compounds can be assigned to a particular fraction according to their chain length and functional groups (Swietlik et al., 2004; Buchanan et al., 2005). The hypothetical molecular structure of HA is presented in Fig. 1.1.

NOM has a significant impact on many aspects of water treatment, including the performance of unit processes, necessity for and application of water treatment chemicals, and the biological stability of the water. As discussed earlier, the current issue related to the increase of NOM concentrations as well as a change in NOM composition from lower to higher molecular weight emphasizes the need of NOM removal at drinking water treatment plants and effective water treatment procedures. NOM also

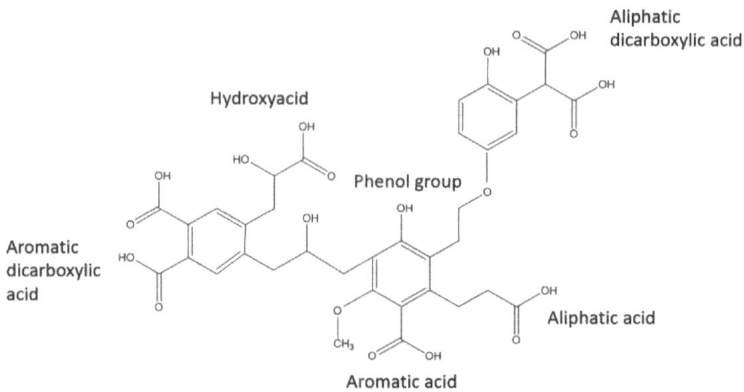

Figure 1.1 Hypothetical molecular structure of humic acid (HA). *(Adapted from Duan and Gregory (2003).)*

affects potable water quality as a carrier of metals and hydrophobic organic chemicals and by contributing to undesirable color, taste, and odor problems. In addition, NOM necessitates the majority of the coagulant and disinfectant used in water treatment. It tends to interfere with the performance of unit operations, such as biofilm growth on media, causing rapid filter clogging and fast saturation of activated carbon beds. NOM is also responsible for the fouling of membranes. NOM contributes to corrosion, is a source of nutrients for heterotrophic bacteria, and acts as a substrate for bacterial growth in distribution systems (Jacangelo et al., 1995). Moreover, NOM has been found to be the major contributor to disinfection by-product (DBP) formation (Trang et al., 2012). NOM also forms stable complexes with metal ions.

Thus, the removal of NOM from water is an emerging issue, and a robust and efficient treatment technology is needed to address it. The number of publications related to NOM research in indexed journals has been escalating, indicating the scientific community's burgeoning interest in NOM research (Fig. 1.2).

Interest in DBPs of water treatment has grown significantly in the past few decades. More than 700 compounds of DBPs have been confirmed; among them, trihalomethanes (THMs) and haloacetic acids (HAAs) are the two groups found most frequently and in the highest concentrations in drinking waters worldwide (Krasner et al., 2006; Sharma et al., 2014). Chlorine, ozone, chlorine dioxide, and chloramine are the most common disinfectants used today, and each produces its own suite of chemical DBPs in drinking water (Krasner et al., 2006). Although chlorinated DBPs have

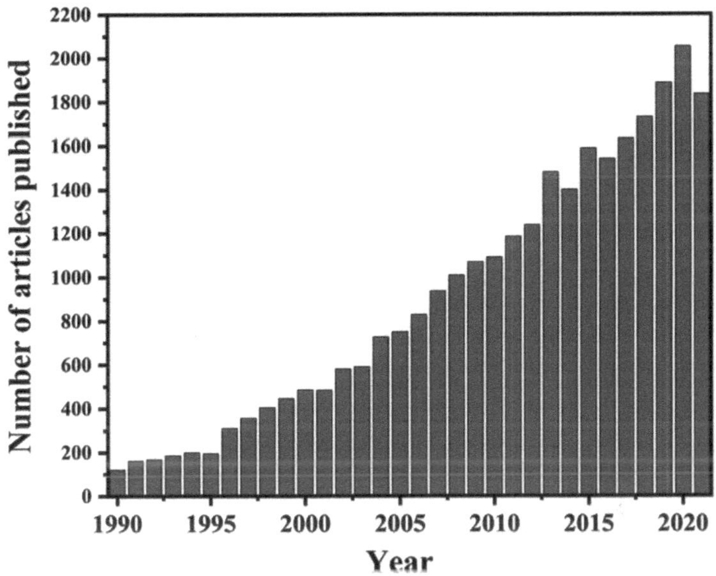

Figure 1.2 Record of the number of publications in indexed journals containing the keyword "natural organic matter" in the abstract between 1990 and 2021. Scopus, October 22, 2021.

been the subject of the greatest concern, brominated DBPs have been considered even more hazardous than their chlorinated counterparts (Singer, 2006), while the formation of iodinated and nitrogen-containing DBPs has been studied with increasing intensity (Krasner et al., 2006; Hua and Reckhow, 2007; Zhao et al., 2008). DBPs have been associated with adverse health effects such as bladder cancer, spontaneous abortions, and birth defects (Singer, 2006; Gough et al., 2014; Diana et al., 2019). Thus, their occurrence in drinking water has been regulated in most countries. NOM has generally been considered the main precursor to DBPs, especially hydrophobic and high molecular mass (HMM) NOM, with its high aromatic carbon content (Hua and Reckhow, 2007; Bond et al., 2009; Liu et al., 2007). It has also been observed that hydrophilic and low molecular mass (LMM) NOM plays a significant role in DBP formation (Hua and Reckhow, 2007; Bond et al., 2009; Chowdhurry et al., 2013). On the one hand, bromine and iodine appear more reactive with hydrophilic and LMM fractions of NOM in the formation of THMs and HAAs. In contrast, chlorine has been shown to react more readily with HMM and hydrophobic NOM compounds (Hua and Reckhow, 2007).

Thus, water treatment should be optimized to remove both hydrophobic and hydrophilic organic compounds, mitigating the formation of DBPs. More efficient, but economical, NOM removal methods are needed to meet stricter drinking water treatment regulations and overcome problems with water quality.

Changes in NOM quantity and quality have a significant influence on the selection, design, and operation of water treatment processes. These changes also cause operational difficulties in water utilities. High seasonal variability and the trend toward elevated levels of NOM concentration pose challenges to water treatment facilities in terms of operational optimization and proper process control. A sustained increase in NOM in raw water will lower the efficiency of water treatment processes and increase the demand for water purification. Changes in the properties of NOM also influence the treatment significantly (Eikebrokk et al., 2004; Sharp et al., 2006a,b). Water treatment facilities thus need to invest in additional NOM removal methods where existing water purification processes become insufficient.

To improve and optimize these processes, it is vital to characterize and quantify NOM at various stages during the purification and treatment process. It is also essential to be able to understand and predict the reactivity of NOM or its fractions during different phases of the treatment. Methods used in the characterization of NOM include resin fractionation, size exclusion chromatography (SEC), nuclear magnetic resonance (NMR) spectroscopy, fluorescence spectroscopy including 3D excitation-emission matrix (EEM) spectroscopy. The amount of NOM in water has been predicted with parameters including ultraviolet and visible (UV–Vis), TOC, and SUVA. More precise methods for determining NOM structures have been developed recently: liquid chromatography or pyrolysis gas chromatography coupled with mass spectrometry, multidimensional NMR techniques, and electrospray ionization (ESI) coupled with Fourier transform ion cyclotron resonance mass spectrometry (FTICR-MS).

Once the composition and quantity of NOM in the water source have been examined, suitable methods for efficient NOM removal can be applied (Nkambule et al., 2012). No single process alone can be used to treat NOM due to its high variability. The development of rapid NOM characterization methods enables the selection of the proper treatment for the water concerned (Nkambule et al., 2012). Thus, NOM characterization and its removal during water treatment are closely connected. Effective selection criteria for precursor removal processes arise from increased knowledge of the nature of the precursor in individual water, facilitating the choice of appropriate technologies for precursor treatment (Bond et al., 2011). One reported variable is the nature of the reactive precursors present. Optimized coagulation treatment may be sufficient in hydrophobic waters, but if the post-coagulation residual remains reactive in DBP formation, technologies such as ion exchange for carboxylic acid precursors and/or granular activated carbon for hydrophobic precursors and/or nanofiltration for hydrophilic precursors are recommended (Bond et al., 2011).

The most common and economically feasible processes available are coagulation and flocculation followed by sedimentation/flotation and

Table 1.2 Advantages and disadvantages of treatment methods for natural organic matter (NOM) removal (Jarvis et al., 2008; Bond et al., 2011; Shestakova and Sillanpää, 2013).

Treatment method	Advantages	Disadvantages
Adsorption	High NOM removal efficiency is achievable. Systems are available for various flow rates and concentrations of pollutants. Easy to implement. Removes hydrophobic NOM fraction.	Requires multiple regenerations and partial replacement of adsorbent. In some cases, adsorbent is not regenerated and needs to be disposed, causing secondary pollution of the environment. Pretreatment of influent from suspended solids is required. Effectiveness depends on the temperature and pH.
Advanced oxidation processes (AOP)	Lower chemical consumption in some AOPs. Complete mineralization of pollutant is achievable. Reactions are often rapid. Unselective oxidants. Harmful to any microorganisms that may be present in the water. Can often be installed in	Requires powdered photocatalyst separation from treated water. Poorer performance for NOM removal than ferric salt coagulation. Hydrogen peroxide residual is toxic. Effectiveness depends on the pH value. Requires additional

Continued

Table 1.2 Advantages and disadvantages of treatment methods for natural organic matter (NOM) removal (Jarvis et al., 2008; Bond et al., 2011; Shestakova and Sillanpää, 2013).—cont'd

Treatment method	Advantages	Disadvantages
	existing water treatment plants. Oxidation of disinfection by-products (DBPs).	reagents. Dependent on the pH value. Radical scavenging. UV irradiation is a high-energy intensive process. Recalcitrant NOM is poorly removed. Ozone is a toxic gas, thus safety monitoring is carefully required. Water is corrosive due to high oxidation power. Short duration of ozone exposure.
Biological	Removes some major NOM fractions that are biologically degradable	Requires a relatively large area. Activated sludge from aerobic treatment is a waste that requires disposal. Requires additional nutrients. Process is sensitive to conditions and concentration variations.
Coagulation	Cost efficient. Conventional method with high levels of NOM removal. Suitable for large molecules.	Sludge production.
Electrochemical methods	Electrocoagulation produces compact flocs. Innovative, inexpensive, and effective. Little or no chemical needed to facilitate NOM removal. Less coagulant is needed in	Energy costs may limit practical use. Electrode materials can be expensive for electrochemical oxidation. Passivation of electrodes due to the presence of oxides and precipitation layers on

Table 1.2 Advantages and disadvantages of treatment methods for natural organic matter (NOM) removal (Jarvis et al., 2008; Bond et al., 2011; Shestakova and Sillanpää, 2013).—cont'd

Treatment method	Advantages	Disadvantages
	electrocoagulation; consequently, less sludge is formed.	the electrode surface, dissolving of the electrodes, and low conductivity of the surface waters. Formation of DBPs.
Ion exchange	Proven technology. Potentially highly efficient. Very low DBP formation. Efficient in treating the transphilic fraction of NOM.	Additional treatment stage required.
Membrane technology	Large, strong flocs formed. Nanofiltration is efficient in the removal of low charged amino acids and carbohydrates. Potential to remove low NOM fraction. Can be integrated into other processes.	Energy-intensive process. Additional treatment stage required.

filtration. Most of the NOM can be removed by coagulation, although the hydrophilic, low molecular weight (LMW) fractions of NOM are apparently removed less efficiently than the hydrophobic, high molecular weight (HMW) compounds (Jacangelo et al., 1995; Matilainen et al., 2010; Sharp et al., 2006a,b). This preference may be due to the more aromatic character, therefore more hydrophobic nature, of the latter (Sharp et al., 2006a,b). Moreover, the hydrophobic fraction generally has a higher specific colloidal charge; more highly charged fractions are more amenable to removal (Sharp et al., 2006a; Bose and Reckhow, 2007). Thus, LMW and hydrophilic NOM dominate the residual organic matter after coagulation (Zhao et al., 2009; Liu et al., 2007).

Other treatment options for NOM removal include magnetic ion exchange resin (MIEX) techniques, activated carbon filtration, membrane filtration methods, and advanced oxidation processes (AOPs) (Jacangelo et al., 1995; Singer and Bilyk 2002; Matilainen et al., 2006a,b; Zularisam et al., 2006; Toor and Mohseni 2007; Matilainen and Sillanpää 2010; Parsons and Byrne, 2004; Pera-Titus et al., 2004; Suty et al., 2004; Agustina et al., 2005; Malato et al., 2007; Comninellis et al., 2008; Klavarioti et al., 2009; Malato et al., 2009). Although the MIEX technique has been reported to remove even hydrophilic NOM, none of the above-named alternative treatment methods successfully remove all the NOM fractions present in raw water. Bioprocesses typically entail the development of a biofilm on a sand or activated carbon filter during water treatment (Bond et al., 2011). Biotreatment is reported to have a significant impact on precursor removal where reactive precursors are readily biodegradable, which is more likely in waters with high amounts of biologically derived NOM (Bond et al., 2011). The pros and cons of some treatment methods of NOM removal are presented in Table 1.2.

This book has three objectives. First, it examines an overview of the impacts of climate change affecting on NOM concentration. Second, it investigates various structural and analytical techniques for NOM characterization, including conventional bulk parameters and advanced instrumental analytical tools. Lastly, it assesses the water treatment methods relevant to NOM removal, such as coagulation, electrochemical methods, membrane technology, AOPs, adsorption, and ion exchange.

References

T. Agustina, H. Ang, V. Vareek, A review of synergistic effect of photocatalysis and ozonation on wastewater treatment, J. Photochem. Photobiol. 6 (2005) 264–273.

T. Bond, E. Goslan, B. Jefferson, F. Roddick, L. Fan, S. Parsons, Chemical and biological oxidation of NOM surrogates and effect on HAA formation, Water Res. 43 (2009) 2615–2622.

T. Bond, E.H. Goslan, S.A. Parsons, B. Jefferson, Treatment of disinfection by-product precursors, Environ. Technol. 32 (2011) 1–21.

P. Bose, D. Reckhow, The effect of ozonation on natural organic matter removal by alum coagulation, Water Res. 41 (2007) 1516–1524.

P. Brezonik, W. Arnold, Water chemistry: fifty years of change and progress, Environ. Sci. Technol. 46 (2012) 5650–5657.

W. Buchanan, F. Roddick, N. Porter, M. Drikas, Fractionation of UV and VUV pretreated natural organic matter from drinking water, Environ. Sci. Technol. 39 (2005) 4647–4654.

L. Bunting, P. Leavitt, P. Weidman, R. Vinebrooke, Regulation of the nitrogen biogeochemistry of mountain lakes by subsidies of terrestrial dissolved organic matter and the implications for climate studies, Limnol. Oceanogr. 55 (1) (2010) 333–345.

U. Chamoli, Disinfection and Self-Sensitized Degradation of Natural Organic Matter (NOM) by TiO_2 Photocatalysis with Visible Light (M.Sc. Thesis), University of Calgary, Calgary: Canada, 2013, p. 129.

X. Chen, X. Chuaia, L. Yanga, H. Zhao, Climatic warming and overgrazing induced the high concentration of organic matter in Lake Hulun, a large shallow eutrophic steppe lake in northern China, Sci. Total Environ. 431 (2012) 332–338.

C. Chow, R. Fabris, M. Drikas, A rapid fractionation technique to characterize natural organic matter for the optimization of water treatment processes, J. Water Supply Res. Technol. - Aqua 53 (2004) 85–92.

S. Chowdhury, Trihalomethanes in drinking water: effect of natural organic matter distribution, WaterSA 39 (1) (2013) 1–8.

W. Clements, M. Brooks, D. Kashian, R. Zuellig, Changes in dissolved organic material determine exposure of stream benthic communities to UV-B radiation and heavy metals: implications for climate change, Global Change Biol. 14 (2008) 2201–2214.

B. Collin, E. Oostveen, O.V. Tsyusko, J.M. Unrine, Influence of natural organic matter and surface charge on the toxicity and bioaccumulation of functionalized ceria nanoparticles in *Caenorhabditis elegans*, Environ. Sci. Technol. 48 (2014) 1280–1289.

C. Comninellis, A. Kapalka, S. Malato, S. Parsons, I. Poulios, D. Mantzavinos, Advanced oxidation processes for water treatment: advances and trends for R&D, J. Chem. Technol. Biotechnol. 83 (2008) 769–776.

S. Couture, D. Houle, C. Gagnon, Increases of dissolved organic carbon in temperate and boreal lakes in Quebec, Canada, Environ. Sci. Pollut. Res. 19 (2012) 361–371.

H.A. De Wit, J. Mulder, A. Hindar, L. Hole, Long-term increase in dissolved organic carbon in streamwaters in Norway is response to reduced acid deposition, Environ. Sci. Technol. 41 (2007) 7706–7713.

H.A. De Wit, S. Valinia, G.A. Weyhenmeyer, M.N. Futter, P. Kortelainen, K. Austnes, J. Vuorenmaa, Current browning of surface waters will be further promoted by wetter climate, Environ. Sci. Technol. Lett. 3 (12) (2016) 430–435.

I. Delpla, A.-V. Jung, E. Baures, M. Clement, O. Thomas, Impacts of climate change on surface water quality in relation to drinking water production, Environ. Int. 35 (2009) 1225–1233.

M. Diana, M. Felipe-Sotelo, T. Bond, Disinfection byproducts potentially responsible for the association between chlorinated drinking water and bladder cancer: a review, Water Res. 162 (2019) 492–504.

S. Diem, M. von Rohra, J. Heringa, H.-P. Kohlera, M. Schirmera, U. von Gunten, NOM degradation during river infiltration: effects of the climate variables temperature and discharge, Water Res. 47 (2013) 6585–6595.

J. Duan, J. Gregory, Coagulation by hydrolysing metal salts, Adv. Colloid Interface Sci. 100–102 (2003) 475–502.

B. Eikebrokk, R. Vogt, H. Liltved, NOM increase in Northern European source waters: discussion of possible causes and impacts on coagulation/contact filtration processes, Water Sci. Technol. Water Supply 4 (4) (2004) 47–54.

C. Evans, D. Monteith, M. Cooper, Long-term increases in surface water dissolved organic carbon: observations, possible causes and environmental impacts, Environ. Pollut. 137 (2005) 55–71.

R. Fabris, C. Chow, M. Drikas, B. Eikebrokk, Comparison of NOM character in selected Australian and Norwegian drinking waters, Water Res. 42 (2008) 4188–4196.

A. Ferretto, R. Brooker, R. Matthews, P. Smith, Climate change and drinking water from Scottish peatlands: where increasing DOC is an issue? J. Environ. Manag. 300 (2021) 113688.

M. Filella, Freshwaters: which NOM matters? Environ. Chem. Lett. 7 (2009) 21–35.

A.G. Finstad, T. Andersen, S. Larsen, K. Tominaga, S. Blumentrath, H.A. De Wit, D.O. Hessen, From greening to browning: catchment vegetation development and reduced S-deposition promote organic carbon load on decadal time scales in Nordic lakes, Sci. Rep. 6 (1) (2016) 1–8.

Ø.A. Garmo, B.L. Skjelkvåle, H.A. de Wit, L. Colombo, C. Curtis, J. Fölster, A. Worsztynowicz, Trends in surface water chemistry in acidified areas in Europe and North America from 1990 to 2008, Water, Air, & Soil Pollut. 225 (3) (2014) 1–14.

E. Gjessing, Physical and Chemical Characteristics of Aquatic Humus, Ann Arbor Science Publishers, Michigan, 1976.

R. Gough, P. Holliman, N. Willis, C. Freeman, Dissolved organic carbon and trihalomethane precursor removal at a UK upland water treatment works, Sci. Total Environ. 468–469 (2014) 228–239.

M. Granskog, R. Macdonald, C.-J. Mundy, D. Barber, Distribution, characteristics and potential impacts of chromophoric dissolved organic matter (CDOM) in Hudson Strait and Hudson Bay, Canada, Continent. Shelf Res. 27 (2007) 2032–2050.

J. Hejzlar, M. Dubrovskýb, J. Buchtelec, M. Růžičkac, The apparent and potential effects of climate change on the inferred concentration of dissolved organic matter in a temperate stream (the Malše River, South Bohemia), Sci. Total Environ. 310 (2003) 143–152.

D. Houle, M. Khadra, C. Marty, S. Couture, Influence of hydro-morphologic variables of forested catchments on the increase in DOC concentration in 36 temperate lakes of eastern Canada, Sci. Total Environ. 747 (2020) 141539.

G. Hua, D. Reckhow, Characterization of disinfection byproduct precursors based on hydrophobicity and molecular size, Environ. Sci. Technol. 41 (2007) 3309–3315.

J. Jacangelo, J. DeMarco, D. Owen, S. Randtke, Selected processes for removing NOM: an overview, J. Am. Water Works Assoc. 87 (1) (1995) 64–77.

P. Jarvis, J. Banks, R. Molinder, T. Stephenson, S. Parsons, B. Jefferson, Processes for enhanced NOM removal: beyond Fe and Al coagulation, Water Sci. Technol. Water Supply 8 (2008) 709–716.

P. Jarvis, E. Sharp, M. Pidou, R. Molinder, S. Parsons, B. Jefferson, Comparison of coagulation performance and floc properties using a novel zirconium coagulant against traditional ferric and alum coagulants, Water Res. 46 (2012) 4179–4187.

F.C. Kent, K.R. Montreuil, A.K. Stoddart, V.A. Reed, G.A. Gagnon, Combined use of resin fractionation and high performance size exclusion chromatography for characterization of natural organic matter, J. Environ. Sci. Health, Part A 49 (14) (2014) 1615–1622.

M. Klavarioti, D. Mantzavinos, D. Kassinos, Removal of residual pharmaceuticals from aqueous systems by advanced oxidation processes, Environ. Int. 35 (2009) 402–417.

A. Korth, C. Fiebiger, K. Bornmann, W. Schmidt, NOM increase in drinking water reservoirs – relevance for drinking water production, Water Sci. Technol. Water Supply 4 (4) (2004) 55–60.

S. Krasner, H. Weinberg, S. Richardson, S. Pastor, R. Chinn, M. Sclimenti, G. Onstad, A. Thruston Jr., Occurrence of a new generation of disinfection by-products, Environ. Sci. Technol. 40 (2006) 7175–7185.

J. Leenheer, J.-P. Croué, Peer reviewed: characterizing aquatic dissolved organic matter, Environ. Sci. Technol. 37 (2003) 18A–26A.

J. Leenher, Comprehensive assessment of precursors, diagenesis, and reactivity to water treatment of dissolved and colloidal organic matter, Water Sci. Technol. Water Supply 4 (4) (2004) 1–9.

S. Liu, M. Lim, K. Chiang, R. Amal, R. Fabris, C. Chow, M. Drikas, A study on the removal of humic acid using advanced oxidation process, Separ. Sci. Technol. 42 (2007) 1391–1404.

S. Louie, R. Tilton, G. Lowry, Effects of molecular weight distribution and chemical properties of natural organic matter on gold nanoparticle aggregation, Environ. Sci. Technol. 47 (2013) 4245–4254.

S. Malato, J. Blanco, D. Alarcón, M. Maldonado, P. Fernández-Ibáñez, W. Gernjak, Photocatalytic decontamination and disinfection of water with solar collectors, Catal. Today 122 (2007) 137–149.

S. Malato, P. Fernández-Ibáñez, M. Maldonado, J. Blanco, W. Gernjak, Decontamination and disinfection of water by solar photocatalysis: recent overview and trends, Catal. Today 147 (2009) 1−59.
A. Matilainen, M. Sillanpää, Removal of natural organic matter from drinking water by advanced oxidation processes: a review, Chemosphere 80 (2010) 351−365.
A. Matilainen, P. Iivari, J. Sallanko, E. Heiska, T. Tuhkanen, The role of ozonation and activated carbon filtration in the natural organic matter removal from drinking water, Environ. Technol. 27 (2006a) 1171−1180.
A. Matilainen, N. Vieno, T. Tuhkanen, Efficiency of the activated carbon filtration in the natural organic matter removal, Environ. Int. 32 (2006b) 324−331.
A. Matilainen, M. Vepsäläinen, M. Sillanpää, Natural organic matter removal by coagulation during drinking water treatment: a review, Adv. Colloid Interface Sci. 159 (2010) 189−197.
P. Maurice, S. Cabaniss, J. Drummond, E. Ito, Hydrogeochemical controls on the variations in chemical characteristics of natural organic matter at a small freshwater wetland, Chem. Geol. 187 (2002) 59−77.
L.K. McDonough, I.R. Santos, M.S. Andersen, D.M. O'Carroll, H. Rutlidge, K. Meredith, A. Baker, Changes in global groundwater organic carbon driven by climate change and urbanization, Nat. Commun. 11 (1) (2020) 1−10.
D.T. Monteith, J.L. Stoddard, C.D. Evans, H.A. De Wit, M. Forsius, T. Høgåsen, J. Vesely, Dissolved organic carbon trends resulting from changes in atmospheric deposition chemistry, Nature 450 (7169) (2007) 537−540.
T. Nkambule, R. Krause, J. Haarhoff, B. Mamba, A three step approach for removing organic matter from South African water sources and treatment plants, Phys. Chem. Earth 50−52 (2012) (2012) 132−139.
C. O'Driscoll, J.L. Ledesma, J. Coll, J.G. Murnane, P. Nolan, E.M. Mockler, L.W. Xiao, Minimal climate change impacts on natural organic matter forecasted for a potable water supply in Ireland, Sci. Total Environ. 630 (2018) 869−877.
J.-H. Park, J.-H. Lee, S.-Y. Kang, S.-Y. Kim, Hydroclimatic controls on dissolved organic matter (DOM) characteristics and implications for trace metal transport in Hwangryong River Watershed, Korea, during a summer monsoon period, Hydrol. Process. 21 (2007) 3025−3034.
S. Parsons, A. Byrne, Water treatment applications, in: S. Parsons (Ed.), Advanced Oxidation Processes for Water and Wastewater Treatment, IWA Publishing, London, 2004, pp. 329−346.
B. Pautler, G. Woods, A. Dubnick, A. Simpson, M. Sharp, S. Fitzimons, M. Simpson, Molecular characterization of dissolved organic matter in glacial ice: coupling natural abundance1H NMR and fluorescence spectroscopy, Environ. Sci. Technol. 46 (2012) 3753−3761.
M. Pera-Titus, V. Garcia-Molina, M. Baños, J. Giménes, S. Esplugas, Degradation of chlorophenols by means of advanced oxidation processes: a general review, Appl. Catal. B Environ. 47 (2004) 219−256.
P. Porcal, J.-F. Koprivnjak, L. Molot, P. Dillon, Humic substances—part 7: the biogeochemistry of dissolved organic carbon and its interactions with climate change, Environ. Sci. Pollut. Res. 16 (2009) 714−726.
E. Rigobello, A. Dantas, L. Di Bernardo, E. Vieira, Influence of the apparent molecular size of aquatic humic substances on colour removal by coagulation and filtration, Environ. Technol. 32 (2011) 1767−1777.
J.P. Ritson, N.J.D. Grahamb, M.R. Templetonb, J.M. Clark, R. Gough, C. Freemand, The impact of climate change on the treatability of dissolved organic matter (DOM) in upland water supplies: a UK perspective, Sci. Total Environ. 473−474 (2014) 714−730.

K. Sawicka, E.C. Rowe, C.D. Evans, D.T. Monteith, E.I. Vanguelova, A.J. Wade, J.M. Clark, Modelling impacts of atmospheric deposition and temperature on long-term DOC trends, Sci. Total Environ. 578 (2017) 323–336.
V.K. Sharma, R. Zboril, T.J. McDonald, Formation and toxicity of brominated disinfection byproducts during chlorination and chloramination of water: a review, J. Environ. Sci. Health, Part B 49 (3) (2014) 212–228.
E. Sharp, P. Jarvis, S. Parsons, B. Jefferson, Impact of fractional character on the coagulation of NOM, Colloids Surf. A Physicochem. Eng. Aspects 286 (2006a) 104–111.
E. Sharp, S. Parsons, B. Jefferson, Seasonal variations in natural organic matter and its impact on coagulation in water treatment, Sci. Total Environ. 363 (2006b) 183–194.
C.M. Sharpless, M. Aeschbacher, S.E. Page, J. Wenk, M. Sander, K. McNeill, Photooxidation-induced changes in optical, electrochemical and photochemical properties of humic substances, Environ. Sci. Technol. 48 (2014) 2688–2696.
M. Shestakova, M. Sillanpää, Removal of dichloromethane from ground and wastewater: a review, Chemosphere 93 (2013) 1258–1267.
P. Singer, DBPs in drinking water: additional scientific and policy considerations for public health protection, J. Am. Water Works Assoc. 98 (10) (2006) 73–79.
P. Singer, K. Bilyk, Enhanced coagulation using a magnetic ion exchange resin, Water Res. 36 (2002) 4009–4022.
E. Smith, Y. Kamal, Optimizing treatment for reduction of disinfection by-product (DBP) formation, Water Sci. Technol. Water Supply 9 (2009) 191–198.
V.L. Snoeyink, D. Jenkins, Water Chemistry, John Wiley & Sons, New York, 1980.
Y. Soh, F. Roddick, J. van Leeuwen, The future of water in Australia: the potential effects of climate change and ozone depletion on Australian water quality, quantity and treatability, Environmentalist 28 (2008) 158–165.
B. Sulzberger, E. Durisch-Kaiser, Chemical characterization of dissolved organic matter (DOM): a prerequisite for understanding UV-induced changes of DOM absorption properties and bioavailability, Aquat. Sci. 71 (2009) 104–126.
H. Suty, C. De Traversay, M. Cost, Applications of advanced oxidation processes: present and future, Water Sci. Technol. 49 (4) (2004) 227–233.
J. Swietlik, A. Dabrowska, U. Raczyk-Stanislawiak, J. Nawrocki, Reactivity of natural organic matter fractions with chlorine dioxide and ozone, Water Res. 38 (2004) 547–558.
W.-W. Tang, G.-M. Zeng, J.-L. Gong, J. Liang, P. Xu, C. Zhang, B.-B. Huang, Impact of humic/fulvic acid on the removal of heavy metals from aqueous solutions using nanomaterials: a review, Sci. Total Environ. 468–469 (2014) 1014–1027.
E. Thurman, Organic Geochemistry of Natural Waters, Martinus Nijhoff/Dr W. Junk Publ., Dordrecht, The Netherlands, 1985.
R. Toor, M. Mohseni, UV-H_2O_2 based AOP and its integration with biological activated carbon treatment for DBP reduction in drinking water, Chemosphere 66 (2007) 2087–2095.
V. Trang, L. Phuong, N. Dan, B. Thanh, C. Visvanathan, Assessment on the trihalomethanes formation potential of tan hiep water treatment plant, J. Water Sustain. 2 (2012) 43–53.
C. Uyguner-Demirel, M. Bekbolet, Significance of analytical parameters for the understanding of natural organic matter in relation to photocatalytic oxidation, Chemosphere 84 (2011) 1009–1031.
M. Winterdahl, Intra-annual Variability of Natural Organic Matter in Boreal Streams Patterns and Controls, Faculty of Natural Resources and Agricultural Sciences, Department of Aquatic Sciences and Assessment, Uppsala (Doctoral Thesis), Swedish University of Agricultural Sciences, Uppsala, Sweden, 2013.

M. Winterdahl, M. Erlandsson, M.N. Futter, G.A. Weyhenmeyer, K. Bishop, Intra-annual variability of organic carbon concentrations in running waters: drivers along a climatic gradient, Global Biogeochem. Cycles 28 (4) (2014) 451–464.

F. Worrall, T. Burt, Trends in DOC concentrations in great britain, J. Hydrol. 346 (2007) 81–92.

F. Worrall, T.P. Burt, Changes in DOC treatability: indications of compositional changes in DOC trends, J. Hydrol. 366 (2009) 1–8.

Y. Xing, Characterization of Dissolved Organic Carbon in Prairie Surface Waters Using Fourier Transform Infrared Spectroscopy (M.Sc. Thesis), University of Saskatchewan, Saskatoon, Saskatchewan, Canada, 2010.

Y.-Y. Zhao, J. Boyd, M. Woodbeck, R. Andrews, F. Qin, S. Hrudey, X.-F. Li, Formation of *N*-nitrosamines from eleven disinfection treatments of seven different surface waters, Environ. Sci. Technol. 42 (2008) 4857–4862.

Z.-Y. Zhao, D.-J. Gu, H.-B. Li, X.-Y. Li, K. Leung, Disinfection characteristics of the dissolved organic fractions at several stages of a conventional drinking water treatment plant in South China, J. Hazard Mater. 172 (2009) 1093–1099.

A. Zularisam, A. Ismail, R. Salim, Behaviours of natural organic matter in membrane filtration for surface water treatment – a review, Desalination 194 (2006) 211–231.

CHAPTER 2

Impact of climate and atmospheric pressures on natural organic matter concentration and surface water treatment practices

Lindsay Anderson[1], Amina Stoddart[1], Graham A. Gagnon[1], Yuri Park[2] and Mika Sillanpää[3,4,5,6]

[1]Center for Water Resources Studies, Dalhousie University, Halifax, Nova Scotia, Canada; [2]Institute of Environmental Technology, Department of Environmental Engineering, Seoul National University of Science and Technology, Seoul, South Korea; [3]Department of Chemical Engineering, School of Mining, Metallurgy and Chemical Engineering, University of Johannesburg, Doornfontein, South Africa; [4]Chemistry Department, College of Science, King Saud University, Riyadh, Saudi Arabia; [5]Department of Applied Physics, Faculty of Science and Technology, Universiti Kebangsaan Malaysia Bangi, Selangor, Malaysia; [6]International Research Centre of Nanotechnology for Himalayan Sustainability (IRCNHS), Shoolini University, Solan, Himachal Pradesh, India

Abstract

For decades, there have been reported widespread increases in the concentration of natural organic matter (NOM) measured as dissolved organic carbon (DOC) throughout the Northern hemisphere, a phenomenon commonly referred to as brownification. Some of the main drivers of brownification are associated with anthropogenic pressures including the reversal of atmospheric acid (e.g., sulfate) deposition as a result of more stringent air emissions policies, and climate pressures (e.g., increasing frequency, intensity and duration of precipitation events, as well as warming temperature). Many studies have predicted that browning will continue considering climate pressures, even after sulfate deposition returns to background levels. Furthermore, browning of surface waters is expected to have a profound impact on drinking water treatment practices. For example, drinking water providers drawing from browning water supplies can expect to experience increased coagulant demand, reduced filter hydraulic performance and potential for elevated disinfection by-products (DBPs) if NOM removal is unoptimized. Inadequate removal of NOM during treatment may also result in elevated trace metal concentration in the distribution system, which can increase the ability of treated drinking water to transport contaminants such as lead. Drinking water treatment processes should be designed to accommodate for fluctuations in both NOM concentration and quality and a buffer between treated water quality and regulatory limits is critical.

Natural Organic Matter in Water
ISBN 978-0-12-824274-2
https://doi.org/10.1016/B978-0-12-824274-2.00004-1

Keywords: Brownification; Climate change; Drinking water treatment; Natural organic matter (NOM); Recovery from acidification.

Abbreviations

DBP	Disinfection by-product
DOC	Dissolved organic carbon
DOM	Dissolved organic matter
INCA-C	Integrated Catchments Model for Carbon
LRTAP	Convention on Long-Range Transboundary Air Pollution
NOM	Natural organic matter
NO_x	Nitrogen oxides
SO_x	Sulfur oxides
THM	Trihalomethane
TOC	Total organic carbon
TTHM	Total trihalomethane
UK-AWMN	United Kingdom Acid Water Monitoring Network
UK-UWMN	United Kingdom Upland Waters Monitoring Network
UNECE	United Nations Economic Commission for Europe

2.1 Introduction

For decades, there have been reported widespread increases in the concentration of natural organic matter (NOM) measured as dissolved organic carbon (DOC) throughout the northern hemisphere, a phenomenon commonly referred to as brownification. Some of the earliest reports of browning have been observed in the UK. As an example, long-term (15-year) monitoring of 22 UK surface waters demonstrated that DOC concentrations increased by 91% (Evans et al., 2005). Several others (Evans and Monteith, 2001; Davies et al., 2005; Broadmeadow et al., 2019) have also observed increasing DOC in lakes and streams throughout the UK. Similar trends in rising DOC have been observed in Scandinavian countries particularly in Norway (Kopacek et al., 2006; De Wit et al., 2007; Haaland et al., 2010; Finstad et al., 2016), as well as Finland and Sweden (e.g., Skjelkvåle et al., 2001; DeWit et al., 2016). In the United States, there have been documented increases in DOC concentration, particularly in the northeastern states including New England (Stoddard et al., 2002; Strock et al., 2014), Maine (SanClements et al., 2012; Gavin et al., 2018), and the Adirondack region of New York (Stoddard et al., 2002; Driscoll et al., 2003, 2016; Burns et al., 2006; Lawrence et al., 2013; Strock et al., 2014; Leach et al., 2019), generally starting in the 1990s–2000s. In Canada, increases in DOC have been occurring predominantly in Ontario (Keller et al., 2019; Imtiazy et al., 2020; Watmough and Eimers, 2020; Meyer-Jacob et al., 2020), Quebec (Houle et al., 2003; Couture et al., 2012), and Atlantic Canada (Clair et al., 2011; Anderson et al., 2017, 2021; Redden et al., 2021). Recently, there have been reports of increasing DOC in Northern China (Chen et al., 2012; Zhang et al., 2016; Guo et al., 2017; Wang et al., 2020) as well as Russia (Kalinkina et al., 2019, 2020). Many other studies have also documented increasing DOC concentration in surface waters throughout the Northern hemisphere (e.g., Europe and North America) (Skjelkvåle et al., 2005; Monteith et al., 2007; Garmo et al., 2014, 2020). An increasing number of peer reviewed publications and citations on the subject of brownifing and rising DOC in surface water (Fig. 2.1) demonstrates the interest gained by the topic of brownification over the past 20 years.

Some of the drivers for the increases in NOM concentration are associated with climate and anthropogenic processes. For example, Imtiazy et al. (2020) observed increasing DOC in over 40 lakes in Ontario and Nova Scotia, Canada, and found that sulfate deposition and precipitation explained 24%–54% and 21%–49%, respectively, of variance in DOC concentration. Others have shown the difficulty in isolating the drivers—Evans et al. (2005) noted that increases in DOC may be driven by climate

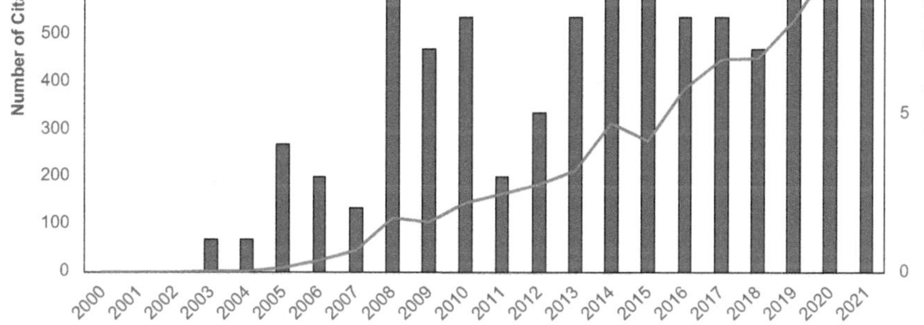

Figure 2.1 Yearly publications (2000–2021) citations for published literature that mentions "brownification and/or browning," "organic carbon," "climate change," and "recovery" in the title, abstract and keywords. Note that irrelevant articles were filtered and omitted, for a total of 134 publications.

factors such as increasing temperature or reductions in atmospheric acid deposition; however, mechanisms could not be isolated (Evans et al., 2005). Other evidence suggests that reductions in acid deposition may drive browning over the short-term, and that climate factors will enhance browning over the long term (Meyer-Jacobs et al., 2019; Kritzberg et al., 2020).

The phenomenon of increasing DOC will undoubtedly have an impact on drinking water treatment facilities that draw from surface waters, particularly those with aging infrastructure or unit processes that were not designed to handle high DOC loads (e.g., direct filtration). This chapter presents an overview of the impacts of climate change and atmospheric/anthropogenic pressures on NOM concentration as well as its impacts on drinking water treatment processes. The following sections provide detailed information on the predominant drivers of increased DOC concentration in the northern hemisphere, with an emphasis on acid deposition and climate change, as well the possible impacts to commonly used water treatment processes.

2.2 Drivers of increasing DOC

Some of the drivers for increasing DOC in surface waters in the Northern hemisphere include changes in land use/cover and land management (Finstad et al., 2016; Kritzberg et al., 2017), although climate change (e.g., increasing frequency/intensity and duration of precipitation events, and warming) and the reversal of acidification (e.g., decreased atmospheric acid deposition) have recently gained popularity. Fig. 2.2 depicts a schematic showing the increasing number of publications (total 134) and citations on the topic of freshwater browning and its relation to climate change as well as reversal of acidification over the past 20 years. Out of the 134 publications, the majority were from Scandinavia (~30%), Canada (25%), the United States (24%), and England (20%). The following sections will provide additional detail on climate and reversal of acidification and their relation to brownification of surface waters in the Northern hemisphere.

2.2.1 Reductions in atmospheric acid deposition

Atmospheric acid (e.g., sulfate) deposition through acid rain, which generally originates from transboundary air pollution, has caused the acidification of surface waters throughout globally since the late 1960s and early 1970s (Schindler, 1988; Jeffries et al., 2003a; Wright et al., 2005). As

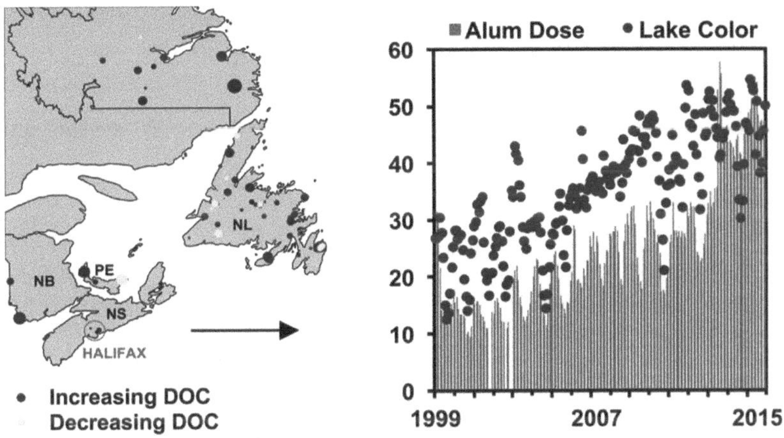

Figure 2.2 Example of the impact of increasing NOM concentration on coagulation practices. *(Adapted from Anderson et al. (2017).)*

of 2019, Canadian SO_x emissions are nearly 80% lower than they were compared to 1990 levels (Canada, 2020), while in the United States over a 90% reduction was achieved for the same period (EPA, 2020). Similar reductions were observed in European countries (EEA, 2020). The resulting impact to freshwater aquatic species was pronounced—particularly in terms of ecosystem diversity as well as population loss of certain fish species such as Atlantic Salmon and trout (Watt, 1987; Schindler, 1988).

To control acidic deposition and acid rain, controls on sulfur and nitrogen oxides (SO_x and NO_x) were implemented through various policies and regulatory frameworks since the 1970s. For example, amendments to the Clean Air Act in the United States, the Canada–United States air quality agreement, the Eastern Canada Acid Rain Control Program, and the United Nations Economic Commission for Europe (UNECE) Convention on Long-Range Transboundary Air Pollution (LRTAP) were all implemented to minimize air emissions. Canada, the United States, and Europe have had massive reductions in emissions followed by reductions in atmospheric acid deposition (Driscoll et al., 2016; Garmo et al., 2020; Stoddard et al., 1999). As a result of reduced atmospheric acid deposition, there have been notable responses in terms of surface water chemistry: improvements in acid-base status (e.g., increasing pH, alkalinity, and increased buffering capacity), reductions in sulfate concentration, and in many cases, increases in NOM (e.g., DOC, color) concentration (e.g., Skjelkvåle et al., 2005; Monteith et al., 2007; Garmo et al., 2014).

Many study sites in Europe and North America have exhibited a correlation between increasing NOM concentration and reductions in atmospheric acid deposition (namely sulfate) or sulfate concentration. For example, Skjelkvale et al. (2005) calculated trends in water quality (1990–2001) for 12 regions in North America and Europe encompassing 189 surface waters to understand the impact of acidification on water quality. With exception to one region, all sites showed decreasing surface water sulfate and increasing DOC concentration. Soon after, Monteith et al. (2007) evaluated water chemistry from over 500 surface waters in northern Europe and North America over the period 1990 to 2004 and found that DOC concentrations were rising in proportion to declines in atmospheric sulfur deposition. They also noted that declining surface water acidity (attributed to decreasing acid deposition) is being partially offset by increasing levels of organic acidity. Garmo et al. (2014) analyzed trends in surface water chemistry from over 170 undisturbed acid-sensitive sites in Europe and North America from 1990 to 2008. It was found that 87% of the 173 sites studied showed significant decrease in sulfate concentration and increasing trends in DOC were more widespread than decreasing. More recently, Garmo et al. (2020) studied nearly 500 surface waters from 13 regions in North America and Europe and found that all regions except for one exhibited increasing DOC concentration which was inversely correlated to deposition chemistry. Most recently, de Wit et al. (2021) studied >400 lakes in Europe and North America between 1990 and 2016 and found that regional browning is occurring at rates of up to 2% annually. It was also noted that DOC export is still dependent on rates of acidic deposition but is becoming more sensitive to climate change. As rates of deposition approach background levels, de Wit et al. (2021) highlighted the importance of understanding the impact of climate on DOC production.

Studies investigating surface waters in the UK Acid Waters Monitoring Network sites (UK-AWMN) have also revealed strong evidence for browning linked to precipitation chemistry. Evans and Monteith (2001) conducted trend analysis for 22 surface waters during the period 1988–2000 in the UK and noted that reductions in anthropogenic sulfur deposition occurred during the latter portion of the monitoring period, which also coincided with widespread increases in DOC. Similarly, Evans et al. (2005) evaluated numerous drivers for increasing DOC concentration in 22 UK-AWMN surface waters over a 15-year period (1988–2003) and found that average DOC concentration increased by more than 90%, which was partially explained by reductions in sulfate deposition. Using experimental data paired with reanalysis of the UK-AWMN dataset, Evans et al. (2006) investigated alternative hypotheses to help understand the driver(s) responsible for the 91% increase in DOC concentration observed in the UK from 1998 to 2003. Many of the trends in DOC concentration observed in this study could be explained by decreasing sulfur deposition. More recently, as part of the UK Upland Waters Monitoring Network (UK-UWMN), Broadmeadow et al. (2019) also observed reductions in excess sulfate concentration as well as increases in DOC concentration in 12 streams over the period 1991–2012.

Similar findings have been observed in Scandinavian surface waters, although in some cases the trends were not as evident as in the UK. This may be attributed to study period, a lagged chemical response to reductions in deposition, or the number of surface waters considered. Skjelkvale et al. (2001) analyzed trends in surface water chemistry over a nine-year period (1990–99) for 344 lakes in Finland (163 lakes), Norway (100 lakes), and Sweden (81 lakes). Trend analysis demonstrated that sulfate concentration decreased significantly in nearly 70% of the lakes monitored over the study period, while only 12% of the lakes showed significant increases in TOC concentration. However, it should be noted that an earlier study by Skjelkvale et al. (1998) presented data from 485 acid sensitive lakes in Norway that were surveyed from 1986 to 95. Overall, all lakes surveyed showed a significant decrease in sulfate concentration, but at the time of this study, there was no evidence for changes in NOM concentration in response to changes in acid deposition. De wit et al. (2007) investigated the role of acid deposition and climate on long-term variation in natural organic matter (DOC) between 1985 and 2003 in three acid sensitive surface waters in Norway. This study demonstrated that there was an overall long-term increase in DOC concentration, which was related to reductions in acid deposition. Using long-term time series of color measurements of four adjacent lakes in Oslo, Norway, Haaland et al. (2010) assessed, how temporal trends in colored organic matter were related to changes in both the quality and quantity of precipitation. Overall, it was determined that the long-term increase in NOM content was related to the reductions in sulfate and chloride concentrations in precipitation in the region. In Finland, Forsius et al. (2003) studied the response of several lakes to reductions in acidic deposition from 1990 to 99 and reported that between 64% and 85% of lakes had significant decline in sulfate concentration. However, only up to 10% of lakes studied showed significant increasing TOC concentration (Forsius et al., 2003). Monteith et al. (2007) observed increases in DOC concentration in many Nordic surface waters and noted that this could be explained by changes in deposition chemistry and catchment sensitivity.

Another highly studied area is the Adirondack region of New York, as well as other areas in the Northeastern US. Driscoll et al. (2003) assessed the response of acid-sensitive lakes in the Adirondack region of New York to declines in atmospheric acidic deposition and found that all lakes studied showed a significant decrease in sulfate concentration since 1982, while a portion of the lakes showed increasing DOC concentration. Burns et al. (2006) evaluated trends in surface water chemistry over the period 1992–2001 in Adirondack and Catskill regions of New York. Decreasing sulfate concentrations were observed in all streams and lakes in the Catskill and Adirondack regions, while 75% and 80% of the Adirondack 80% of the Catskill showed significant increases in DOC concentration, respectively. Similarly, Waller et al. (2012) conducted time-series analysis on lake chemistry data for 42 lakes in the Adirondack region of New York over the period 1991–2007 and found that nearly all the lakes showed reduced sulfate concentration and nearly 70% of the study lakes studied showed increasing DOC concentration. Stoddard et al. (2003) assessed changes in surface water chemistry from 1990 to 2000 in response to changes in acid deposition throughout the northern and eastern portions of the United States in response to the Clean Air Act Amendments of 1990. Overall, it was determined that DOC increased in New England lakes, Adirondack lakes, Northern Appalachian streams, Upper Midwest lakes (Stoddard et al., 2003). SanClements et al. (2012) analyzed water quality data from nine acid-sensitive lakes in the Northeastern United States over the period 1982–2010 and found that DOC concentrations increased in five of the nine lakes studied, three of which showed significant negative correlations between sulfate and DOC concentration.

In Eastern Canada, increases in DOC have been lagged in comparison to other regions. Jeffries et al. (2003b) assessed the recovery of lakes in Southeastern Canada from the effects of acidic deposition throughout the 1990s and found that study lakes in Atlantic Canada, Ontario, and Quebec all experienced declining sulfate concentrations. However, most of these lakes showed no trend for DOC concentration. Clair et al. (2011) analyzed water chemistry trends for 66 lakes in Atlantic Canada throughout 1984–2007 and observed a general downwards trend in sulfate deposition and lake concentration; however, increases in DOC were only observed during the period 2000–07. More recently, Anderson et al. (2017) studied sulfate deposition in Atlantic Canada and noted that in response to decreasing sulfate deposition in the region, lake color increased by nearly fourfold over a ~15-year period. Redden et al. (2021) evaluated a large dataset of 81 lakes and rivers in Nova Scotia over a period of 41 years and found that in contrast to previous reports, surface waters were exhibiting decreases in sulfate and increases in color. Anderson et al. (2021) also observed browning in six lakes over a 30-year period in Nova Scotia, a region where acid deposition has been reduced significantly.

It has been hypothesized that reductions in sulfate (SO_4^{2-}) leads to a reduction in soil acidity, resulting in more soluble organic matter (DeWit et al., 2007; Haaland et al., 2010; Evans et al., 2012; SanClements et al., 2012). Others have noted that a reduction of ionic strength in soil solutions caused by lower sulfate concentration causes an increase in DOC flux (Monteith et al., 2007). As an example, Clark et al. (2011) performed experiments to evaluate DOC release from soils because of changes in ionic strength and acidity through sulfate addition. The release of DOC was hindered by between 21% and 60% as a result of the addition of sulfuric acid to soil solutions (Clark et al., 2011).

Based on this, increasing DOC in the Northern hemisphere is likely in part attributed to reductions in atmospheric acid deposition, although additional drivers such as climate pressures (e.g., temperature, precipitation) need to be considered.

2.2.2 Climate change pressures

In addition to atmospheric deposition, climate change processes can also impact NOM concentration in surface waters. As reductions in acid deposition become more stable, it is expected that climate change will be the dominant driver of NOM increase. In addition, many lakes that are not considered to be acidified have had increases in DOC concentration and in some regions, climate has been the dominant driver since the 2000s (Riise et al., 2018; Imtiazy et al., 2020).

Many researchers have investigated the impact of climate factors on DOC trends. Lipczynka-Kochany (2018) noted that increasing temperature will increase the biodegradation of NOM, while runoff will increase export, both leading to increased NOM concentration in freshwaters. In addition to sulfate concentration, Couture et al. (2012) identified short-term precipitation, long-term temperature, and radiation as possible climate variables that contributed significantly to increasing DOC in 30

Quebec surface waters between 1989 and 2006. Finstad et al. (2016) noted that increased temperature and runoff, in addition to reduced sulfate deposition, adds to freshwater DOC concentration and that climate mediated changes in vegetation cover may increase browning in northern surface waters over the long term. Meyer-Jacob et al. (2020) assessed whether DOC levels were still influenced by acid deposition in Ontario (1981–2018) and found that weakening correlations between DOC and acid variables indicated a reduced impact of acid deposition and that climate changes are becoming the dominant controls in DOC concentration. Meyer Jacob et al. (2019) also noted that climate change is forcing lakes to new ecological states as DOC concentrations exceed pre-industrial levels. Xiao et al. (2020) estimated lake color over a long period (1800–2015) based on DOM from sediment and found that lake color was four times higher than during 1800–1915, and that impacts of anthropogenic sulfate deposition, mainly the interplay between different drivers including increased temperature and precipitation is responsible for lake browning. Others have noted that DOC is in fact a response variable for monitoring climate change. Adrian et al. (2009) described DOC as a measurable response parameter for climate change in lakes as it integrates transparency and metabolism, although its main disadvantage was that it can be confounded by other factors such as acid deposition. Hongve et al. (2004) also found significant correlation between color and accumulated precipitation in Norwegian lakes, while temperature variations were insignificant.

2.2.2.1 Precipitation

Precipitation has been identified as one of the main drivers of DOC content in surface waters. Precipitation mobilizes DOC and therefore increased rainfall amounts associated with climate change may result in larger DOC export from soil. As an example, De Wit et al. (2016) evaluated 474 surface waters in Scandinavia over 1990 and 2013 and showed that in dry regions, precipitation was a strong driver of TOC concentration, while in wetter regions the correlation was weaker. As precipitation increases, browning trends will continue (De Wit et al., 2016). Hongve et al. (2004) studied a number of lakes and drinking water supplies in Norway and found that increases in DOC and water color were significantly correlated with precipitation accumulation. Strock et al. (2016) investigated the effects of extreme weather over 30 years on 84 lakes in northeastern US and found that brownification was observed during wet years while acidification was observed during dry years. Using 30 years of measurements, de Wit et al.

(2018) calculated that concentrations and fluxes of DOC in a Norwegian lake would rise significantly with declining sulfate deposition and increased precipitation levels and noted that precipitation matters most in DOC controls for small northern lakes. Therefore, as the frequency of extreme precipitation events increases, it is likely that DOC concentration will continue to increase.

Periods of climate induced drought can also impact DOC export. For example, Strock et al. (2016) showed that dry periods caused episodic acidification resulting in DOC suppression (2.5 mg L^{-1}). It has been hypothesized that drought causes acidification, as a result of the oxidation of legacy sulfur in soils (Clark et al., 2005; Watmough et al., 2016; Strock et al., 2016). In addition, precipitation patterns (e.g., extreme wet or dry weather) can also alter the characteristics of NOM. For example, Baek et al. (2019) investigated the change in NOM composition before and after rainfall and found that after rainfall, lateral flows transported large molecular weight, hydrophobic NOM that was comprised majorly of lignin (as measured using fluorescence excitation emission, molecular composition hydrophobicity and molecular size distribution). On the other hand, more hydrophilic NOM is typically expected during extreme drought events. Fabris et al. (2015) observed distinct water quality conditions resulting from extreme climate (e.g., drought) and noted that drought-impacted NOM had low specific UV-absorbance (SUVA) and apparent molecular weight distribution. Accordingly, as extreme weather events continue because of changing climate, there will likely be episodic shifts in NOM quality (hydrophobic during wet events, hydrophilic during dry events) which will impact the effectiveness of water treatment practices.

2.2.2.2 Temperature

Increasing temperatures associated with climate change are also known to impact the concentration of DOC, although the effects may be lesser than those observed by increasing precipitation. Freeman et al. (2001) found a 65% increase in DOC concentrations in UK freshwaters over a 65-year period and attributed increases in part to rising temperatures because of global warming. Further, DOC trends were observed at both sites that were unacidified as well as those recovering from acidification, suggesting that temperature could have influenced all study sites (Freeman et al., 2001). Evans et al. (2005) noted that increasing DOC may be associated with both reductions in atmospheric acid deposition as well as rising temperatures, although it is difficult to isolate mechanisms. Warming of surface waters also causes increased microbial activity, which can also contribute to increased production of NOM (Xiao et al., 2020). Dieleman et al. (2016) also investigated decomposition rates of NOM experimentally through increased temperature, CO_2 and water table position and found that under elevated temperature conditions, DOC decomposition rates increased significantly, and the quality of DOC was also variable (lability and recalcitrance).

2.3 Predicting future DOC concentration under climate scenarios

Given the widespread increases in DOC throughout the northern hemisphere, many have modeled the effect of climate change on freshwater DOC concentrations. Heizlar et al. (2003) used a linear regression model and climate-based scenarios (e.g., temperature and precipitation projections) to predict long term NOM concentrations for a river in the Czech Republic. The climate scenario with a doubling of atmospheric CO_2 concentration suggested a 7% increase in NOM concentration in the study stream, and multiple linear regression models showed that temperature and interflow could explain nearly 70% of NOM variability (Heizlar et al., 2003). Futter et al. (2008) used the integrated catchments model for carbon (INCA-C) and artificial neural networks to model the effects and mechanisms of climate effects on surface water DOC concentration on a Finnish catchment between 1990 and 2003. It was found that the INCA-C could simulate some of the long-term trends in DOC; however, factors other than climate played a role. Production within the catchment and mineralization within the lake were found to control DOC concentrations in the study lake (Futter et al., 2008). Similarly, Futter et al. (2009) also used the INCA-C to simulate long term DOC at four monitoring sites in Sweden, and found that long-term variations in DOC caused by factors including runoff, temperature and sulfate deposition were observed at all monitoring sites; short-term variability was attributed to runoff, while acid deposition could explain some of the long term variability. Ledesma et al. (2012) evaluated long-term dynamics of TOC for a drinking water supply in Sweden also using the INCA-C model. The INCA-C model could successfully predict TOC variation with exception to an increase during 2009, and therefore, it was concluded that there may be unmeasured processes controlling TOC in surface waters (Ledesma et al., 2012). Weyhenmeyer et al. (2016) also predicted a420 (absorbance at 420 nm, a proxy for water color) would increase substantially in Sweden; median a420 would rise by a factor between 1.1 and 7.6 by 2030 given a 32% increase in precipitation levels. More recently, Lee et al. (2020) evaluated the impact of droughts and acid deposition on long-term DOC concentration in a Welsh catchment using the INCA-C model. Results suggested that DOC increases would likely continue in the near future, stabilizing in the far future with seasonal DOC concentrations shifting toward a post-drought surge of DOC in the fall (Lee et al., 2020). Xu et al. (2020) also used INCA-C to predict discharge and DOC concentration in nine peatland derived drinking water catchments in the UK under climate and sulfate deposition scenarios. Simulations revealed that mean annual discharge would decrease, and mean annual DOC concentrations would increase by up to 53% under all future climate and emissions scenarios by the 2090s. DOC increases were particularly notable in the fall and winter, when treatment plants were already operating at DOC removal capacity and have difficulty coping with elevated DOC; large capital investments would likely be required to maintain safe drinking water quality in the future (Xu et al., 2020). Accordingly, models such as INCA-C can be useful in predicting long-term patterns in surface water DOC under various climate scenarios, which can be beneficial for projecting drinking water treatment plant design needs.

2.4 Impact on water treatment

Since the onset of brownification, many studies have begun to document the impact on drinking water supplies and treatment facilities. Delpla et al. (2009) reviewed the impacts of climate change on surface water quality in relation to drinking water production and noted that there is a degradation trend for drinking water quality, among the parameters reported were DOM. A need for the incorporation of water quality monitoring (e.g., online TOC measurements, fluorescence analysis) and predictive models and decision support systems was also highlighted as a means to adapt to climate change impacts (Delpla et al., 2009). Ritson et al. (2014) evaluated the impact of climate change on the treatability of DOM in UK water supplies and found that hydrophobic DOM may become more prominent, and that utilities may see increased intra-annual variation in both quantity and quality with climate change (Ritson et al., 2014). A need for more robust treatment processes that can handle variation in NOM hydrophobicity, concentration and molecular weight distribution are needed, for example employing the use of enhanced coagulation, switching coagulants, or alternative methods of NOM removal such as magnetic ion exchange (MIEX) or activated carbon (Ritson et al., 2014). Online process monitoring tools such as UV-absorbance and fluorescence were identified as viable methods as they can rapidly detect changes in NOM properties and can predict removal efficiencies and inform DBP formation (Ritson et al., 2014).

Coagulation and direct filtration processes are some of the most impacted aspects of drinking water treatment considering increasing DOC concentration. Anderson et al. (2017) documented the impact of browning surface waters in Atlantic Canada on drinking water treatment processes and found that monthly color concentrations increased by 1.7–3.8-fold over a 16-year period, with coagulant (alum) increases occurring at the same rate. Fig. 2.2 below depicts an example of how alum dose can be impacted by brownification as described by Anderson et al. (2017). Further, filter performance was also compromised in the direct filtration plant supplied by browning water because of increased coagulant dosing (Anderson et al., 2017). DeMont et al. (2021) assessed strategies to improve the efficacy and efficiency of direct filtration plants facing changes in source water quality from anthropogenic and climatic pressures. This was conducted as an immediate strategy to manage increasing organic loading while more long-term, sustainable solutions are developed. The impact of coagulant and polyelectrolyte combinations, as well as increasing the effective size of filter media, on NOM removal and unit filter run volumes were investigated. Although the partial substitution of alum with polyelectrolyte did not improve NOM removal (consistent removals of 35%–40% were observed in all test conditions), increasing the effective size of filter media from 0.9 to 1.3 mm did improve unit filter run volumes without impacting finished water quality (DeMont et al., 2021). This strategy could be used as a short-term solution to improve filter operations for plants needing to increase chemical dosing to overcome increased organic load.

If not removed adequately, increasing NOM concentration will likely result in exceedances of regulated disinfection by-products such as total trihalomethanes (TTHMs). Cool et al. (2019) predicted trihalomethane occurrence for >100 drinking water utilities in Quebec, Canada impacted by climate change (e.g., temperature and precipitation) and found a significant increase in the probability of TTHM threshold exceedances (80 ug/L threshold) for the 2050 and 2080 prediction periods. Further, it was noted that there were differences between treatment type and season, with higher values for summer and for utilities using chlorination alone as treatment (e.g., small systems)—utilities using more advanced treatment were more resilient to climate change (Cool et al., 2019). Similarly, Valdivia-Garcia et al. (2019) estimated the impact of climate change on THM formation in drinking water treatment using monitoring data from five Scottish drinking water treatment plants. Significant correlations were observed between THMs, temperature, and DOC concentration, and the impact of global warming on THM formation was estimated—a mean summer temperature increase of 1.8 °C would yield a 39% increase in THMs by 2050 (Valdivia-Garcia et al., 2019). Thus, utilities are faced with an increased risk of THM formation or increased treatment costs to maintain equivalent treated water quality (Valdivia-Garcia et al., 2019). DeMont et al. (2021) also noted a lack of improved treatment efficacy through coagulation and direct filtration resulted consistently high DBP concentrations in treated water, and it was attributed to NOM that was recalcitrant to coagulation as evidenced by low specific UV-absorbance. Additional treatment plant upgrades consisting of biofiltration, membrane filtration, activated carbon, or ion exchange may be required to target DBP precursor material remaining after coagulation (DeMont et al., 2021).

Poor NOM removal during treatment also has implications for distribution system water quality, particularly in terms of trace metals. Many studies have demonstrated increased trace metals cycling in conjunction with brownification. For example, Neal et al. (2008) found increasing iron concentrations in upland waters in the UK—iron concentrations have doubled over a 20-year period and the greatest rates of iron increase were associated with DOC. Similarly, Kritzberg and Ekstrom (2012) investigated trends in color, NOM, and iron in 30 Swedish rivers and found that iron explained a significant portion of water color, and that iron and NOM are controlled by similar processes. Bjorneras et al. (2017) found widespread increases in iron concentration in North American and European freshwaters and noted that regions with increasing iron coincided with increasing organic carbon as well. In addition to iron, aluminum content has also been increasing. As an example, Redden et al. (2021) studied brownification in 81 lakes and rivers in Atlantic Canada and found widespread increases in color and aluminum in study lakes, where aluminum was determined to be bound to both iron and organic colloidal material. Anderson et al. (2021) showed similar trends in Atlantic Canadian surface waters. Accordingly, NOM is likely in part responsible for maintaining trace metals in solution and this has substantial implications for distributed water quality. Inadequate removal of NOM during treatment may result in elevated trace metal concentration, such as iron and/or aluminum in the distribution system. This also has relevance for controlling lead in distribution systems—if NOM-stabilized trace metals are persistent in treated water, they can increase lead leaching from plumbing (Anderson et al., 2021). These colloidal interactions have been demonstrated by Trueman et al. (2019) who characterized colloidal metals in drinking water by field flow fractionation and found that organic (e.g., humic substances) and inorganic colloidal fractions can present additional capacity for drinking water to transport lead. NOM itself can also reduce lead oxides—A study by Winning et al. (2015) investigated the impact of TOC and humic substances on the occurrence of lead at the tap and found that a 50% reduction of total NOM concentration resulted in a 75% reduction in lead release at the tap. This further demonstrates the critical importance of adequate NOM removal during drinking water treatment.

2.5 Conclusions

The quality of surface water supplies is expected to worsen due to increases in NOM (i.e., brownification) resulting from climate and anthropogenic pressures. Some of the drivers for the increases in NOM concentration are associated with climate and anthropogenic processes such declining acid (e.g., sulfate) deposition; however, the mechanisms are difficult to isolate. A substantial amount of research has demonstrated that reductions in atmospheric sulfate deposition have contributed to increasing DOC concentration in the Northern hemisphere, particularly in the UK, Scandinavia, Northeastern US, and Canada.

As reductions in acid deposition become more stable, it is expected that climate change processes will be the dominant driver of NOM. Many researchers have investigated the impact of climate factors on DOC trends—increasing temperature may increase the biodegradation of NOM, while increased runoff can increase export, both leading to increased NOM concentration in freshwaters. Weakening correlations between DOC and acid deposition have also indicated that climate changes may become the dominant controls in DOC concentration, which is forcing lakes to new ecological states as DOC concentrations exceed pre-industrial levels. As rates of deposition approach background levels, it will be increasingly important to understand the impact of climate on DOC concentration.

Precipitation has been identified as one of the main drivers of increasing DOC in surface waters and increased rainfall caused by climate change may result in greater DOC export from soil. As the frequency of extreme precipitation events increases, it is likely that DOC export will continue to increase resulting in further browning. Periods of climate induced drought also impact DOC export, namely resulting from episodic acidification corresponding with DOC suppression. The quality of NOM can also vary with precipitation, with large MW hydrophobic NOM after rainfall, and more hydrophilic NOM following periods of drought. As climate change continues, there will likely be episodic shifts in NOM quality. Increasing temperatures associated with climate change are also known to impact the concentration of DOC, although the effects may be lesser than those observed by increasing precipitation. Research has demonstrated that INCA-C models can be useful in predicting long-term patterns in surface water DOC under different climate scenarios, which may be helpful for projecting drinking water treatment needs.

The impacts of increasing DOC on drinking water treatment practices are profound. Specifically, water utilities may be faced with increased coagulant demand exhibited by elevated DOC. Without effective clarification processes, filter performance may also be compromised. Further, if not removed adequately, increasing NOM concentration will likely result in exceedances of regulated DBPs. Additional treatment plant upgrades may be required to target DBP precursor material remaining after coagulation. Inadequate removal of DOC during treatment may result in elevated trace metal concentration (e.g., iron and/or aluminum) in the distribution system, which can increase the ability of treated drinking water to transport lead.

Accordingly, treatment plants that are already operating at DOC removal capacity and have difficulty coping with current elevated DOC content may be faced with large capital investments to maintain safe drinking water quality in the future. This will likely result in drinking water production becoming more expensive, and water utilities or providers should begin implementing treatment processes that are more robust. In the future, processes should be designed to be able to handle fluctuations in both NOM concentration and quality caused by extreme weather events, and a greater buffer between treated water quality and regulatory limits is critical.

References

R. Adrian, C.M. O'Reilly, H. Zagarese, S.B. Baines, D.O. Hessen, W. Keller, M. Winder, Lakes as sentinels of climate change, Limnol. Oceanogr. 54 (6part2) (2009) 2283–2297.

L.E. Anderson, W.H. Krkošek, A.K. Stoddart, B.F. Trueman, G.A. Gagnon, Lake recovery through reduced sulfate deposition: a new paradigm for drinking water treatment,

Environ. Sci. Technol. 51 (2017) 1414–1422, https://doi.org/10.1021/acs.est.6b04889.
L.E. Anderson, B.F. Trueman, D.W. Dunnington, G.A. Gagnon, Relative importance of organic- and iron-based colloids in six Nova Scotian lakes, NPJ Clean Water (2021) 1–10, https://doi.org/10.1038/s41545-021-00115-4.
S. Baek, H. Lee, J. Park, K.H. Cho, Investigating influence of hydrological regime on organic matters characteristic in a Korean Watershed, Water 11 (3) (2019) 512.
C. Björnerås, G.A. Weyhenmeyer, C.D. Evans, M.O. Gessner, H.P. Grossart, K. Kangur, I. Kokorite, P. Kortelainen, H. Laudon, J. Lehtoranta, N. Lottig, D.T. Monteith, P. Nõges, T. Nõges, F. Oulehle, G. Riise, J.A. Rusak, A. Räike, J. Sire, E.S. Kritzberg, Widespread increases in iron concentration in European and North American freshwaters, Global Biogeochem. Cycles 31 (10) (2017) 1488–1500, https://doi.org/10.1002/2017GB005749.
S.B. Broadmeadow, T.R. Nisbet, J. Forster, Trends in surface water chemistry in afforested Welsh catchments recovering from acidification, 1991–2012, Environ. Pollut. 247 (2019) 27–38, https://doi.org/10.1016/j.envpol.2018.12.048.
D.A. Burns, M.R. McHale, C.T. Driscoll, K.M. Roy, Response of surface water chemistry to reduced levels of acid precipitation: comparison of trends in two regions of New York, USA, Hydrol. Process. 20 (7) (2006) 1611–1627, https://doi.org/10.1002/hyp.5961.
Canada, Air Pollutant Emissions, 2021. Retrieved from, https://www.canada.ca/en/environment-climate-change/services/environmental-indicators/air-pollutant-emissions.html.
X. Chen, X. Chuai, L. Yang, H. Zhao, Climatic warming and overgrazing induced the high concentration of organic matter in Lake Hulun, a large shallow eutrophic steppe lake in northern China, Sci. Total Environ. 431 (2012) 332–338.
T.A. Clair, I.F. Dennis, R. Vet, Water chemistry and dissolved organic carbon trends in lakes from Canada's Atlantic Provinces: no recovery from acidification measured after 25 years of lake monitoring, Can. J. Fish. Aquat. Sci. 68 (2011) 663–674, https://doi.org/10.1139/f2011-013.
J.M. Clark, P.J. Chapman, J.K. Adamson, S.N. Lane, Influence of drought-induced acidification on the mobility of dissolved organic carbon in peat soils, Global Change Biol. 11 (5) (2005) 791–809, https://doi.org/10.1111/j.1365-2486.2005.00937.x.
G. Cool, I. Delpla, P. Gagnon, A. Lebel, R. Sadiq, M.J. Rodriguez, Climate change and drinking water quality: predicting high trihalomethane occurrence in water utilities supplied by surface water, Environ. Model. Software 120 (2019) 104479.
S. Couture, D. Houle, C. Gagnon, Increases of dissolved organic carbon in temperate and boreal lakes in Quebec, Canada, Environ. Sci. Pollut. Control Ser. 19 (2) (2012) 361–371, https://doi.org/10.1007/s11356-011-0565-6.
J.J.L. Davies, A. Jenkins, D.T. Monteith, C.D. Evans, D.M. Cooper, Trends in surface water chemistry of acidified UK Freshwaters, 1988-2002, Environ. Pollut. 137 (1) (2005) 27–39, https://doi.org/10.1016/j.envpol.2004.12.029.
H.A. de Wit, J. Mulder, A. Hindar, L. Hole, Long-term increase in dissolved organic carbon in streamwaters in Norway is response to reduced acid deposition, Environ. Sci. Technol. 41 (22) (2007) 7706–7713, https://doi.org/10.1021/es070557f.
H.A. de Wit, S. Valinia, G.A. Weyhenmeyer, M.N. Futter, P. Kortelainen, K. Austnes, D.O. Hessen, A. Räike, H. Laudon, J. Vuorenmaa, Current browning of surface waters will Be further promoted by wetter climate, Environ. Sci. Technol. Lett. 3 (12) (2016) 430–435.
H.A. de Wit, R.M. Couture, L. Jackson-Blake, M.N. Futter, S. Valinia, K. Austnes, Y. Lin, Pipes or chimneys? For carbon cycling in small boreal lakes, precipitation matters most, Limnol. Oceanography Lett. 3 (3) (2018) 275–284.

H.A. de Wit, J.L. Stoddard, D.T. Monteith, J.E. Sample, K. Austnes, S. Couture, C.D. Evans, Cleaner air reveals growing influence of climate on dissolved organic carbon trends in northern headwaters, Environ. Res. Lett. 16 (10) (2021) 104009.

I. Delpla, A.V. Jung, E. Baures, M. Clement, O. Thomas, Impacts of climate change on surface water quality in relation to drinking water production, Environ. Int. 35 (8) (2009) 1225−1233.

I. DeMont, A.K. Stoddart, G.A. Gagnon, Assessing strategies to improve the efficacy and efficiency of direct filtration plants facing changes in source water quality from anthropogenic and climatic pressures, J. Water Proc. Eng. 39 (October 2020) (2021) 101689, https://doi.org/10.1016/j.jwpe.2020.101689.

C.T. Driscoll, K.M. Driscoll, K.M. Roy, M.J. Mitchell, Chemical response of lakes in the Adirondack Region of New York to declines in acidic deposition, Environ. Sci. Technol. 37 (10) (2003) 2036−2042, https://doi.org/10.1021/es020924h.

C.M. Dieleman, Z. Lindo, J.W. McLaughlin, A.E. Craig, B.A. Branfireun, Climate change effects on peatland decomposition and porewater dissolved organic carbon biogeochemistry. Biogeochemistry 128 (3) (2016) 385−396, https://doi.org/10.1007/s10533-016-0214-8.

C.T. Driscoll, K.M. Driscoll, H. Fakhraei, K. Civerolo, Long-term temporal trends and spatial patterns in the acid-base chemistry of lakes in the Adirondack region of New York in response to decreases in acidic deposition, Atmos. Environ. 146 (x) (2016) 5−14, https://doi.org/10.1016/j.atmosenv.2016.08.034.

EEA, Sulphur Dioxide (SO_2) Emissions, 2020. Retrieved from: https://www.eea.europa.eu/data-and-maps/indicators/eea-32-sulphur-dioxide-so2-emissions-1/assessment-3.

EPA, EPA's 2019 Power Plant Emissions Data Demonstrate Significant Progress, 2020. Retrieved from: https://www.epa.gov/newsreleases/epas-2019-power-plant-emissions-data-demonstrate-significant-progress.

C.D. Evans, T.G. Jones, A. Burden, N. Ostle, P. Zieliński, M.D.A. Cooper, M. Peacock, J.M. Clark, F. Oulehle, D. Cooper, C. Freeman, Acidity controls on dissolved organic carbon mobility in organic soils. Global Change Biology 18 (11) (2012) 3317−3331, https://doi.org/10.1111/j.1365-2486.2012.02794.x.

C.D. Evans, D.T. Monteith, Chemical trends at lakes and streams in the UK Acid Waters Monitoring Network, 1988-2000: evidence for recent recovery at a national scale, Hydrol. Earth Syst. Sci. 5 (3) (2001) 351−366, https://doi.org/10.5194/hess-5-351-2001.

C.D. Evans, D.T. Monteith, D.M. Cooper, Long-term increases in surface water dissolved organic carbon: observations, possible causes and environmental impacts, Environ. Pollut. 137 (1) (2005) 55−71, https://doi.org/10.1016/j.envpol.2004.12.031.

C.D. Evans, P.J. Chapman, J.M. Clark, D.T. Monteith, M.S. Cresser, Alternative explanations for rising dissolved organic carbon export from organic soils, Global Change Biol. 12 (11) (2006) 2044−2053.

R. Fabris, K. Braun, J.Y. Morran, L. Ho, M. Drikas, The evolution of organic character in a drought-and flood-impacted water source and the relationship with drinking water treatment, J. Water Climate Change 6 (3) (2015) 401−413.

A.G. Finstad, T. Andersen, S. Larsen, K. Tominaga, S. Blumentrath, H.A. De Wit, H. Tømmervik, D.O. Hessen, From greening to browning: catchment vegetation development and reduced S-deposition promote organic carbon load on decadal time scales in Nordic lakes, Sci. Rep. 6 (7485) (2016) 1−8, https://doi.org/10.1038/srep31944.

M. Forsius, J. Vuorenmaa, J. Mannio, S. Syri, Recovery from acidification of Finnish lakes: regional patterns and relations to emission reduction policy, Sci. Total Environ. 310 (1−3) (2003) 121−132.

C. Freeman, C.D. Evans, D.T. Monteith, B. Reynolds, N. Fenner, Export of organic carbon from peat soils, Nature 412 (6849) (2001), 785-785.

M.N. Futter, M. Starr, M. Forsius, M. Holmberg, Modelling the effects of climate on long-term patterns of dissolved organic carbon concentrations in the surface waters of a boreal catchment, Hydrol. Earth Syst. Sci. 12 (2) (2008) 437−447.

M.N. Futter, M. Forsius, M. Holmberg, M. Starr, A long-term simulation of the effects of acidic deposition and climate change on surface water dissolved organic carbon concentrations in a boreal catchment, Nord. Hydrol 40 (2−3) (2009) 291−305.

Ø.A. Garmo, B.L. Skjelkvåle, H.A. De Wit, L. Colombo, C. Curtis, J. Fölster, A. Hoffmann, J. Hruška, T. Høgåsen, D.S. Jeffries, W.B. Keller, P. Krám, V. Majer, D.T. Monteith, A.M. Paterson, M. Rogora, D. Rzychon, S. Steingruber, J.L. Stoddard, A. Worsztynowicz, Trends in surface water chemistry in acidified areas in Europe and North America from 1990 to 2008, Water Air Soil Pollut. 225 (3) (2014), https://doi.org/10.1007/s11270-014-1880-6.

Ø.A. Garmo, Ø. Kaste, J. Arle, K. Austnes, H. De Wit, J. Fölster, D. Houle, J. Hruška, I. Indriksone, D. Monteith, M. Rogora, J. Sample, J.L. Stoddard, R. Talkop, W. Trodd, R. Ulańczyk, J. Vuorenmaa, ICP Waters Report 142/2020, 2020.

A.L. Gavin, S.J. Nelson, A.J. Klemmer, I.J. Fernandez, K.E. Strock, W.H. McDowell, Acidification and climate linkages to increased dissolved organic carbon in high-elevation lakes, Water Resour. Res. 54 (8) (2018) 5376−5393, https://doi.org/10.1029/2017WR020963.

W. Guo, F. Yang, Y. Li, S. Wang, New insights into the source of decadal increase in chemical oxygen demand associated with dissolved organic carbon in Dianchi Lake, Sci. Total Environ. 603 (2017) 699−708.

S. Haaland, D. Hongve, H. Laudon, G. Riise, R.D. Vogt, Quantifying the drivers of the increasing colored organic matter in boreal surface waters, Environ. Sci. Technol. 44 (8) (2010) 2975−2980, https://doi.org/10.1021/es903179j.

J. Hejzlar, M. Dubrovský, J. Buchtele, M. Růžička, The apparent and potential effects of climate change on the inferred concentration of dissolved organic matter in a temperate stream (the Malše River, South Bohemia), Sci. Total Environ. 310 (1−3) (2003) 143−152.

D. Hongve, G. Riise, J.F. Kristiansen, Increased colour and organic acid concentrations in Norwegian forest lakes and drinking water - a result of increased precipitation? Aquat. Sci. 66 (2) (2004) 231−238, https://doi.org/10.1007/s00027-004-0708-7.

D. Houle, C. Gagnon, S. Couture, A. Kemp, Recent recovery of lake water auality in Southern Quebec following reductions in sulfur emissions, Water Air Soil Pollut. 4 (2) (2003) 247−261.

M.N. Imtiazy, A.M. Paterson, S.N. Higgins, H. Yao, S. Couture, J.J. Hudson, Dissolved organic carbon in eastern Canadian lakes: novel patterns and relationships with regional and global factors, Sci. Total Environ. 726 (2020) 138400, https://doi.org/10.1016/j.scitotenv.2020.138400.

D.S. Jeffries, T.G. Brydges, P.J. Dillon, W. Keller, Monitoring the results of Canada/USA acid rain control programs: some lake responses, Environ. Monit. Assess. 88 (1) (2003a) 3−19.

D.S. Jeffries, T.A. Clair, S. Couture, P.J. Dillon, J. Dupont, W.B. Keller, R. Weeber, Assessing the recovery of lakes in southeastern Canada from the effects of acidic deposition, AMBIO A J. Hum. Environ. 32 (3) (2003b) 176−183.

N.M. Kalinkina, E.V. Tekanova, A.V. Sabylina, A.V. Ryzhakov, Changes in the hydrochemical regime of Onego Lake since the early 1990s, Izvestiya Rossiiskoi akademii nauk. Seriya geograficheskaya (1) (2019) 62−72.

N. Kalinkina, E. Tekanova, A. Korosov, M. Zobkov, A. Ryzhakov, What is the extent of water brownification in Lake Onego, Russia? J. Great Lake. Res. 46 (4) (2020) 850–861.
W.B. Keller, J. Heneberry, B.A. Edwards, Recovery of acidified Sudbury, Ontario, Canada, lakes: a multi-decade synthesis and update, Environ. Rev. 27 (1) (2019) 1–16, https://doi.org/10.1139/er-2018-0018.
J. Kopáček, E. Stuchlík, D. Hardekopf, Chemical composition of the tatra mountain lakes: recovery from acidification, Biologia 61 (18) (2006), https://doi.org/10.2478/s11756-006-0117-6.
E.S. Kritzberg, Centennial-long trends of lake browning show major effect of afforestation, Limnol. Oceanography Lett. 2 (4) (2017) 105–112.
E.S. Kritzberg, S.M. Ekström, Increasing iron concentrations in surface waters - a factor behind brownification? Biogeosciences 9 (4) (2012) 1465–1478, https://doi.org/10.5194/bg-9-1465-2012.
E.S. Kritzberg, E.M. Hasselquist, M. Škerlep, S. Löfgren, O. Olsson, J. Stadmark, S. Valinia, L.A. Hansson, H. Laudon, Browning of freshwaters: consequences to ecosystem services, underlying drivers, and potential mitigation measures, Ambio 49 (2) (2020) 375–390, https://doi.org/10.1007/s13280-019-01227-5.
G.B. Lawrence, J.E. Dukett, N. Houck, P. Snyder, S. Capone, Increases in dissolved organic carbon accelerate loss of toxic Al in adirondack lakes recovering from acidification, Environ. Sci. Technol. 47 (13) (2013) 7095–7100, https://doi.org/10.1021/es4004763.
T.H. Leach, L.A. Winslow, N.M. Hayes, K.C. Rose, Decoupled trophic responses to long-term recovery from acidification and associated browning in lakes, Global Change Biol. 25 (5) (2019) 1779–1792, https://doi.org/10.1111/gcb.14580.
J.L.J. Ledesma, S.J. Köhler, M.N. Futter, Long-term dynamics of dissolved organic carbon: implications for drinking water supply, Sci. Total Environ. 432 (2012) 1–11, https://doi.org/10.1016/j.scitotenv.2012.05.071.
J. Lee, P.G. Whitehead, M. Futter, J.W. Hall, Impacts of droughts and acidic deposition on long-term surface water dissolved organic carbon concentrations in upland catchments in Wales, Front. Environ. Sci. 8 (2020) 578611, https://doi.org/10.3389/fenvs.2020.578611.
E. Lipczynska-Kochany, Effect of climate change on humic substances and associated impacts on the quality of surface water and groundwater: a review, Sci. Total Environ. 640 (2018) 1548–1565.
C. Meyer-Jacob, N. Michelutti, A.M. Paterson, B.F. Cumming, W. Keller, J.P. Smol, The browning and re-browning of lakes: divergent lake-water organic carbon trends linked to acid deposition and climate change, Sci. Rep. 9 (1) (2019) 1–10, https://doi.org/10.1038/s41598-019-52912-0.
C. Meyer-Jacob, A.L. Labaj, A.M. Paterson, B.A. Edwards, W. Keller, B.F. Cumming, J.P. Smol, Re-browning of Sudbury (Ontario, Canada) lakes now approaches pre-acid deposition lake-water dissolved organic carbon levels, Sci. Total Environ. 725 (2020) 138347.
D.T. Monteith, J.L. Stoddard, C.D. Evans, H.A. De Wit, M. Forsius, T. Høgåsen, A. Wilander, B.L. Skjelkvåle, D.S. Jeffries, J. Vuorenmaa, B. Keller, J. Kopécek, J. Vesely, Dissolved organic carbon trends resulting from changes in atmospheric deposition chemistry, Nature 450 (7169) (2007) 537–540, https://doi.org/10.1038/nature06316.
C. Neal, S. Lofts, C.D. Evans, B. Reynolds, E. Tipping, M. Neal, Increasing iron concentrations in UK upland waters, Aquat. Geochem. 14 (3) (2008) 263–288, https://doi.org/10.1007/s10498-008-9036-1.

D.J. Redden, B.F. Trueman, D.W. Dunnington, L.E. Anderson, G. Gagnon, Chemical recovery and browning of Nova Scotia surface waters in response to declining acid deposition, Environ. Sci. Process. Impacts (2021) 12–14, https://doi.org/10.1039/d0em00425a.

G. Riise, R.A. Müller, S. Haaland, G.A. Weyhenmeyer, Acid rain — a strong external driver that has suppressed water colour variability between lakes, Boreal Environ. Res. 23 (February) (2018) 69–81.

J.P. Ritson, N.J.D. Graham, M.R. Templeton, J.M. Clark, R. Gough, C. Freeman, The impact of climate change on the treatability of dissolved organic matter (DOM) in upland water supplies: a UK perspective, Sci. Total Environ. 473–474 (2014) 714–730, https://doi.org/10.1016/j.scitotenv.2013.12.095.

M.D. SanClements, G.P. Oelsner, D.M. McKnight, J.L. Stoddard, S.J. Nelson, New insights into the source of decadal increases of dissolved organic matter in acid-sensitive lakes of the Northeastern United States, Environ. Sci. Technol. 46 (6) (2012) 3212–3219, https://doi.org/10.1021/es204321x.

D.W. Schindler, Effects of acid rain on freshwater ecosystems, Science 239 (4836) (1988) 149–157.

B.L. Skjelkvåle, R.F. Wright, Mountain Lakes; Sensitivity to Acid Deposition and Global Climate Change, Ambio, 1998, pp. 280–286.

B.L. Skjelkvåle, J. Mannio, A. Wilander, T. Andersen, Recovery from acidification of lakes in Finland, Norway and Sweden 1990–1999, Hydrol. Earth Syst. Sci. 5 (3) (2001) 327–338, https://doi.org/10.5194/hess-5-327-2001.

B.L. Skjelkvåle, J.L. Stoddard, D.S. Jeffries, K. Tørseth, T. Høgåsen, J. Bowman, J. Mannio, D.T. Monteith, R. Mosello, M. Rogora, D. Rzychon, J. Vesely, J. Wieting, A. Wilander, A. Worsztynowicz, Regional scale evidence for improvements in surface water chemistry 1990-2001, Environ. Pollut. 137 (1) (2005) 165–176, https://doi.org/10.1016/j.envpol.2004.12.023.

J.L. Stoddard, D.S. Jeffries, A. Lükewille, T.A. Clair, P.J. Dillon, C.T. Driscoll, M. Forsius, M. Johannessen, J.S. Kahl, J.H. Kellogg, A. Kemp, J. Mannlo, D.T. Monteith, P.S. Murdoch, S. Patrick, A. Rebsdorl, B.L. Skjelkvale, M.P. Stainton, T. Traaen, A. Wllander, Regional trends in aquatic recovery from acidification in North America and Europe, Nature 401 (6753) (1999) 575–578, https://doi.org/10.1038/44114.

J.L. Stoddard, J.S. Kahl, F.A. Deviney, D.R. Dewalle, C.T. Driscoll, A.T. Herlihy, J.H. Kellogg, P.S. Murdoch, J.R. Webb, K.E. Webster, Response of Surface Water Chemistry to the Clean Air Act Amendments of 1990, United States Environmental Protection Agency, 2002, pp. 1–7.

J.L. Stoddard, J.S. Kahl, F.A. Deviney, D.R. DeWalle, C.T. Driscoll, A.T. Herlihy, K.E. Webster, Response of Surface Water Chemistry to the Clean Air Act Amendments of 1990, US Environmental Protection Agency, Research Triangle Park (NC), 2003.

K.E. Strock, S.J. Nelson, J.S. Kahl, J.E. Saros, W.H. McDowell, Decadal trends reveal recent acceleration in the rate of recovery from acidification in the northeastern U.S, Environ. Sci. Technol. 48 (9) (2014) 4681–4689, https://doi.org/10.1021/es404772n.

K.E. Strock, J.E. Saros, S.J. Nelson, S.D. Birkel, J.S. Kahl, W.H. McDowell, Extreme weather years drive episodic changes in lake chemistry: implications for recovery from sulfate deposition and long-term trends in dissolved organic carbon, Biogeochemistry 127 (2–3) (2016) 353–365, https://doi.org/10.1007/s10533-016-0185-9.

B.F. Trueman, T. Anaviapik-Soucie, V. L'Hérault, G.A. Gagnon, Characterizing colloidal metals in drinking water by field flow fractionation, Environ. Sci. Water Res. Technol. 5 (12) (2019) 2202–2209.

M. Valdivia-Garcia, P. Weir, D.W. Graham, D. Werner, Predicted impact of climate change on trihalomethanes formation in drinking water treatment, Sci. Rep. 9 (1) (2019) 1–10.

K. Waller, C. Driscoll, J. Lynch, D. Newcomb, K. Roy, Long-term recovery of lakes in the Adirondack region of New York to decreases in acidic deposition, Atmos. Environ. 46 (2012) 56–64.

W. Wang, B. Zheng, X. Jiang, J. Chen, S. Wang, Characteristics and source of dissolved organic matter in lake hulun, A large shallow eutrophic steppe lake in northern China, Water 12 (4) (2020) 953.

S.A. Watmough, M.C. Eimers, Rapid recent recovery from acidic deposition in central Ontario lakes, Soil Syst. 4 (1) (2020) 10, https://doi.org/10.3390/soilsystems4010010.

S.A. Watmough, C. Eimers, S. Baker, Impediments to recovery from acid deposition, Atmos. Environ. 146 (2016) 15–27.

W.D. Watt, A summary of the impact of acid rain on Atlantic salmon (*Salmo salar*) in Canada, Water Air Soil Pollut. 35 (1) (1987) 27–35.

G.A. Weyhenmeyer, R.A. Müller, M. Norman, L.J. Tranvik, Sensitivity of freshwaters to browning in response to future climate change, Climatic Change 134 (1–2) (2016) 225–239, https://doi.org/10.1007/s10584-015-1514-z.

L.D. Winning, B. Gorczyca, K. Brezinski, Effect of total organic carbon and aquatic humic substances on the occurrence of lead at the tap, Water Qual. Res. J. 52 (1) (2015) 2–10.

R.F. Wright, T. Larssen, L. Camarero, B.J. Cosby, R.C. Ferrier, R. Helliwell, W. Schöpp, Recovery of acidified European surface waters, Environ. Sci. Technol. 39 (3) (2005) 64A–72A.

Y. Xiao, T. Rohrlack, G. Riise, Unraveling long-term changes in lake color based on optical properties of lake sediment, Sci. Total Environ. 699 (2020) 134388, https://doi.org/10.1016/j.scitotenv.2019.134388.

J. Xu, P.J. Morris, J. Liu, J.L. Ledesma, J. Holden, Increased dissolved organic carbon concentrations in peat-fed UK water supplies under future climate and sulfate deposition scenarios, Water Resour. Res. 56 (1) (2020) e2019WR025592.

L.J. Zhang, S. Zhang, B. Sun, S.N. Zhao, L.E. Liang, Spatial and temporal variation of water quality in Hulun Lake, J. Anhui Agric. Univ. 43 (2016) 568–575.

CHAPTER 3

Characterization of natural organic matter

Yuri Park[1], Mika Sillanpää[2,3,4,5], Lindsay Anderson[6], Graham A. Gagnon[6] and Anu Matilainen[7]

[1]Institute of Environmental Technology, Department of Environmental Engineering, Seoul National University of Science and Technology, Seoul, South Korea; [2]Department of Chemical Engineering, School of Mining, Metallurgy and Chemical Engineering, University of Johannesburg, Johannesburg, South Africa; [3]Chemistry Department, College of Science, King Saud University, Riyadh, Saudi Arabia; [4]Department of Applied Physics, Faculty of Science and Technology, Universiti Kebangsaan Malaysia, Bangi, Selangor, Malaysia; [5]International Research Centre of Nanotechnology for Himalayan Sustainability (IRCNHS), Shoolini University, Solan, Himachal Pradesh, India; [6]Center of Water Resource Studies, Dalhousie University, Halifax, NS, Canada; [7]Finnish Safety and Chemicals Agency, Tampere, Finland

Abstract

Worldwide reports over the last few decades have shown that the amount of natural organic matter (NOM) in surface water is continuously increasing, which has an adverse effect on drinking water purification. For many practical and hygienic reasons, the presence of NOM in drinking water is undesirable. Various technologies have been proposed for NOM removal with varying degrees of success. The properties and amount of NOM, however, can significantly affect the process efficiency. To improve and optimize these processes, it is essential to characterize and quantify NOM at various points during purification and treatment. It is also important to be able to understand and predict the reactivity of NOM or its fractions at different stages of the process. Methods used in the characterization of NOM include resin adsorption, size exclusion chromatography (SEC), nuclear magnetic resonance (NMR) spectroscopy, and fluorescence spectroscopy. The NOM in water has been quantified with parameters including ultraviolet and visible, total organic carbon, and specific UV-absorbance. More comprehensive analytical methods for determining NOM structures have been developed recently: liquid chromatography-mass spectrometry (LC-MS), pyrolysis gas chromatography-mass spectrometry (Py-GC-MS), multidimensional NMR techniques, and Fourier transform ion cyclotron resonance mass spectrometry (FTICR-MS). This chapter focuses on the methods used for the characterization and quantification of NOM in relation to drinking water treatment.

Keywords: Characterization; Chromatography; Fractionation; FTICR-MS; Natural organic matter (NOM); NMR; Spectroscopy.

Natural Organic Matter in Water
ISBN 978-0-12-824274-2
https://doi.org/10.1016/B978-0-12-824274-2.00002-8

Abbreviations

AFM	Atomic force microscopy
AOC	Assimilable organic carbon
AOM	Algogenic organic matter
AOPs	Advanced oxidation processes
ATR	Attenuated total reflectance
BDOC	Biodegradable dissolved organic carbon
COD	Chemical oxygen demand
DBP	Disinfection by-product
DBPFP	Disinfection by-product formation potential
DOC	Dissolved organic carbon
EEM	Excitation-emission matrix
ESI	Electron spray ionization
FA	Fulvic acid
FIFFF	Flow field-flow fractionation
FTICR-MS	Fourier transform ion cyclotron resonance-mass spectrometry
FTIR	Fourier transform infrared
GAC	Granular activated carbon
GC	Gas chromatography
HA	Humic acid
HMBC	Heteronuclear multiple bond correlation
HMW	High molecular weight
HPLC	High-performance liquid chromatography
HPSEC	High-performance size exclusion chromatography
HR-MAS	High-resolution magic-angle spinning
ICP AES	Inductively coupled plasma atomic emission spectroscopy
IR	Infrared
LC-MS	Liquid chromatography-mass spectrometry
LC-OCD	Liquid chromatography-organic carbon detection
MM	Molar mass
MWD	Molecular weight distribution
NMR	Nuclear magnetic resonance
OCD	Organic carbon detection
PARAFAC	Parallel factor
PSS	Polystyrene sulfonate
Py-GC-MS	Pyrolysis gas chromatography-mass spectrometry
RO	Reverse osmosis
RPHPLC	Reversed-phase high performance liquid chromatography
SEC	Size exclusion chromatography
SEM	Scanning electron microscopy
SHA	Slightly hydrophobic acid
SUVA	Specific UV-absorbance
TEM	Transmission electron microscopy
THM	Trihalomethane
TOC	Total organic carbon
UV-Vis	ultraviolet and visible
VHA	Very hydrophobic acid

3.1 Introduction

Natural organic matter (NOM) is a complex mixture of organic compounds. Some of this organic matter is negatively charged, and it can possess a wide variety of chemical compositions and molecular sizes (Thurman, 1985; Swietlik et al., 2004).

All disinfection methods (chlorine, ozone, chlorine dioxide, chloramines, and UV radiation) reportedly produce their own suite of disinfection by-products (DBPs) and bioactive compounds in drinking water (Richardson et al., 2007). Present knowledge and experience show that the hydrophobic and high molecular weight (HMW) compounds of NOM are the most significant precursors to DBP formation (Hua and Reckhow, 2007; Chen et al., 2008). Hydrophilic organic matter, which can generally be poorly removed during conventional water treatment processes may also play a key role in the formation of new compounds during disinfection, especially in waters with low humic components.

More efficient removal of NOM requires more knowledge of the organic matter present in raw water, and novel methods have been developed for the characterization of these organic compounds. At the same time, existing methods and techniques have been improved. These characterization methods are used to study the composition of NOM prior to treatment and during various stages of the treatment process (Chen et al., 2007; Sarathy and Mohseni, 2007; Her et al., 2008a; Liu et al., 2008, 2010; Tercero Espinoza et al., 2009; Zhao et al., 2009). The diversity of molecules that constitute NOM and the relatively low concentrations of NOM in water often make characterization difficult. There is, therefore, a significant need for methods that can either accurately characterize NOM in these dilute solutions or isolated or concentrated NOM.

The NOM in raw water must be characterized to understand its role in water treatment (Matilainen et al., 2011). Ideally, once the various NOM components and fractions of a raw water source have been determined, the treatment processes that will eliminate the most dominant NOM fractions could be selected. Yet, NOM contains literally thousands of chemical constituents. Thus, it is not realistic to characterize NOM on the basis of individual components, as discussed in the introduction, but it is more practical to identify groups of chemicals with similar properties (i.e., hydrophobicity, molecular weight, etc.) (Croue et al., 2000).

An alternative approach to the characterization of NOM is to study how it reacts to DBP formation and the occurrence of different DBPs in drinking water (Culea et al., 2006; Kanokkantapong et al., 2006b; Chen et al., 2008; Cooper et al., 2008; Richardson et al., 2008; Blodau et al., 2009). The binding potential of NOM with inorganic and organic micropollutants may also be relevant to drinking water treatment (Gjessing et al., 2007; Laborda et al., 2009; Park, 2009).

Different unit processes remove different NOM fractions during water treatment (Haarhoff et al., 2010). Therefore, more sophisticated characterization techniques are required (Matilainen et al., 2011). Several complementary methods can provide definitive structural or functional information about NOM, and have been found to correlate well, allowing for comprehensive characterization (Jaouadi et al., 2012; Penru et al., 2013). These analytical methods also provide better information for process design and optimization than the conventional dissolved organic carbon (DOC) measurements.

DOC, chemical oxygen demand (COD), UV_{254}, pH, turbidity, and color are common water quality parameters assessed by water treatment facilities in their quality control. Assessment of these parameters does not require sophisticated equipment, is simple and fast to perform, and can be automated. Yet, such assessments offer no information about the characteristics of NOM such as molar mass (MM) or hydrophobicity.

Traditionally, humic substances have been classified into three categories based on their solubility: humic acid (HA), fulvic acid (FA), and humin. This is an operational division; however, the terms do not refer to single compounds but to a wide range of compounds of similar origin (Uyguner-Demirel and Bekbolet, 2011). NOM may be characterized by high performance size exclusion chromatography (HPSEC) analysis, which determines the molecular weight distribution (MWD) of NOM (Chow et al., 2008; Korshin et al., 2009), or by fractionation techniques that use resins to divide the mixture of organic compounds of NOM into hydrophilic and hydrophobic fractions (Sharp et al., 2006a,b). Although these parameters provide useful information regarding the change in organic characteristics during treatment and impact on DBP formation, it has been suggested that their use alone may not be efficient for predicting DBP formation (Sanly et al., 2007). Recently, excitation-emission matrix (EEM) spectroscopy has shown promise for predicting DBP formation (Trueman, 2018).

The hydrophilic fractions of NOM are composed mostly of aliphatic carbon nitrogenous compounds such as carboxylic acids, carbohydrates, and proteins. Hydrophobic NOM primarily consists of HA and FA (humic substances) and is rich in aromatic carbon, phenolic structures, and conjugated double bonds (Swietlik et al., 2004). Humic substances generally account for over half of the total organic carbon (TOC) content in water (Thurman, 1985). Humic substances can be regarded as natural anionic polyelectrolytes, with an anionic charge at pH values higher than 4. Another way to define hydrophobicity is to determine the specific UV-absorbance (SUVA) value (i.e., UV_{254} absorbance divided by the TOC concentration). A high SUVA value indicates that the organic matter is largely composed of hydrophobic, high MW organic material. A low SUVA value indicates that the water contains organic compounds that are mainly hydrophilic, with a low MW and charge density (Sharp et al., 2006a, b; Edzwald and Tobiason, 1999). Due to their diverse properties, different NOM fractions respond differently to treatment by coagulation, coagulant demand, chlorine and ozone reactivity, and disinfection by-product formation potential (DBPFP) (Fabris et al., 2008; Sharp et al., 2006 a,b; Archer and Singer, 2006).

3.2 General parameters

In practice, NOM is usually characterized to understand its physico-chemical characteristics by the measurement of its TOC, DOC, absorption of UV light (UV_{254}), or COD. NOM is also the major contributor to the brownish-yellow color in water; thus, measurement of color can provide some indication of the amount of NOM that the water contains (Uyguner et al., 2007). All these tests are fast and do not require sophisticated sample pretreatment or analytical equipment. However, these analyses provide information mostly about the quantity of NOM while offering limited information on the characteristics of NOM.

3.2.1 Total organic carbon/dissolved organic carbon

TOC is the sum of the particulate and DOC when existing inorganic carbon is removed by acidification. A widely accepted operational definition of DOC is the organic carbon in the water sample filtered through a 0.45 μm filter (Danielsson, 1982). TOC and DOC are the most convenient parameters for analyzing the NOM removal of treatment processes. Essentially, all methods for quantifying organic carbon in water involve oxidation. Before TOC analyzers, an oxidizing agent such as potassium permanganate ($KMnO_4$) was added to the water and to indirectly measure the overall amount of organic compounds in water, indicating the COD index (mg L^{-1}). Currently, off-line COD test kits with potassium dichromate ($K_2Cr_2O_7$) as oxidizing agent are the most popular and practical for routine applications. In the recent years, some new green COD methods such as photoelectrochemical oxygen demand (peCOD) have been developed which use TiO_2 and no toxic reagents. pPeCOD offers rapid measurement (i.e., <15 min) of COD which is correlated with NOM and organic compounds (Stoddart and Gagnon, 2014). The peCOD measurement has also shown to have higher resolution compared to TOC (Stoddart and Gagnon, 2014). A variety of oxidation techniques and instruments are currently in use, including burning, radiation, and oxidizing agents. The resulting CO_2 is measured mostly by infrared spectroscopy (IR). TOC and DOC measurements alone do not provide much insight into the behavior of NOM during water treatment (Haarhoff et al., 2010).

3.2.2 Specific UV-absorbance

Specific UV-absorbance (SUVA) is defined as the UV absorbance of a given sample at 254 nm divided by the DOC concentration of the sample. This ratio describes the hydrophobicity and hydrophilicity of NOM in water; a SUVA > 4 indicates mainly hydrophobic and especially aromatic material, whereas a SUVA < 2 illustrates mainly hydrophilic material (Edzwald and Tobiason, 1999). Similarly, $SUVA_{365}$ and $SUVA_{436}$ specific absorbance parameters are associated with increasing molecular size and chromophoric groups in NOM, respectively (Valencia et al., 2013). Several researchers have emphasized that NOM removal by coagulation is likely to have a high SUVA and absorption ratios at low wavelength (A_{254}/A_{204}) correlate relatively well because both are associated with the amount of aromatic carbon present in NOM (Hur et al., 2006; Al-Juboori et al., 2016a,b). Jung and Son (2008) also reported a clear correlation between the SUVA value and trihalomethane (THM) formation potential. However, the latter relationship has not been confirmed by others (Ates et al., 2007; Wei et al., 2008). Methods used to characterize different features of NOM are presented in Table 3.1. On the other hand, the absorption ratios at relatively long wavelengths (>600 nm) provides poor signal to noise ratio, which is not recommended to use.

3.2.3 Polarity

Polarity of NOM is another important parameter to measure of the change in chemical characteristics in drinking water treatment processes. Conventionally, NOM polarity can be accessed by fractionation with

Table 3.1 Methods used to characterize different features of natural organic matter (NOM), with their advantages and disadvantages.

Bulk parameters				
Methods	**Detected features**	**Advantages**	**Disadvantages**	**References**
Total Organic Carbon (TOC) Dissolved Organic Carbon (DOC)	Total organic carbon content in water Dissolved organic carbon in water, filtered through a 0.45 μm filter	Easy to use; analytical equipment relatively inexpensive; can be used as on-line method	Only offers information on the quantity of NOM but limited to access the quality of NOM	Koprivnjak et al. (2006), Liu et al. (2007), Seredynska-Sobecka et al. (2007), and Spencer et al. (2007)
Specific UV-Absorbance (SUVA)	UV-absorption $A_{254nm\ (1cm)} mgC^{-1}L^{-1}$; High SUVA value > 4 refers to hydrophobic, aromatic compounds SUVA value < 3 indicates mainly hydrophilic material	Easy to determine; simple instrumentation	High nitrate content in low-DOC waters may interfere with measurement	Edzwald and Tobiason (1999), Archer and Singer (2006), Bose and Reckhow (2007), Ates et al. (2007), Jung and Son (2008), and Wei et al. (2008)

macroeticular resins (Thurman and Malcolm, 1981) and the NOM polarity can be recently evaluated by a new technique (i.e., the polarity rapid assessment method, PRAM), which was developed by Rosario-Oriz (2007a, b). This technique quantifies the amount of material adsorbed onto various solid-phase extraction (SPE) sorbents under ambient conditions without pretreatment steps and allows to measure the polarity of NOM as it exists in the environment (Pan et al., 2016).

3.3 Biological tests

Several biological tests have been developed to assess the level of biodegradable organic matter in water. These bioassays are generally based on two concepts: (1) the assimilable organic carbon (AOC), which measures the growth of bacteria inoculum in response to the amount of nutrients in the water and (2) the biodegradable dissolved organic carbon (BDOC), which measures the fraction of DOC assimilated and mineralized by heterotrophic microbes (Escobar and Randall, 2001).

AOC is believed to be composed mostly of LMW compounds, which are the most difficult to remove. These are also readily degradable, thus indicating the potential for bacterial growth in the distribution system (Escobar and Randall, 2001; Kang et al., 2006; Matilainen et al., 2006; Chen et al., 2007; Chow et al., 2009a). BDOC methods have been developed and used to determine the biological stability of raw and treated water (Escobar and Randall, 2001; Chen et al., 2007; Soh et al., 2008; Zappia et al., 2008; Yapsakli and Cecen, 2009; Haarhoff et al., 2010). BDOC fractionation provides further insight into existing oxidation and biofiltration processes (Haarhoff et al., 2010).

Another biological aspect of organic matter that is gaining considerable attention is algogenic organic matter (AOM), particularly due to the rise of eutrophication of aquatic environments and promulgation of harmful algal blooms. AOM is considered to have a significant impact on NOM composition and is released into water as a result of algal and cyanobacterial photosynthesis and secondary metabolism (Pivokonsky et al., 2015), and is known to have a substantial impact on DBP formation (Hua et al., 2019). AOM is comprised of extracellular organic matter (EOM) and cellular organic matter (COM). An entire review was devoted to AOM and its impact on treatment plant operation and water quality (Pivokonsky et al., 2015). In a review article by Hua et al. (2019), the character of AOM is generally quantified via the organic carbon per organic nitrogen content

(DOC/DON ratio), hydrophilicity, MW fractionation, and optical properties (UV absorbance and fluorescence). In addition, Her et al. (2004) also noted HPSEC-UV-fluorescence-DOC, FTIR and UV absorbance ratio index (URI) as means to analyze AOM. More detail on some of these measures are provided in the sections below.

3.4 Isolation and concentration

Isolation of NOM can increase ionic strength and concentration levels and a variety of NOM isolation and concentration methods are currently available, of which the most commonly used are reverse osmosis (RO), evaporation under reduced pressure (vacuum distillation). Freeze-dry, solid-phase extraction (SPE), and dialysis are often employed (Moody, 2020). Some sorption methods exist (Aiken, 1985; Aiken et al., 1992; Croue, 2004). The disadvantage of most of these techniques is that they also concentrate the salts that are always present in natural waters (Croue, 2004). RO has been used in many NOM isolation studies (Croue, 2004; Koprivnjak et al., 2006; Song et al., 2009), since it has a relatively small alternation effect on the physicochemical properties of NOM (Song et al., 2009; Alasonati et al., 2010). In the RO method, the water is passed through a cation exchanger (mostly Na- or H- saturated) prior to RO filtration. Recovery of NOM from waters with a high SUVA value (mass recoveries 80%–99%) is more efficient than for waters with low SUVA. Recovery is also affected by the ambient pH (Song et al., 2009).

3.5 Fractionation

NOM molecules are generally unique but share many common properties. One of the major analytical difficulties in the characterization of aqueous NOM is the lack of a suitable fractionation procedure to yield pure components that can subsequently be characterized by standard techniques (Leenher, 1985). To characterize NOM effectively, it must first be isolated and fractionated into separate groups of molecules possessing properties. Commonly used chemical fractionation methods include precipitation, solvent extraction, and adsorption chromatography. Available physical fractionation methods include electrophoresis, ultrafiltration, size exclusion chromatography (SEC), reversed-phase high performance liquid chromatography (RPHPLC), field-flow fractionation (FFF), and ultracentrifugation

(Leenher, 1985; Leenher and Croue, 2003; Chow et al., 2005; Matilainen et al., 2011).

3.5.1 Resin and membrane fractionation

3.5.1.1 Resin fractionation

The most common way of distinguishing between hydrophobic and hydrophilic NOM is to determine whether or not it is adsorbed onto Amberlite XAD resin. This is a commercially available synthetic resin introduced by (Malcolm and MacCarthy, 1992) at the U.S. Geological Survey in Denver for the isolation humic substances from water. Many researchers have developed and modified resin adsorption methods (Leenher, 1981; Thurman and Malcolm, 1981; Malcolm and MacCarthy, 1992; Marhaba et al., 2003), and these are still used in NOM fractionation studies (Swietlik et al., 2004; Kanokkantapong et al., 2006a,b; Sharp et al., 2006a, Her et al., 2004; Jung and Son, 2008; Wei et al., 2008; Zhang et al., 2008). This technique was adopted by the International Humic Substances Society (IHSS) as a standard method for isolating FA and HA. Yet, FA and HA do not correspond to Svein Oden's classical definition (Oden, 1919). The XAD method involves the Amberlite resins XAD-8 and XAD-4. The XAD-8 resin favors the isolation of hydrophobic, HMW, and aromatic NOM. The hydrophilic fraction does not adsorb onto either of the resins. XAD-4 resin adsorbs weakly hydrophobic acid fractions, commonly defined as transphilic NOM (Sharp et al., 2006a; Croue, 2004; Wei et al., 2008). The IHSS Standard Method for FA is based on the matter adsorbed onto XAD-8 after reducing the pH to 2 and prefiltering with a 0.45 μm filter. The hydrophilic fraction can be separated further, for example, AG-MP-50 cationic resin and WA-10 a weak anionic resin (Marhaba et al., 2003).

The rapid resin fraction technique, based on XADs (Chow et al., 2004), separates DOC into four fractions based on character and molecular weight (MW). Fractions are defined as very hydrophobic acids (VHA), slightly hydrophobic acids (SHA), charged hydrophilics, and neutral hydrophilics (NEU). According to Chow et al. (2004), VHA is composed of HMW HAs, whereas FAs constitute SHA, and the NEU fraction containing substances that do not adsorb onto any of the applied resins. A variety of studies use this method (Fabris et al., 2008; Liu et al., 2008, 2010; Soh et al., 2008). A new fractionation technique based on NOM polarity in combination with an XAD-8 adsorption chromatography technique involving gradient elution has also been shown to separate different hydrophobic and hydrophilic moieties (Lie et al., 2009). Another rapid technique by Ratpukdi et al. (2009) uses SPE cartridges (Bond Elute ENV, Strata XC and Strata XAW), as little as 1 L sample volume and takes approximately 6 h. It has also been demonstrated that the rapid SPE method (Ratpukdi et al., 2009) can provide comparable results to the method described by Marhaba et al. (2003).

Although fractionation methods based on various resins have been extensively used and accepted, they still have disadvantages. These discrepancies include possible chemical or physical alternations of NOM due to extreme pH levels and changes in pH during fractionation, irreversible adsorption of NOM compounds onto the resin, contamination from resin bleeding, and size-exclusion effects. Different operational conditions during fractionation may also influence the results (Song et al., 2009). Further, many of the fractionation methods can be time intensive which is a drawback when considering their practicality as a way to inform water treatment decision making. The proposed composition of NOM fractions separated by fractionation are presented in Table 3.2.

3.5.1.2 Membrane filtration

Filtration is prevalent in water research. The current international definition of dissolved matter in water is "the matter that passes a filter with a cut-off of 0.45 μm" and RO is the most used research method for isolating NOM from water. As early as 1970, it was demonstrated that NOM (aquatic humus) could be fractionated by membranes with different cut-off ranges (Gjessing, 1970). Further development of this technique confirms that NOM is a mixture of organic matter with different sized molecules (Aiken, 1984; Liu et al., 2007; Lankes et al., 2008; Wei et al., 2008; Zhao et al., 2009).

Separation by membranes may not only be dependent on the size of NOM, but also on its molecular structures (Assemi et al., 2004). Methodological conditions also influence the separation of molecules according to size and shape by membrane filtration. One problem associated with this method is the adherence of macromolecules to the sides of membrane pores, affecting permeability. Results may also be affected by flow restriction, also known as concentration polarization, in which the deposition of macromolecules results in a gel layer that becomes the dominant resistance reasonably well with SEC in terms of size and MW (Gjessing, 1973). It should be emphasized, however, that the reproducibility of the characterization method is more important than accuracy regarding size and shape.

Table 3.2 Proposed compositions of natural organic matter (NOM) fractions separated using fractionation techniques.

Fraction	Organic compound class	References
Humic acid	Portion of humic substances precipitated at pH 1	Peuravuori et al. (1997)
Hydrophobic acid	Soil fulvic acids, C5−C9 aliphatic carboxulic acids, 1- and 2-ring aromatic carboxylic acids, 1- and 2-ring phenols	Leenher (1981), Aiken et al. (1992), Marhaba et al. (2000), and Barber et al. (2001)
Hydrophobic base	1-and 2-ring aromatics (except pyridine), proteinaceous substances	Leenher (1981), Marhaba et al. (2000), and Barber et al. (2001)
Hydrophobic neutral	Mixture of hydrocarbons, >C5 aliphatic alcohols, amides, aldehydes, ketones, esters, >C9 aliphatic carboxylic acids and amines, >3-ring aromatic carboxylic acids and amines	Leenher (1981), Marhaba et al. (2000), and Barber et al. (2001)
Hydrophilic acid	Mixtures of hydroxy acids, <C5 aliphatic carboxylic acids, polyfunctional carboxylic acids	Leenher (1981), Aiken et al. (1992), Marhaba et al. (2000), and Barber et al. (2001)
Hydrophilic base	Pyridine, amphoteric proteinaceous material (i.e., aliphatic amino acids, amino sugars, <C9 aliphatic amines, peptides, and proteins)	Leenher (1981), Marhaba et al. (2000), and Barber et al. (2001)
Hydrophilic neutral	<C5 aliphatic alcohols, polyfunctional alcohols short-chain aliphatic amines, amides, aldehydes, ketones, esters, cyclid amides, polysaccharides, and carbohydrates	Leenher (1981), Marhaba et al. (2000), Barber et al. (2001)

Adapted from Swietlik, J., Dabrowska, A., Raczyk-Stanislawiak, U., Nawrocki, J., 2004. Reactivity of natural organic matter fractions with chlorine dioxide and ozone. Water Res. 38, 547−558.

3.6 Spectroscopic methods

3.6.1 Ultraviolet and visible

UV-visible absorption spectroscopy is a semiquantitative method to determine humic substances in natural waters. It is based on the measurement of the attenuation of a beam of light after it passes through a sample or after its reflection from a sample surface. The concentration of an analyte in a solution can be determined by measuring the absorbance at a certain wavelength, applying the Beer—Lambert law. Absorption measurements can be taken at a single wavelength or over an extended spectral range. The UV absorbance spectra of NOM decreases monotonically with increasing wavelength (Hur et al., 2006). Any wavelength from 220 to 280 nm is appropriate for NOM measurements, although the molar absorptivity will vary due to the range of chromophores in the structure of NOM. The different wavelengths are believed to identify different chromophores: absorbance at 220 nm is associated with both the carboxylic and aromatic chromophores, whereas absorbance at 254 nm is typical for aromatic groups with varying degrees of activation (Korshin et al., 2009). UV_{254} has been identified as a potential surrogate measure for DOC despite its tendency to represent only the aromatic character. Yet, due to the presence of a high number of overlapping groups, humic substances display featureless and monotonic absorbance spectra (Uyguner-Demirel and Bekbolet, 2011 and references therein).

The ratio of absorbance/absorption coefficients at two wavelengths, such as A_{250}/A_{365}, A_{265}/A_{465}, A_{465}/A_{665}, A_{254}/A_{204}, A_{280}/A_{350}, or A_{254}/A_{436}, can be used to monitor the chemical compositions and properties of NOM including the molecular weight distribution, aromaticity, polarity, humification, and hydrophobicity of NOM (Hur et al., 2006; Spencer et al., 2007; Li et al., 2009; Rodriguez et al., 2016). For instance, the ratios of A_{280}/A_{350} and A_{250}/A_{365} along with $SUVA_{254}$, $SUVA_{280}$ show the best correlations with aromaticity while the best correlations with molecular weight are for A_{280}/A_{350} ratio, $SUVA_{254}$, and $SUVA_{280}$ (Rodriguez et al., 2016). The ratio of A_{254}/A_{203} correlates with the formation of DBPs (Kim and Yu, 2007; Al-Juboori et al., 2016b). Yet, there is a significant possibility of errors due to interference at 200 nm through 230 nm by inorganic ions such as nitrates or sulfates (Her et al., 2008a, b).

3.6.2 Differential absorption

An alternative approach is to measure differential absorbance, defined as the change in the absorbance of a sample in response to any forcing function, such as chlorination (Korshini et al., 1999; Ates et al., 2007). Differential absorbance spectroscopy (DAS) can detect subtle changes in NOM structure (Uyguner-Demirel and Bekbolet, 2011 and references therein, Chen et al., 2018). The differential absorbance at 272 nm has been proposed for the evaluation of DBP formation. This differential absorbance correlates strongly with total THM and HAA concentrations (Roccaro and Vagliasindi, 2009). However, differential absorbance at 254 nm (ΔA_{254}) has no correlation with total THM and HAA formations (Ates et al., 2007).

3.6.3 Fluorescence

Fluorescence spectroscopy excites the analyte molecules by irradiation at a certain wavelength, and the emitted radiation is measured at another wavelength. This method has caught the attention of the water industry due to its potential as a monitoring technique: recent research has shown its applicability for long-term monitoring (Sanchez et al., 2013). Fluorescence can provide insight into the chemical characteristics of NOM, because it is a function of molecular structure and functional groups (Hudson et al., 2007). The main advantages of fluorescence over traditional methods such as UV-Vis are increased sensitivity and selectivity (Bieroza et al., 2009; Peiris et al., 2010). The method is also simple, relatively inexpensive, and requires little sample pretreatment (Baghoth et al., 2011; Carstea et al., 2020). Yet, fluorescence measurements are dependent on environmental factors such as pH, ionic strengths, and temperature (Markechova et al., 2013). A particular molecular conformation called fluorophore is characterized by specific excitation and emission wavelengths. These fluorophores are helpful in describing the structural composition of humic materials (Datta et al., 1971; Uyguner et al., 2007; Wu et al., 2007b; Baker et al., 2008; Zhang et al., 2008; Bieroza et al., 2009). Fluorescence has also been recommended as a tool for estimating the presence of biodegradable NOM, since strong correlations between fluorescence and biological oxygen demand values have been demonstrated (Hudson et al., 2008).

In the last decade, three-dimensional fluorescence excitation-emission matrix (EEM) spectrophotometry has been increasingly studied for the assessment of NOM (Carstea et al., 2020). An entire review was recently devoted to EEM and parallel factor (PARAFAC) drinking water treatment

(Markechova et al., 2013). EEM is a 3D spectrum in which fluorescence intensity can be presented as a function of excitation and emission wavelength (Valencia et al., 2014; Park et al., 2021). The EEM spectrum visualizes a range of different fluorophores with an excitation and emission wavelength ranging from ~200 to ~500 nm, which is more revealing than the traditional single-scan technique (Spencer et al., 2007). EEM fluorescence provides valuable information on the removal of different NOM fractions in water treatment (Jeong et al., 2013; Lee et al., 2013; Sanchez et al., 2014; Valencia et al., 2014). Chen et al. (2003) classified NOM into five different fractions to interpret EEM fluorescence spectra: aromatic proteins (regions I and II); fulvic acid-like and humic acid-like compounds (regions III and V); and microbial by-products (region IV). Typical raw water may contain two major fluorescence peaks, described as humic-like and protein-like fluorescence maxima (Wu et al., 2007b; Baghoth et al., 2009) or three peaks referred to as tryptophan-like, fulvic-like, and humic-like fluorophores (Liu et al., 2007; Seredynska-Sobecka et al., 2007; Spencer et al., 2007; Baker et al., 2008; Hudson et al., 2008). The fluorescence intensity peaks for different NOM fractions of natural waters are listed in Table 3.3 (Coble, 1996; Leenher and Croue,

Table 3.3 Major fluorescence peaks for water samples.

Range of excitation	Range of emission	Component type	References
270–280 nm	310–320 nm	Tyrosine-like, protein-like	Coble (1996) and Baghoth et al. (2009)
270–285 (220–235) nm	340–360 nm	Tryptophan-like, protein-like	Coble (1996), Spencer et al. (2007), Baker et al. (2008), and Baghoth et al. (2009)
320–350 nm	400–450 nm	Fulvic-like	Spencer et al. (2007) and Baker et al. (2008)
310–320 nm	380–420 nm	Humic-like (Marine humic-like)	Coble (1996) and Baghoth et al. (2009)
320–390 nm	410–500 nm	Humic-like	Coble (1996), Stedmon et al. (2003), Stedmon and Markager (2005), Spencer et al. (2007), Murphy et al. (2008), Yamashita and Jaffe (2008), and Baghoth et al. (2009)

2003; Spencer et al., 2007; Baghoth et al., 2009). To interpret the fluorescence data in order to access the structural information of NOM deeper, various methods have been developed. Marhaba (2000) developed the spectral fluorescent signature technique to model these characteristics of NOM. Also, peak-picking, fluorescence regional integration (FRI), parallel factor (PARAFAC), multivariate data analysis such as principal component analysis (PCA), self-organizing map analysis, 2DCOS, and prior linear decomposition (PLD) are widely used for the evaluation of the NOM characteristics in various environments (Chen and Yu, 2020 and references there in).

The EEMs of NOM can be decomposed into regionally-specific EEM volumes through FRI method, which integrates the volume beneath each EEM region, enabling semi-quantitative analysis of the configuration and heterogeneity of NOM (Xu et al., 2018). EEM coupled with FRI analysis can also be used to track the DBP precursors originated from DOM (Deng et al., 2018). Whereas PCA coupled with EEM allows to extract common features from datasets, but it lacks chemical significance (Stedmon and Bro, 2008; Park et al., 2021). Hence, PCA combined with other methods or parameters to reflect the difference between NOM samples and identify the predominant influential factors. In comparison to PCA, PARAFAC analysis deconvolutes EEMs into chemically meaningful components based on a mathematical tri-linear to process EEM data (Ishii and Boyer, 2012), and shows superior capacity on the differentiation and characterization of fluorescent NOM indicators (He and Hur, 2015). PARAFAC models have been found to fit on excitation-emission matrices (Sanchez et al., 2013). According to Markechova et al. (2013), PARAFAC models can be applied by initializing the algorithm, establishing the number of components, identifying the fluorophores, and calibrating the model. They may be utilized to determine the component most easily removed from a treatment process, establish the order of preferred removal, and identify persistent components to be removed (Pifer and Fairey, 2012; Markechova et al., 2013; Kothawala et al., 2013; Shutova et al., 2014). Thus, PARAFAC constitutes a valuable tool for modeling NOM behavior during water treatment. Furthermore, strong linear correlations have been demonstrated between EEMs and DOC, SUVA, and their organic precursors (Valencia et al., 2013).

Several studies have developed models using EEM data to predict DBP formation (Johnstone and Miller, 2009; Hua et al., 2010; Peleato and Andrews, 2015; Peleato et al., 2018) and other aspects of source water monitoring and drinking water treatment. As an example, Hua et al. (2010) used PARAFAC models to identify THM precursors in 55 US lakes and found that factors 1 and 2 were THM precursors and provided better surrogates for DBPFP compared to SUVA. Trueman et al. (2016) compared several methods (FRI and principal components regression, supervised principal components regression, lasso and boosted regression tree ensembles) for the prediction of THM and HAAs as well as DBPFP in drinking water samples and found that boosted regression tree ensembles and the lasso yielded the most accurate predictions for full-scale conventional treatment. Recently, Park et al. (2021) used EEM fluorescence spectroscopy coupled with PARAFAC and PCA analysis identify different NOM "fingerprints" in a source water that receives wastewater discharge and also acts as the source for drinking water treatment plant.

3.6.4 Fourier transform infrared spectroscopy and Raman spectroscopy

Since NOM is known to be a complex heterogeneous mixture, Fourier transform infrared spectroscopy (FT-IR) provides specific information on NOM characteristics (e.g., chemical bonding, molecular environment, and molecular structure, see Table 3.4) (Her et al., 2008a; Yang et al., 2013; Pan et al., 2016). Among the IR range, mid-infrared (2.5−25 μm,

Table 3.4 Main characteristics of the Fourier transform infrared spectra (FTIR) of humic substances.

Wavenumbers (cm^{-1})	Assignment
3400	Associated O-H stretch (alcohols, phenols, and carboxylic groups)
2850−2960	C-H stretch (CH_3 and CH_2)
2620	O-H stretch (hydrogen-bonded carboxylic groups)
1720	C=O stretch (carboxylic groups)
1630	C=C stretch (alkenes and aromatic rings)
1540	N-H bend (N−H structures)
1455	C-H bend (CH_3 and CH_2)
1410	O-H bend (carboxylic groups)
1375	C-H bend (CH_3)
1260 and 1220	C-O stretch (carboxylic groups, phenols, aromatic/unsaturated ethers)
1095 and 1030	C-O stretch (alcohols, aliphatic esthers)
805	C-H bend (tri- and tetra-substituted aromatic rings)

400–4000 cm^{-1} wavenumbers) is the most popularly used technique for NOM studies as it contains two regions, the functional group region (4000–1350 cm^{-1}) and the unique fingerprint region (1350–400 cm^{-1}) to identify organic and inorganic functional groups in NOM. However, direct interpretation of the obtained absorption spectrum may be challenging due to the heavily overlapped spectral features of NOM as a result of the high polyfunctionality and complexity (Pan et al., 2016). Hence, FTIR spectroscopy is mostly used as a qualitative method rather than quantitative technique.

Another vibrational spectroscopic technique that offers similar but complementary structural information compared to IR spectra is Raman spectroscopy. Since the interference of water signal does not exist with Raman spectroscopy, Raman spectroscopy is used for real time and semi-quantitative monitoring of membrane fouling during crow flow ultrafiltration and online sensing of biofouling in drinking water membrane filtration (Kogler et al., 2016; Virtanen et al., 2017; Chen et al., 2018). Due to the practical implementation this technique is not yet widely applied for NOM characterization, and there are relatively few works reported in the literature on the study of the structural characteristics of NOM by means of Raman spectra.

3.6.5 Nuclear magnetic resonance

NOM is the environmental matrix that is frequently found in our environment but its molecular complexity and heterogeneity, and by the variation in NOM with origin, climate, and season, NOM structure is poorly understood (Hayes et al., 1989). Nuclear magnetic resonance (NMR) spectroscopy is one of the most useful spectrometric methods to offer qualitative and quantitative information on NOM characterization with close range molecular order (Hertkorn et al., 2016) and functional groups in NOM structures (Barron and Wilson, 1981). Both qualitative and quantitative organic structure for NOM in solid and liquid states can be generated by NMR.

However, NMR spectroscopy is not as sensitive as infrared or UV-Vis spectroscopy, thus, generated NMR spectra in combination with IR spectra, elemental composition, and apparent MW can be effectively utilized, particularly for fractionated NOM (Gjessing et al., 1998; Assemi et al., 2004; Templier et al., 2005; Deshmukh et al., 2007; Mao et al., 2007a,b; Lankes et al., 2008; Ritchie and Perdue, 2008; Li et al., 2009; Jeong et al.,

2013). To improve the practical availability of NMR information, various fraction methods (SEC, resin fractionation) have been used, although in some cases the pre-fractionation of NOM has simplified spectra (Wong et al., 2002). Additionally, NMR-based NOM research often requires pre-concentration via methods including RO coupled with electrodialysis (ED) (to remove salts), SPE, or ultrafiltration followed by rotary evaporation or freeze drying in order to obtain solid material prior to analysis (Mao et al., 2017). According to Minor et al. (2014), pre-concentration with RO and ED provide additional peaks in ^{13}C-NMR and ^{1}H-NMR spectra compared to that with SPE or ultrafiltration.

Both solid- and liquid-state NMR spectroscopy have extensively been used to investigate NOM structure. Liquid-state NMR is useful for characterization of soluble components in NOM (particularly with low molecular weight), which still limits for the analysis of some insoluble fractions of NOM. Solid-state NMR has also been employed to examine NOM structure, at semi-quantitative ^{13}C cross polarization/magic angle spinning (CP/MAS) technique, but this has some known limitations such as non-quantitative peak intensities, spinning sidebands, baseline distortion, and ambiguous assignments (Botto et al., 1987; Mao et al., 2007a, 2017; Lankes et al., 2008). The development of the high-resolution magic angle spinning (HRMAS) probe has led to the examination organic materials that are poorly soluble (Zhong et al., 2011) using liquid NMR methods, which also allows to understand NOM in the soil-water interface (Simpson et al., 2006). Two dimensional ^{1}H-^{13}C heteronuclear correlation NMR can be used to identify spatial proximity of different specific functional groups and thus, it can be useful in determination of complex NOM structures (Deshmukh et al., 2007; Mao et al., 2007a,b; Cook, 2004). In addition, ^{15}N NMR has also been used to identify peptides and other nitrogen containing structures of NOM (Thorn and Cox, 2009).

3.6.6 ^{1}H NMR

The obtained structural features of NOM by ^{1}H NMR is presented in Table 3.5 (Ma et al., 2001; Kim et al., 2006a; Kim and Yu, 2007). In application, ^{1}H NMR can be used for rapid screening of NOM for quantification of aromaticity and humification degree (Kim et al., 2006a, b; Lam and Simpson, 2008). ^{1}H NMR spectra have been used in previous studies to conduct characterization of NOM in aquatic environments (e.g., Lam and Simpson, 2008). As an example, Hertkorn et al. (2016)

Table 3.5 General assignments of the 1H NMR spectra of humic substances.

Chemical shift (ppm)	General assignment	References
0.0–1.6	Protons on methyl and methylene carbons directly bonded to other carbons	Ma et al. (2001), Kim et al. (2006a), and Kim and Yu (2007)
1.6–3.2	Protons of the methyl and methylene carbons alpha to aromatic rings, carboxyl and carbonyl groups	
3.2–4.3	Protons on carbon of hydroxyl, ester and ether and protons on methyl, methylene and methylene carbons directly bonded to oxygen and nitrogen	
6.0–8.5	Protons attached to unsaturated carbons. Aromatic protons including quinones, phenols, oxygen containing heteroaromatics	

characterized DOM from subtropical wetlands using ^{1}H NMR and found clear distinctions across samples, confirming wetland specific molecular signatures. They also used 2-D NMR spectroscopy to reveal aliphatic and unsaturated substructures which indicated that NOM was derived from microbial sources. However, ^{1}H NMR has been used to a lesser extent in drinking water monitoring and treatment applications. Rodriguez et al. (2016) did use NMR to track changes in the structure and properties of humic substances following ozonation during drinking water treatment and found that a decrease in aromaticity post-ozone could be monitored with ^{1}H NMR spectra. Gonsior et al. (2014) used NMR to monitor changes in NOM quality and formation of unknown DBPs during the water treatment process. In this work, ^{1}H NMR spectra of chlorinated samples revealed a molecular variance of DOM, indicating that distinct molecules were formed during chlorination. Accordingly, ^{1}H NMR could be useful for understanding the transformations of NOM during treatment.

3.6.7 ^{13}C NMR

Solid-state ^{13}C NMR is the most used NMR method for the characterization of NOM via cross-polarization-magic-angle spinning (CP-MAS) and enables the determination of the carbon functional groups in NOM (Chen et al., 2002; Wong et al., 2002; Mao et al., 2007b). This technique is rather qualitative analysis for obtaining relative compositions of NOM and specific compounds are hardly obtained (Peuravuori et al., 2003). The general assignments for structural features of NOM obtained by solid-state ^{13}C NMR are shown in Table 3.6 (Chen et al., 2002; Mao et al., 2007b).

Like ^{1}H NMR, ^{13}C NMR has been widely used to study spatiotemporal variations in NOM composition in environmental samples (e.g., Benner et al., 1992; Bianchi et al., 2004; Fimmen et al., 2007; Abdulla et al., 2010; Mao et al., 2012), to understand sources of NOM (e.g., Wilson et al., 1983; Engelhaupt and Bianchi, 2001), to understand photochemical changes in (e.g., NOM (Wetzel et al., 1995; Kulovaara et al., 1996; Osburn et al., 2001), and to characterize NOM fractions (e.g., Templier et al., 2005; Koprivniak et al., 2009; Nwosu and Cook, 2015). In drinking water treatment applications, Yamamura et al. (2014) used ^{13}C NMR to characterize isolated NOM fractions obtained from four source waters and were able to distinguish different properties (aromatic, carbohydrate, aliphatic, and protein) of the individual fractions. Ratpukdi et al. (2009) also applied ^{13}C NMR to characterize the functional and carbon structural groups in isolated NOM fractions from a water treatment context.

Table 3.6 Chemical shift regions for the integration of peak areas in solid-state 13C NMR spectra of NOM and their respective assignments.

Chemical shift (ppm)	General assignment	References
0–50	Aliphatic or paraffinic carbon chains	(Chen et al., 2002; Mao et al., 2007b)
50–75	Methoxyl groups	
75–112	Carbohydrate RCs-OH or RC-OR functional groups	
112–145	Aromatic carbons	
145–163	Phenolic groups	
163–180	Carboxylic, carbonyl, amine, and ester carbons	

3.7 Chromatographic methods

3.7.1 Size exclusion chromatography

3.7.1.1 Fractionation of NOM by size exclusion chromatography

Size exclusion chromatography (SEC) is a fractionation method based on molecular size: the larger the molecules, the shorter the retention time. Yet, molecular shape and some interaction characteristics may influence the results (Lankes et al., 2009). SEC fractionation of NOM has been utilized since the early 1960s (Posner, 1963; Gjessing, 1965); however, the soft gels used had some major disadvantages (Gjessing, 1973), such as poor separation. To overcome this, SEC column fractionation was applied to high-performance liquid chromatography (HPLC), developing a technique known as HPSEC (Fukano et al., 1978). HPSEC has become particularly useful for NOM characterization in terms of NOM weight fractionation (average and weight-average MW) during various stages of drinking water treatment (Matilainen et al., 2006; Allpike et al., 2007; Fabris et al., 2008; Chow et al., 2009a,b; Tercero Espinoza et al., 2009; Zhao et al., 2009).

3.7.1.2 Eluents and columns used for HPSEC measurements

In NOM studies using HPSEC, the choice of a proper eluent is important, since its ionic strength and pH significantly affect results. The surface charge characteristics of the gel, the charge and structure of the NOM, and NOM-gel interactions are all affected by the effluent (Specht and Frimmel, 2000). Studies of NOM characterization by HPSEC have used various eluents as a mobile phase, including a phosphate buffer solution with ionic strength adjustment using sodium chloride and sodium acetate solution. A newer application of HPSEC is established by Chon et al. (2017) that separation of organic matter can be obtained with de-ionized (DI) water as mobile phase. The columns have different silica-based and polymer-based supports (Swietlik et al., 2004; Matilainen et al., 2006; Sarathy and Mohseni, 2007; Wu et al., 2007a,b; Fabris et al., 2008; Her et al., 2008a; Liu et al., 2008, 2010; Soh et al., 2008; Zhang et al., 2008; Chow et al., 2009b; Korshini et al., 2009; Zhao et al., 2009). According to Specht and Frimmel (2000), however, both polymer-based and silica-based columns have specific interactions with NOM.

3.7.1.3 Detectors used for HPSEC measurements

Detectors used in different HPSEC systems include Fourier transform infrared spectroscopy (FTIR) (Allpike et al., 2007), on-line DOC analyzers

(Her et al., 2008a), and excitation emission fluorescence detection (Wu et al., 2007a). The detectors most widely used in HPSEC fractionation are the single and variable wavelength ultraviolet and visible light (UV-Vis) detectors or diode array detectors (DAD; Matilainen et al., 2006; Ates et al., 2007; Sarathy and Mohseni, 2007; Wu et al., 2007b; Fabris et al., 2008; Her et al., 2008a,b; Liu et al., 2008, 2010; Zhang et al., 2008; Chow et al., 2009b; Korshin et al., 2009). UV-Vis detectors are simple and quick to use and are available in most laboratories. However, they only respond to or detect analytes that absorb on the wavelength at which they are operating. Since NOM structures contain a range of chromophores with varying molar absorptivities, the MW calculated may not involve all the NOM compounds present (O'Loughlin and Chin, 2001; Wu et al., 2007b; Soh et al., 2008; Tercero Espinoza et al., 2009). Reducing the wavelength results in an increase in the absorbance (O'Loughlin and Chin, 2001; Lankes et al., 2009). On the other hand, low molecular weight (LMW) compounds reportedly are not easily detected by UV-Vis due to their lack of conjugated double bonds. HPSEC-UV coupling with nondispersive infrared (NDIR) detector enables to detect low molecular weight DOM (~2000 Da) with low UV absorbance, leading to determine the chemical and physical properties of DOM (Kawasaki et al., 2011).

Despite of the success of on-line specific Ex/Em fluorescence detection in the characterization of NOM fractions, this only provides limited information and somewhat poorer separation resolution compared to UV-Vis absorbance detection (Wu et al., 2007a). Yet, in contrast to UV-Vis, online 3D excitation-emission matrix (EEM) fluorescence detection gives more insight into the chemical qualitative and structural features of NOM besides molecular size (Wu et al., 2003, 2007a; Li et al., 2013; Ignatev and Tuhkanen, 2019).

Online organic carbon detectors (OCD) may offer better quantification than UV-Vis detectors (Her et al., 2002; Soh et al., 2008; Lankes et al., 2009; Tercero Espinoza et al., 2009; Simon et al., 2013). Liquid chromatography-organic carbon detection (LC-OCD) is a combination of SEC separation and continuous analysis to quantify organic carbon, nitrogen, and UV absorbance (Penru et al., 2013). During the last 20 years, an LC-OCD combination has been develop that both quantifies NOM and characterizes its compounds (Huber and Frimmel, 1992a,b). Using LC-OCD, it is also possible to evaluate the content of HMW polysaccharides and biopolymers in the sample, which are not generally visible or traceable by UV detectors. Recently, LC-OCD has been successfully visible or traceable by UV detectors. Recently, LC-OCD has been successfully used to characterize NOM fractions and assess the impact of pretreatments on them (Simon et al., 2013). Huber and Frimmel (1994) used an HPSEC coupled with an LC-OCD to fractionate NOM into five parts: (1) biopolymers (such as polypeptides, polysaccharides, proteins, and amino sugars), (2) humic substances (fulvic and humic acids), (3) building blocks (hydrolysates of humic substances), (4) Low MW humic substances and acids, and (5) Low MW neutrals (such as alcohols, aldehydes, ketones, and amino acids). High-resolution mass spectrometry (MS) has also been coupled with HPSEC to provide more information about the molecules in complex mixtures of NOM fractions.

To achieve better quantitative and qualitative NOM evaluation for separation and determination of heterogeneous MW distribution of NOM and its structural identification, an increasing number of reports couple multiple detectors on-line, such as UV-Vis and spectrofluorometer (Wu et al., 2007a, 2020; Wünsch et al., 2017), or UV-Vis and OCD (Her et al., 2002; Allpike et al., 2007; Soh et al., 2008; Baghoth et al., 2009; Lankes et al., 2009; Tercero Espinoza et al., 2009; Huber et al., 2011). Accordingly, an SEC setup that uses a UV detector followed by a TOC module and finally an MS as the detector has been described (Warton et al., 2008). The results of these combined detector systems look promising.

3.7.1.4 Determination of MW by HPSEC

The size of the MW of the NOM may be specified by comparison with known molecular mass standards. The selection of proper standards to characterize the MWs of organic substances in NOM using HPSEC is based mostly on the hypothesized structure of NOM. Polystyrene sulfonate (PSS) as MW standards have been widely used (Zhou et al., 2000; Swietlik et al., 2004; Ates et al., 2007; Wu et al., 2007a; Fabris et al., 2008; Chow et al., 2009b; Korshin et al., 2009; Zhao et al., 2009; Liu et al., 2010). Yet, NOM in water can be more branched and cross linked than PSS. Therefore, the resulting MWs are approximations. Polyethylene glycols are also used to determine the MWD of NOM (Her et al., 2008a; Zhang et al., 2008). The MW magnitudes determined by HPSEC reportedly correlate well with other MW measuring techniques (Egeberg et al., 2002; Assemi et al., 2004).

3.7.1.5 HPSEC determinations in water treatment monitoring

HPSEC can be used as a tool to characterize the MW profile of NOM as well as to model and predict NOM removal (Xing et al., 2012). SEC profiles provide valuable information in the study of the removal mechanism of HMW NOM during coagulation. The advantages of SEC include ease of operation, minimal sample pretreatment, and rapid analysis (Kristiana et al., 2010). Several studies compare changes in both the quality of NOM during different stages of the treatment process (Matilainen et al., 2006; Fabris et al., 2008; Baghoth et al., 2009; Chow et al., 2009b; Zhao et al., 2009) and the actual degree of its removal (Wu et al., 2007b). These SEC profiles do not, however, contain sufficient information to evaluate the DBPFP of the remaining NOM molecules. Nevertheless, Korshin et al. (2009) reported a good correlation between DBPFP and HPSEC multi-wavelength UV results based on spectroscopic index calculations (Korshin et al., 2009).

SEC determinations by Haahorff et al. (2010) revealed that rapid sand filtration targeted the low and high molecular fractions of NOM, granular activated carbon targeted the low to intermediate fractions, and ultrafiltration (UF) targeted the high molecular weight (HMW) fraction.

3.7.2 Flow field-flow fractionation

Flow field-flow fractionation (FIFFF) is a flow-based fractionation method for separating macromolecules, colloids, and particles without a stationary phase, like the column packing material (Messaud et al., 2009). Hence, this is a chromatography-like technique to determine size or the molecular weight distribution of NOM collected from different environments (Beckett et al., 1987; Moon et al., 2006; Rosario-Ortiz et al., 2007a; Alasonati et al., 2010; Baalousha et al., 2011; Pifer and Fairey, 2012). Smaller particles have higher diffusion rates than larger particles and build up diffusional clouds which extend higher into the channel than those of the larger particles (Hassellov et al., 2007; Neubauer et al., 2013) FIFFF is a sub-technique of field-flow fractionation (FFF). Among variants of FIFFF, asymmetrical flow field-flow fractionation (AF4) has been widely used and in combination with online fluorescence and UV-Vis detectors, chromophoric and fluorophoric organic components of various size and origin are characterized (Lee et al., 2015; Yamashita and Saito, 2015). Detectors used in FIFFF include UV-Vis, fluorescence, inductively coupled plasma atomic emission spectroscopy (ICP-AES), and inductively coupled plasma mass spectrometry (ICP-MS) (Moon et al., 2006; Alasonati et al., 2010). The results obtained by FIFFF are comparable to those obtained by other techniques such as HPSEC (Assemi et al., 2004). Although the technique as compared with HPSEC has not yet matured for characterization of molecular weight distribution of NOM (Messaud et al., 2009), FlFFF is becoming increasingly popular, with regard to the criticisms of HPSEC (Zhou and Guo, 2015). Size distributions determined by FlFFF under varying pH and ionic strength should be treated with caution, because the sample recovery of humic substances decreases rapidly with increasing cross flow, lower pH, and greater FlFFF carrier ionic strength, and varies according to the material of the accumulation wall membrane (Baalousha et al., 2011).

FFF has been used in conjunction with detectors including UV-absorbance and ICP-MS to characterize the interactions between inorganic and organic based colloids in freshwater sources (Lyvén et al., 2003; Stolpe et al., 2010; Anderson et al., 2021) as well as treated drinking water (Trueman et al., 2019). For example, Fu et al. (2020) showed that FFF can be used to quantify the MW distribution of chromophoric and fluorescent components of DOM from natural waters. Stolpe et al. (2010) coupled FFF to detectors including UV-absorbance, fluorescence, and ICP-MS and identified colloidal spectra including 0.5—4 nm colloidal DOM which was found to bind with most elements. Anderson et al. (2021) studied the relative importance of organic and iron based colloids in six lakes undergoing brownification via FFF-UV-ICPMS and found a 1 nm size fraction containing the primary DOM signal which was primarily associated with colloidal iron and aluminum. Further, it was noted that source water monitoring should include FFF-UV-ICPMS characterization of DOM-inorganic colloids to better inform treatment process operation and design. Trueman et al. (2019) used FFF coupled with UV and ICP-MS detection to separate drinking water samples into ionic species, colloids and metal-organic complexes and noted that during long stagnation periods in plumbing, high MW organic and colloidal fractions represent capacity to transport lead in drinking water.

3.7.3 Reverse-phase high-performance liquid chromatography

RPHPLC is rather limited in application to NOM fractionation, especially compared to HPSEC (Stenson, 2008; Reemtsma, 2009). RPHPLC is based on a nonpolar stationary phase and an aqueous, moderately polar mobile phase, thus fractionating NOM into hydrophobic or hydrophilic properties according to its polarity. Xing et al. (2012) investigated RPHPLC as a rapid assessment of the hydrophobicity or hydrophilicity of NOM. It was found that the reduction of total RPHPLC peak area correlated well with DOC and UV absorbance removal efficiency. The ratio between the hydrophobic and hydrophilic peak areas could be used to quantify the traceability of NOM (Xing et al., 2012).

3.7.3.1 Liquid chromatography-mass spectrometry

Liquid chromatography-mass spectrometry (LC-MS or alternatively HPLC-MS) is a technique that combines the physical separation capabilities of liquid chromatography with the mass analysis capabilities of MS. LC-MS is a powerful technique with high sensitivity and it is one of the main recent innovations in NOM structure research (Pan et al., 2016). Coupling LC with MS by an advanced technique called electrospray ionization (ESI) made the laborious derivatization of NOM molecules unnecessary (Reemtsma, 2009), leading to observing direct mass-to-charge ratio (m/z) values of NOM molecular ions. The LC-MS system is used in combination with SEC in describing low MW FAs (Reemtsma and These, 2005) and ESI-MS is more sensitive to characterizing low MW FAs than high MW humics (Reemtsma and These, 2005). Furthermore, ESI-tandem MS (ESI-MS/MS) such as ion trap, triple quadrupoles, or quadrupole time-of-flight equipped with ion mobility can be used to obtain NOM structural information (Reemtsma and These, 2005; Mawhinney et al., 2009). High resolution mass spectrometry has been utilized to study marine NOM evolution through a desalination process on a molecular scale (Cortes-Francisco and Caixach, 2013).

3.7.3.2 Fourier transform ion cyclotron resonance mass spectrometry

Fourier transform ion cyclone resonance mass spectrometry (FTICR-MS) is currently the only technique that can separate NOM into individual isobars and determine molecular formulae with reasonable accuracy. The introduction of MS (particularly ESI method) by Reemtsma (2009) attributes significantly to NOM research, which allowed the infusion of aqueous solutions into the MS directly. Another discovery was ultra-resolution in the form of Fourier transform ion cyclotron resonance. Ultrahigh-resolution mass spectrometry (UHRMS), utilizing an FTICR-MS has emerged as an important analytical technique to obtain molecular-level characteristics of NOM (Brown et al., 2016; Zark and Dittmar, 2018). Coupling ESI with FTICR-MS (ESI-FTICR-MS) has recently been employed to investigate NOM structures (Reemtsma, 2009;

Zhang et al., 2012a,b; Zhang et al., 2020) as well as to detect elemental composition (Kim et al., 2006a,b). However, only polar or semi-polar NOM molecules can be ionized by ESI, which limits the detection of nonpolar NOM molecules. Recently, a combination of ESI and atmospheric pressure photospray ionization (APPI) technique has been employed to extend MS capacities to less polar molecules (D'Andrilli et al., 2010; Hockaday et al., 2009). Additionally, the laser desorption/ionization (LDI) and matrix-assisted laser desorption/ionization (MALDI) ionization techniques can be powerful complementary ionization methods to ESI prior to FTICR-MS analysis of NOM (Blackburn et al., 2017; Solihat et al., 2019; Zhang et al., 2020). More than 4000 NOM components could be resolved in one sample using this technique. Interestingly, in practical applications, NOM molecules with a low degree of oxidation were preferentially removed during coagulation by inorganic polymer flocculants (e.g., polyaluminum chloride) (Zhang et al., 2012b). FTICR-MS is further expected to the understanding of NOM transformations on a molecular level during water treatment processes and provides useful evidence for improving their performance (Zhang et al., 2012b). Although FTICR-MS enhances the knowledge on the nature of DOM at the molecular level, more specific information (for those ions with same elemental composition and both enter the interface at the same time) is obtained by coupling with ESI-FTICR-MS with SEC (Reemtsma et al., 2006, 2008; Reemtsma, 2009) and RPHPLC (Stenson, 2008; Reemtsma, 2009), thereby producing different fractions of NOM, which are easier to be determined by ESI-FTICR-MS. However, the challenges still exist in molecular formula assignment, which can be improved by consideration of C-13 isotopes (Herzsprung et al., 2015) as well as group-wise decisions based on the frequency (Herzsprung et al., 2014). An application of the universal algorithm for automatic treatment of mass spectra, based on total mass difference statistics, has been also developed and implemented to support the interpretation of data on complex NOM structures (Kunenkov et al., 2009).(Solihat et al., 2019)

3.7.4 Gas chromatography

3.7.4.1 Pyrolysis gas chromatography-mass spectrometry

In pyrolysis, the large complex molecules in NOM are broken down into more analytically available fragments by the application of heat and under anoxic conditions. These fragments are then swept into the analytical column for gas chromatography (GC) and detected by MS. Pyrolysis combined with gas chromatography-mass spectrometry (Py-GC-MS) is a useful technique to clarify both volatile and non-volatile NOM compositions, and through the Py-GC-MS, both quantitative and qualitative analysis for producing structural information on the molecular building blocks of NOM can be determined (Bruchet et al., 1990; Schulten, 1993; Fabbri et al., 1998; Christy et al., 1999; Leenher and Croue, 2003; Templiere et al., 2005; Chow et al., 2009b; Derenne et al., 2014; Li et al., 2017). The analysis of complex oxygen-containing macromolecules can be challenging (Parsi et al., 2007). Polar pyrolysis products are poorly GC amenable and undergo undesirable thermal reactions (Templier et al., 2005), thus providing only little information on NOM chemical structure. Development in the pyrolysis technique has made it possible to overcome some of these analytical limitations (Parsi et al., 2007).

Microscale sealed vessel pyrolysis is a promising approach to NOM characterization. This technique promotes the reduction of polar NOM moieties and significantly increases the yield of GC amenable products (Greenwood et al., 2006; Berwick et al., 2007, 2010). This analytical method represents an excellent complement to conventional flash pyrolysis techniques, which may have a limited capacity for chromatographic resolution of polar biochemicals. However, the technique needs to be developed further to enable more advanced NOM characterization.

Major pyrolysis fragments, most commonly yielded from aquatic NOM pyrochromatograms, have been listed by Leenher and Croue (2003) and are summarized in Table 3.7. Besides pyrolysis-MS, the molecular-level structural information can also be obtained by chemical degradation methods such as using a reducing agent (e.g., n-butylsilane) in combination with MS (Nimmagadda and McRae, 2007). Main key methods for characterization and quantification of NOM in relation to drinking water treatment processes have been evaluated and discussed in Table 3.8.

3.8 Other characterization methods

Many other NOM characterization methods are rather complex and time-consuming, requiring access to sophisticated analytical instrumentation. Even though these techniques are not used routinely, they may provide further insight into the quality and structural features of NOM at various stages in the water treatment process, promoting a better understanding of its behavior.

Table 3.7 The predominant pyrolysis by-products from aquatic natural organic matter (NOM).

Type	Common fragments
Polysaccharides	Methylfuran, furfural, acetylfuran, methylfurfural, levoglucosenone, hydroxypropanone, cyclopentanone, methylcyclopentenone, acetic acid
Amino sugars	Acetamide, *N*-methylacetamide, propionamide, acetic acid
Proteins	Acetonitrile, benzonitrile, phenylacetonitrile, pyridine, methylpyridine, pyroole, methylpyrrole, indole, methylindole (from tryptophan), toluene, styrene (from phenylalanine), phenol, *p*-cresol (from tyrosine)
Polyphenolic compounds	Phenol, o-, m-, p-cresol; methylphenols, dimethylphenols
Lignins	Methoxyphenols
Tannins	Catechol
DNA	Furfuryl alcohol
Polyhydroxybutyrates	Butenoic acid

The elemental composition of NOM is often used to calculate atomic ratios such as O/C, H/C, or N/C. In addition to the classic Elemental Analyzer (Li et al., 2009; Zhao et al., 2009), elemental composition can be determined by ICP-AES (Blodau et al., 2009). Another structural analysis method is to measure the size and conformation of NOM under different environmental conditions with photon correlation spectroscopy, transmission electron microscopy (TEM), or atomic force microscopy (AFM) for obtaining better chemical and structural information of NOM (Lead et al., 2000; Baalousha et al., 2006; Abu-Lail et al., 2007). Recent advances in AFM measures adhesion forces between the foulant (colloidal NOM) and the membrane, and between the foulant in bulk solution (Costa et al., 2006). AFM also determines NOM to bacteria-NOM adhesion forces (Abu-Lail et al., 2007). Field emission scanning electron microscopy (SEM) is commonly used to evaluate foulant layers on the surface of membranes used for drinking water treatment (e.g., Frimmel et al., 2004; Kim et al., 2010; Yu et al., 2018), and is sometimes used to understand the structure of NOM (e.g., humic acid) (Redwood et al., 2005). Methods mentioned above to characterize different features of NOM with their pros and cons are presented in Table 3.8.

Table 3.8 Advantages and disadvantages of the spectroscopic and chromatographic methods used to characterize different features of natural organic matter (NOM).

	Methods	Advantages	Disadvantages	References
Spectroscopic methods				
UV-Vis	Quantitative measurement of all compounds in the sample that adsorb UV light. Conjugated C–C multiple bonds, aromatic carbon, -COOH and -OH increase adsorption	Simple; rapid	Not all compounds of NOM can be detected; wavelength-specific adsorption; sensitive to chemical environment (e.g., pH and ionic strength)	Korshin et al. (1999), Hur et al. (2006), Ates et al. (2007), Spencer et al. (2007), Li et al. (2009), and Roccro and Vagliasindi (2009)
Fluorescence	Molecules of the sample are excited by irradiation at a certain wavelength and the emitted radiation is measured at a different wavelength. Conjugated double bonds and aromatic rings, -OH and $-NH_2$ enhance fluorescence, while -COOH diminishes it. Three major groups: Tryptophan-, humic-, and fulvic-like fluorophores	Relatively simple, sensitivity, specificity, rapid	Sensitivity to chemical environment (e.g., pH)	Liu et al. (2007), Seredynska-Sobecka et al. (2007), Spencer et al. (2007), Wu et al. (2007a,b), Baker et al. (2008), Hudson et al. (2008), Zhang et al. (2008), Bieroza et al. (2009), and Peiris et al. (2010)

Continued

Table 3.8 Advantages and disadvantages of the spectroscopic and chromatographic methods used to characterize different features of natural organic matter (NOM).—cont'd

	Methods	Advantages	Disadvantages	References
FTIR	Provide information on aromatic and aliphatic hydrocarbon, different bonds and functional groups (e.g., OH in carboxylic and alcoholic groups)	Sample can be measured both in liquid and solid phases; compact instruments; reliability; ease of use	Interpretation of the analysis can be difficult due to overlapping of characteristic spectral features due to the complexity and polyfunctional of NOM	Kanokkantapong et al. (2006b), Hay and Myneni (2007), and Her et al. (2008a, b)
^{1}H NMR	Determine the structure with respect to hydrogen-1 nuclei within the molecules of a substance	Highly sensitive; powerful tool especially together with ^{13}C NMR in structural analysis	Interpretation of spectra can be limited	Ma et al. (2001), Kim et al. (2006a), Kim and Yu (2007), and Mao et al. (2007a)
^{13}C NMR	Determine a molecule's structure related with regarding to ^{13}C nuclei in a magnetic field absorbing and re-emitting electromagnetic radiation Carboxylic structures of NOM, as well as aliphatic carbon, o-alkyl carbon, aromatic carbon and carbonyl carbon. Can be done either solid or liquid sample	Powerful tool especially together with ^{1}H NMR in structural analysis	Less sensitive to carbon than ^{1}H NMR to hydrogen; NOM sample needs to be isolated in solid-state NMR measurements; parameter sensitive	Chen et al. (2002), Wong et al. (2002), Mao et al. (2007a,b), and Lankes et al. (2008)

^{15}N NMR	Identifies carbon bonded to nitrogen	Broader insight into NOM structure	Rather nuclear insensitive; ^{15}N appears at low percentage in natural	Mao et al. (2007b), Lankes et al. (2008), and Thorn and Cox (2009)
2-D NMR	Spreads the spectrum in two dimensions and allows for the observation of cross peaks. Structural environments of, e.g., carboxyls in detail	Multi-dimensional NMR techniques can be used to resolve contributions that overlap strongly in a 1D NMR experiment	Might not detect very large molecules	Deshmukh et al. (2007) and Mao et al. (2007a,b)
Analytical methods				
HPSEC	Fractionates NOM on basis of molecular sizes of organic compounds present in a sample	Rapid; sensitive; no sample preparation needed; equipment relatively simple; inexpensive	Charge effects during measurement (column/NOM, eluent/NOM), choice of detector and proper standards	Wu et al. (2007a), Chow et al. (2005), Fabris et al. (2008), Her et al. (2008a, b), Zhang et al. (2008), and Korshin et al. (2009)
FIFFF	A separation technique similar to chromatography. Fractionate NOM by size and molecular weight distribution (MWD)	No stationary phase	Choice of proper standards	Moon et al. (2006), Messaud et al. (2009), and Alasonati et al. (2010)
GC-MS	Structural and molecular properties of compounds in NOM	Sensitive; specific; rapid	Unwanted thermal reactions; data interpretation is complicated due to the complexity of NOM	Templier et al. (2005) and Parsi et al. (2007)

Continued

Table 3.8 Advantages and disadvantages of the spectroscopic and chromatographic methods used to characterize different features of natural organic matter (NOM).—cont'd

	Methods	Advantages	Disadvantages	References
FTICR-MS	Molecular species, exact masses, and molecular formulas of compounds on NOM	The full scale of the molecular complexity of NOM can be revealed; can be combined with SEC or RPHPLC	Data interpretation is difficult, due to the complexity of NOM	Reemtsma et al. (2006, 2008), Stenson (2008), Kunenkov et al. (2009), and Reemtsma (2009)

3.9 Conclusions

Surface water is the main drinking water source in many regions of the world. Due to contact with the soil surface, fresh water will always contain some inorganic and organic matter. The quality of all fresh water thus depends on the nature of the catchment. In the northern hemisphere, surface water often has a high NOM content. Knowledge of this matter has considerably increased with the last three or four decades of international research, leading to the general conclusion that for both practical and hygienic reasons, NOM should be removed from drinking water. The nature, character, and amount of NOM in raw water all depend on geology, topography, and climate; and climate change clearly has an effect on these parameters, as is demonstrated by a number of reports from across the globe testifying to a significant increase in the both the amount, and probably also the quality, of NOM in water sources.

Water purification should be optimized by controlling water quality change during the treatment process. Many advanced characterization methods to describe the nature of NOM have been developed over the last few years, which are an asset in optimizing NOM removal from raw water. The main aim is to produce quality water that in turn will create fewer problems at later stages, such as disinfection and alkalization during transport to the tap.

Isolation, concentration, and fractionation of NOM are often implemented before analytical procedures are performed. Resin isolation and fractionation are widely accepted pretreatment methods to HPSEC or FTIR analysis. HPSEC has proved highly popular for size fractionation of NOM in conjunction with a number of different detection procedures. NMR spectroscopy has also been used extensively during the last few decades to investigate the structures and other properties of NOM.

With the development of more and more characterization techniques, the single-throughput complementary high-resolution analytical techniques are recently put forward. It is expected that a more comprehensive combination of current analytical and spectroscopic methods is promising to shed more light on the molecular and even structural features of NOM. Coupling MS with HPLC, allowing direct analysis without prior destruction or derivatization of the NOM molecules, has increased the significance of MS as a technique. More recently, the development of FTICR-MS has facilitated the molecular-level analysis of NOM, enabling the separation of molecular species from very complex NOM structures. By integrating

GCMS with FTICR-MS, both hydrosoluble and organosoluble compounds in NOM can be well characterized. Multidimensional NMR techniques together with FTICR-MS are new methods that appear promising for characterizing the major fractional features of NOM and fractionated NOM. Finally, reproducible characterization methods are more important than chemical structural analysis in promoting better knowledge of the practical, hygienic, and ecological impact of NOM.

References

H.A. Abdulla, E.C. Minor, R.F. Dias, P.G. Hatcher, Changes in the compound classes of dissolved organic matter along an estuarine transect: a study using FTIR and 13C NMR, Geochem. Cosmochim. Acta 74 (13) (2010) 3815–3838.

L. Abu-Lail, Y. Liu, A. Atabek, T. Camesano, Quantifying the adhesion and interaction forces between *Pseudomonas aeruginosa* and natural organic matter, Environ. Sci. Technol. 41 (2007) 8031–8037.

G. Aiken, Evaluation of ultrafiltration for determining molecular weight of fulvic acid, Environ. Sci. Technol. 18 (1984) 978–981.

G. Aiken, Isolation and concentration techniques for aquatic humic substances, in: G. Aiken, D. McKnight, R. Wershaw, P. MacCarthy (Eds.), Humic Substances in Soil, Sediment and Water, John Wiley & Sons, 1985, pp. 363–385.

G. Aiken, D. McKnight, K. Thorn, E. Thurman, Isolation of hydrophilic organic acids from water using non-ionic macroporous resins, Org. Geochem. 18 (4) (1992) 567–573.

R.A. Al-Juboori, T. Yusaf, V. Aravinthan, L. Bowtell, Tracking ultrasonically structural changes of natural aquatic organic carbon: chemical fractionation and spectroscopic approaches, Chemosphere 145 (2016a) 231–248.

R.A. Al-Juboori, T. Yusaf, P.A. Pittaway, Exploring the correlations between common UV measurements and chemical fractionation for natural waters, Desalin. Water Treat. 57 (35) (2016b) 16324–16335.

E. Alasonati, V. Slaveykova, H. Gallard, J.-P. Croué, M. Benedetti, Characterization of the colloidal organic matter from Amazonian basin by asymmetrical flow field-flow fractionation and size exclusion chromatography, Water Res. 44 (2010) 223–231.

B. Allpike, A. Heitz, C. Joll, R. Kagi, A new organic carbon detector for size exclusion chromatography, J. Chromatogr. A 1157 (2007) 472–476.

L.E. Anderson, B.F. Trueman, D.W. Dunnington, G.A. Gagnon, Relative importance of organic-and iron-based colloids in six Nova Scotian lakes, Npj Clean Water 4 (1) (2021) 1–10.

A. Archer, P. Singer, An evaluation of the relationship between SUVA and NOM coagulation using ICR database, J. Am. Water Works Assoc. 98 (7) (2006) 110–123.

S. Assemi, G. Newcombe, C. Hepplewhite, R. Beckett, Characterization of natural organic matter fractions separated by ultrafiltration using flow field-flow fractionation, Water Res. 38 (2004) 1467–1476.

N. Ates, M. Kitis, U. Yetis, Formation of chlorination by-products in waters with low SUVA – correlations with SUVA and differential UV spectroscopy, Water Res. 41 (2007) 4139–4148.

M. Baalousha, M. Motelica-Heino, P. Le Coustumer, Conformation and size of humic substances: effects of major cation concentration and type, pH, salinity, and residence time, Colloids Surf. A 272 (2006) 48–55.

M. Baalousha, B. Stolpe, J.R. Lead, Flow field-flow fractionation for the analysis and characterization of natural colloids and manufactured nanoparticles in environmental systems: a critical review, J. Chromatogr. A 1218 (2011) 4078–4103.
S. Baghoth, M. Dignum, A. Grefte, J. Kroesbergen, G. Amy, Characterization of NOM in a drinking water treatment process train with no disinfectant residual, Water Sci. Technol. 9 (1) (2009) 379–386.
S. Baghoth, S. Sharma, G. Amy, Tracking natural organic matter (NOM) in a drinking water treatment plant using fluorescence excitation–emission matrices and PARAFAC, Water Res. 45 (2011) 797–809.
A. Baker, E. Tipping, S. Thacker, D. Gondar, Relating dissolved organic matter fluorescence and functional properties, Chemosphere 73 (2008) 1765–1772.
L. Barber, J. Leenher, T. Noyes, E. Stiles, Nature and transformation of dissolved organic matter in treatment wetlands, Environ. Sci. Technol. 35 (2001) 4805–4816.
P. Barron, M. Wilson, Humic soil and coal structure study with magic-angle spinning 13C CP NMR, Nature 289 (1981) 275–276.
R. Beckett, Z. Jue, J.C. Giddings, Determination of molecular weight distributions of fulvic and humic acids using flow field-flow fractionation, Environ. Sci. Technol. 21 (1987) 289–295.
R. Benner, J.D. Pakulski, M. McCarthy, J.I. Hedges, P.G. Hatcher, Bulk chemical characteristics of dissolved organic matter in the ocean, Science 255 (5051) (1992) 1561–1564.
L. Berwick, P. Greenwood, R. Kagi, J.-P. Croué, Thermal release of nitrogen organics from natural organic matter using micro scale sealed vessel pyrolysis, Org. Geochem. 38 (2007) 1073–1090.
L. Berwick, P. Greenwood, R. Smernik, The use of MSSV pyrolysis to assist the molecular characterization of aquatic natural organic matter, Water Res. 44 (2010) 3039–3054.
T.S. Bianchi, T. Filley, K. Dria, P.G. Hatcher, Temporal variability in sources of dissolved organic carbon in the lower Mississippi River, Geochem. Cosmochim. Acta 68 (5) (2004) 959–967.
M. Bieroza, A. Baker, J. Bridgeman, Relating freshwater organic matter fluorescence to organic carbon removal efficiency in drinking water treatment, Sci. Total Environ. 407 (2009) 1765–1774.
J.W. Blackburn, W. Kew, M.C. Graham, D. Uhrin, Laser desorption/ionization coupled to FTICR mass spectrometry for studies of natural organic matter, Anal. Chem. 89 (8) (2017) 4382–4386.
C. Blodau, M. Bauer, S. Regenspurg, D. Macalady, Electron accepting capacity of dissolved organic matter as determined by reaction with metallic zinc, Chem. Geol. 260 (2009) 186–195.
P. Bose, D. Reckhow, The effect of ozonation on natural organic matter removal by alum coagulation, Water Res. 41 (2007) 1516–1524.
R.E. Botto, R. Wilson, R.E. Winans, Evaluation of the reliability of solid 13C-NMR spectroscopy for the quantitative analysis of coals: study of whole coals and maceral concentrates, Energy Fuels 1 (2) (1987) 173–181.
T.A. Brown, B.A. Jackson, B.J. Bythell, A.C. Stenson, Benefits of multidimensional fractionation for the study and characterization of natural organic matter, J. Chromatogr. A 1470 (2016) 84–96.
A. Bruchet, C. Rousseau, J. Mallevialle, Pyrolysis-GC-MS for investigating highmolecular-weight THM precursors and other refractory organics, J. Am. Water Works Assoc. 82 (1990) 66–74, 46.
E.M. Carstea, C.L. Popa, A. Baker, J. Bridgeman, In situ fluorescence measurements of dissolved organic matter: a review, Sci. Total Environ. 699 (2020) 134361.

W. Chen, H.Q. Yu, Advances in the characterization and monitoring of natural organic matter using spectroscopic approaches, Water Res. (2020) 116759.

J. Chen, B. Gu, E.J. LeBoef, H. Pan, S. Dai, Spectroscopic characterization of the structural and functional properties of natural organic matter fractions, Chemosphere 48 (2002) 59–68.

W. Chen, P. Westerhoff, J. Leenheer, K. Booksh, Fluoresence excitation-emission matrix regional integration to quantify spectra for dissolved organic matter, Environ. Sci. Technol. 37 (2003) 5701–5710.

C. Chen, X. Zhang, W. He, W. Lu, H. Han, Comparison of seven kinds of drinking water treatment processes to enhance organic material removal: a pilot test, Sci. Total Environ. 382 (2007) 93–102.

C. Chen, X. Zhang, L. Zhu, J. Liu, W. He, H. Han, Disinfection by-products and their precursors in a water treatment plant in North China: seasonal changes and fraction analysis, Sci. Total Environ. 397 (2008) 140–147.

W. Chen, Y. Han-Qing, Advances in the characterization and monitoring of natural organic matter using spectroscopic approaches, Water Res. 189 (2021) 116759, https://doi.org/10.1016/j.watres.2020.116759.

W. Chen, C. Qian, K.G. Zhou, H.Q. Yu, Molecular spectroscopic characterization of membrane fouling: a critical review, Chem 4 (7) (2018) 1492–1509.

K. Chon, K. Chon, J. Cho, Characterization of size fractionated dissolved organic matter from river water and wastewater effluent using preparative high performance size exclusion chromatography, Org. Geochem. 103 (2017) 105–112.

C. Chow, R. Fabris, M. Drikas, A rapid fractionation technique to characterize natural organic matter for the optimization of water treatment processes, J. Water Supply Res. Technol. Aqua 53 (2004) 85–92.

A.T. Chow, R.A. Dahlgren, S. Gao, Physical and chemical fractionation of dissolved organic matter and trihalomethane precursors: a review, J. Water Supply Res. Technol. 54 (2005) 475–507.

C. Chow, R. Fabris, J. van Leeuwen, D. Wang, M. Drikas, Assessing natural organic matter treatability using high performance size exclusion chromatography, Environ. Sci. Technol. 42 (2008) 6683–6689.

C. Chow, P. Kuntke, R. Fabris, M. Drikas, Organic characterization tools for distribution system management, Water Sci. Technol. 9 (2009a) 1–8.

C. Chow, J. von Leeuwen, R. Fabris, M. Drikas, Optimized coagulation using aluminium sulfate for the removal of dissolved organic carbon, Desalination 245 (2009b) 120–134.

A. Christy, A. Bruchet, D. Rybacki, Characterization of natural organic matter by pyrolysis/GC-MS, Environ. Int. 25 (1999) 181–189.

P.G. Coble, Characterization of marine and terrestrial DOM in seawater using excitation–emission matrix spectroscopy, Mar. Chem. 51 (1996) 325–346.

R.L. Cook, Coupling NMR to NOM, Anal. Bioanal. Chem. 376 (2004) 1484–1503.

W. Cooper, W. Song, M. Gonsior, D. Kalnina, B. Peake, S. Mezyk, Recent advances in structure and reactivity of dissolved organic matter in natural waters, Water Sci. Technol. 8 (2008) 615–623.

N. Cortés-Francisco, J. Caixach, Molecular characterization of dissolved organic matter through a desalination process by high resolution mass spectrometry, Environ. Sci. Technol. 47 (2013) 9619–9627.

A.R. Costa, M.N. de Pinho, M. Elimelech, Mechanisms of colloidal natural organic matter fouling in ultrafiltration, J. Membr. Sci. 281 (1–2) (2006) 716–725.

J.-P. Croue, Isolation of humic and non-humic NOM fractions: structural characterization, Environ. Monit. Assess. 92 (2004) 193–207.

J.-P. Croue, G.V. Korshin, M.M. Benjamin, Characterization of Natural Organic Matter in Drinking Water, American Water Works Association, 2000, p. 324.

M. Culea, O. Cozar, D. Ristoiu, Methods validation for the determination of trihalomethanes in drinking water, J. Mass Spectrom. 41 (2006) 1594–1597.
J. D'Andrilli, T. Dittmar, B.P. Koch, J.M. Purcell, A.G. Marshall, W.T. Cooper, Comprehensive characterization of marine dissolved organic matter by Fourier transform ion cyclotron resonance mass spectrometry with electrospray and atmospheric pressure photoionization, Rapid Commun. Mass Spectrom.: Int. J. Devoted Rapid Dissemen. Up-to-the-Minute Res. Mass Spectro. 24 (5) (2010) 643–650.
L. Danielsson, On the use of filters for distinguishing between dissolved and particulate fractions in natural waters, Water Res. 16 (1982) 179–182.
C. Datta, K. Ghosh, S. Mukherjee, Fluorescence excitation spectra of different fractions of humus, J. Indian Chem. Soc. 48 (1971) 279–287.
Y. Deng, C. Jung, R. Zhao, K. Torrens, L. Wu, Adsorption of UV-quenching substances (UVQS) from landfill leachate with activated carbon, Chem. Eng. J. 350 (2018) 739–746.
S. Derenne, T.T.N. Tu, Characterizing the molecular structure of organic matter from natural environments: an analytical challenge, Compt. Rendus Geosci. 346 (3–4) (2014) 53–63.
A. Deshmukh, C. Pacheco, M. Hay, S. Myneni, Structural environments of carboxyl groups in natural organic molecules from terrestrial systems. Part 2: 2D NMR spectroscopy, Geochem. Cosmochim. Acta 71 (2007) 3533–3544.
J. Edzwald, J. Tobiason, Enhanced coagulation: USA requirements and a broader view, Water Sci. Technol. 40 (1999) 63–70, 47.
P. Egeberg, A. Christy, M. Eikenes, The molecular size of natural organic matter (NOM) determined by diffusivimetry and seven other methods, Water Res. 36 (2002) 925–932.
E. Engelhaupt, T.S. Bianchi, Sources and composition of high-molecular-weight dissolved organic carbon in a southern Louisiana tidal stream (Bayou Trepagnier), Limnol. Oceanogr. 46 (4) (2001) 917–926.
I. Escobar, A. Randall, Assimilable organic carbon (AOC) and biodegradable dissolved organic carbon (BDOC): complementary measurements, Water Res. 35 (2001) 4444–4454.
D. Fabbri, M. Mongardi, L. Mintanari, G. Galletti, G. Chiavari, R. Scotti, Comparison between CP/MS 13C-NMR and pyrolysis-GC/MS in the structural characterization of humins and humic acids of soil and sediments, J. Anal. Chem. 362 (1998) 299–306.
R. Fabris, C. Chow, M. Drikas, B. Eikebrokk, Comparison of NOM character in selected Australian and Norwegian drinking waters, Water Res. 42 (2008) 4188–4196.
R.L. Fimmen, R.M. Cory, Y.P. Chin, T.D. Trouts, D.M. McKnight, Probing the oxidation–reduction properties of terrestrially and microbially derived dissolved organic matter, Geochem. Cosmochim. Acta 71 (12) (2007) 3003–3015.
F.H. Frimmel, F. Saravia, A. Gorenflo, NOM removal from different raw waters by membrane filtration, Water Sci. Technol. Water Supply 4 (4) (2004) 165–174.
X. Fu, J. Pan, H. Xu, Characterization and modification of the molecular weight distribution within dissolved organic matter using flow field-flow fractionation, Limnol. Oceanogr. Methods 18 (10) (2020) 560–569.
K. Fukano, K. Komiya, H. Sasaki, T. Hashimoto, Evaluation of new supports for high-pressure aqueous gel permeation chromatography: TSK-GEL SW type columns, J. Chromatogr. A 166 (1978) 47–54.
E. Gjessing, Use of Sephadex gels for estimation of humic substances in natural water, Nature 208 (1965) 1091–1092.
E. Gjessing, Ultrafiltration of aquatic humus, Environ. Sci. Technol. 4 (1970) 437–438.
E.T. Gjessing, Gel- and ultramembrane filtration of aquatic humus: a comparison of the two methods, Schweiz. Z. Hydrol. 35 (1973) 286–294.

E. Gjessing, J. Alberts, A. Bruchet, P. Egeberg, E. Lydersen, L. McGown, J. Mobed, U. Munster, J. Pempkowiak, M. Perdue, H. Ratnawerra, D. Rybacki, M. Takacs, G. Abbt-Braun, Multi-method characterization of natural organic matter isolated from water: characterization of reverse osmosis-isolates from water of two semi-identical dystrophic lakes basins in Norway, Water Res. 32 (1998) 3108–3124.

E. Gjessing, C. Steiro, G. Becher, A. Christy, Reduced analytical availability of polychlorinated biphenyls (PCB's) in colored surface water, Chemosphere 66 (2007) 644–649.

M. Gonsior, P. Schmitt-Kopplin, H. Stavklint, S.D. Richardson, N. Hertkorn, D. Bastviken, Changes in dissolved organic matter during the treatment processes of a drinking water plant in Sweden and formation of previously unknown disinfection byproducts, Environ. Sci. Technol. 48 (21) (2014) 12714–12722.

P. Greenwood, J. Leenher, C. McIntyre, L. Berwick, P. Franzmann, Bacterial biomarkers thermally released from dissolved organic matter, Org. Geochem. 37 (2006) 597–609.

J. Haarhoff, M. Kubare, B. Mamba, R. Krause, T. Nkambule, B. Matsebula, J. Menge, NOM characterization and removal at six Southern African water treatment plants, Drink. Water Eng. Sci. 3 (2010) 53–61.

M.A.R.T.I.N. Hassellov, F. von der Kammer, R. Beckett, Characterisation of aquatic colloids and macromolecules by field-flow fractionation, IUPAC Series Analyt. Phys. Chem. Environ. Systems 10 (2007) 223.

M. Hay, S. Myneni, Structural environments of carboxyl groups in natural organic molecules from terrestrial systems. Part 1: infrared spectroscopy, Geochem. Cosmochim. Acta 71 (2007) 3518–3532.

M.H.B. Hayes, P. MacCarthy, R.L. Malcolm, R.S. Swift, The Search for Structure: Setting the Scene, John Wiley & Sons, New York, 1989, pp. 3–31.

W. He, J. Hur, Conservative behavior of fluorescence EEM-PARAFAC components in resin fractionation processes and its applicability for characterizing dissolved organic matter, Water Res. 83 (2015) 217–226.

N. Her, G. Amy, D. Foss, J. Cho, Y. Yoon, P. Kosenka, Optimization of method for detecting and characterizing NOM by HPLC-size exclusion chromatography with UV and on-line DOC detection, Environ. Sci. Technol. 36 (2002) 1069–1076.

N. Her, G. Amy, H.R. Park, M. Song, Characterizing algogenic organic matter (AOM) and evaluating associated NF membrane fouling, Water Res. 38 (6) (2004) 1427–1438.

N. Her, G. Amy, J. Chung, J. Yoon, Y. Yoon, Characterizing dissolved organic matter and evaluating associated nanofiltration membrane fouling, Chemosphere 70 (2008a) 495–502.

N. Her, G. Amy, J. Sohn, U. Gunten von, UV absorbance ratio index with size exclusion chromatography (URI-SEC) as an NOM property indicator, J. Water Supply Res. Technol. Aqua 57 (2008b) 35–44.

N. Hertkorn, M. Harir, K.M. Cawley, P. Schmitt-Kopplin, R. Jaffé, Molecular characterization of dissolved organic matter from subtropical wetlands: a comparative study through the analysis of optical properties, NMR and FTICR/MS, Biogeosciences 13 (8) (2016) 2257–2277.

P. Herzsprung, N. Hertkorn, W. von Tümpling, M. Harir, K. Friese, P. Schmitt-Kopplin, Understanding molecular formula assignment of Fourier transform ion cyclotron resonance mass spectrometry data of natural organic matter from a chemical point of view, Anal. Bioanal. Chem. 406 (30) (2014) 7977–7987.

P. Herzsprung, W. v.Tümpling, N. Hertkorn, M. Harir, K. Friese, P. Schmitt-Kopplin, High-field FTICR-MS data evaluation of natural organic matter: are CHON5S2 molecular class formulas assigned to 13C isotopic m/z and in reality CHO components? Anal. Chem. 87 (19) (2015) 9563–9566.

W.C. Hockaday, J.M. Purcell, A.G. Marshall, J.A. Baldock, P.G. Hatcher, Electrospray and photoionization mass spectrometry for the characterization of organic matter in natural waters: a qualitative assessment, Limnol Oceanogr. Methods 7 (1) (2009) 81–95.
G. Hua, D. Reckhow, Characterization of disinfection byproduct precursors based on hydrophobicity and molecular size, Environ. Sci. Technol. 41 (2007) 3309–3315.
B. Hua, K. Veum, J. Yang, J. Jones, B. Deng, Parallel factor analysis of fluorescence EEM spectra to identify THM precursors in lake waters, Environ. Monit. Assess. 161 (1) (2010) 71–81.
L.C. Hua, S.J. Chao, C. Huang, Fluorescent and molecular weight dependence of THM and HAA formation from intracellular algogenic organic matter (IOM), Water Res. 148 (2019) 231–238.
S. Huber, F. Frimmel, A liquid chromatographic system with multi-detection for direct analysis of hydrophilic organic compounds in natural waters, J. Anal. Chem. 342 (1992a) 198–200.
S. Huber, F. Frimmel, A new method for the characterization of organic carbon in aquatic systems, Int. J. Environ. Anal. Chem. 49 (1992b) 49–57.
S. Huber, F. Frimmel, Direct gel chromatic characterization and quantification of marine dissolved organic carbon using high-sensitivity DOC detection, Environ. Sci. Technol. 28 (1994) 1194–1197.
S.A. Huber, A. Balz, M. Abert, W. Pronk, Characterisation of aquatic humic and non-humic matter with size-exclusion chromatography—organic carbon detection—organic nitrogen detection (LC-OCD-OND), Water Res. 45 (2) (2011) 879–885.
N. Hudson, A. Baker, D. Reynolds, Fluorescence analysis of dissolved organic matter in natural, waste and polluted waters—a review, River Res. Appl. 23 (2007) 631–649.
N. Hudson, A. Baker, D. Ward, D. Reynolds, C. Brunsdon, C. Carliell-Marquet, S. Browning, Can fluorescence spectrometry be used as a surrogate for the biochemical oxygen demand (BOD) test in water quality assessment? An example from South West England, Sci. Total Environ. 391 (2008) 149–158.
J. Hur, M. Williams, M. Schlautman, Evaluating spectroscopic and chromatographic techniques to resolve dissolved organic matter via end member mixing analysis, Chemosphere 63 (2006) 387–402.
A. Ignatev, T. Tuhkanen, Step-by-step analysis of drinking water treatment trains using size-exclusion chromatography to fingerprint and track protein-like and humic/fulvic-like fractions of dissolved organic matter, Environ. Sci. J. Integr. Environ. Res.: Water Res. Technol. 5 (9) (2019) 1568–1581.
S.K. Ishii, T.H. Boyer, Behavior of reoccurring PARAFAC components in fluorescent dissolved organic matter in natural and engineered systems: a critical review, Environ. Sci. Technol. 46 (4) (2012) 2006–2017.
M. Jaouadi, N. Amdouni, L. Duclaux, Characteristics of natural organic matter extracted from the waters of Medjerda dam (Tunisia), Desalination 305 (2012) 64–71.
S. Jeong, S.-J. Kim, C. Kim, S. Vigneswaran, T. Nguyen, H.-K. Shon, J. Kandasamy, I. Kim, A detailed organic matter characterization of pretreated seawater using low pressure microfiltration hybrid systems, J. Membr. Sci. 428 (2013) 290–300.
D.W. Johnstone, C.M. Miller, Fluorescence excitation—emission matrix regional transformation and chlorine consumption to predict trihalomethane and haloacetic acid formation, Environ. Eng. Sci. 26 (7) (2009) 1163–1170.
C. Jung, H. Son, The relationship between disinfection by-products formation and characteristics of natural organic matter in raw water, Kor. J. Chem. Eng. 25 (2008) 714–720.
M.-G. Kang, Y.-H. Ku, Y.-K. Cho, M.-J. Yu, Variation of dissolved organic matter and microbial regrowth potential through drinking water treatment processes, Water Sci. Technol. 6 (2006) 57–66.

V. Kanokkantapong, T. Marhaba, B. Panyapinyophol, P. Pavasant, FTIR evaluation of functional groups involved in the formation of haloacetic acids during the chlorination of raw water, J. Hazard Mater. B136 (2006a) 188–196.

V. Kanokkantapong, T. Marhaba, P. Pavasant, B. Panyapinyophol, Characterization of haloacetic acid precursors in source water, J. Environ. Manag. 80 (2006b) 214–221.

N. Kawasaki, K. Matsushige, K. Komatsu, A. Kohzu, F.W. Nara, F. Ogishi, A. Imai, Fast and precise method for HPLC–size exclusion chromatography with UV and TOC (NDIR) detection: importance of multiple detectors to evaluate the characteristics of dissolved organic matter, Water Res. 45 (18) (2011) 6240–6248.

H.-C. Kim, M.-J. Yu, Characterization of aquatic humic substances to DBPs formation in advanced treatment processes for conventionally treated water, J. Hazard Mater. 143 (2007) 486–493.

H.C. Kim, M.J. Yu, I. Han, Multi-method study of the characteristic chemical nature of aquatic humic substances isolated from the Han River, Korea, Appl. Geochem. 21 (7) (2006a) 1226–1239.

S. Kim, R.P. Rodgers, A.G. Marshall, Truly "exact" mass: elemental composition can be determined uniquely from molecular mass measurement at ~ 0.1 mDa accuracy for molecules up to ~ 500 Da, Int. J. Mass Spectrom. 251 (2–3) (2006b) 260–265.

J. Kim, Z. Cai, M. Benjamin, NOM fouling mechanisms in a hybrid adsorption/membrane system, J. Membr. Sci. 349 (2010) 35–43.

M. Kögler, B. Zhang, L. Cui, Y. Shi, M. Yliperttula, T. Laaksonen, K. Zhang, Real-time Raman based approach for identification of biofouling, Sensor. Actuator. B Chem. 230 (2016) 411–421.

J. Koprivnjak, E. Perdue, P. Pfromm, Coupling reverse osmosis with electrodialysis to isolate natural organic matter from fresh waters, Water Res. 40 (2006) 3385–3392.

J.F. Koprivnjak, P.H. Pfromm, E. Ingall, T.A. Vetter, P. Schmitt-Kopplin, N. Hertkorn, E.M. Perdue, Chemical and spectroscopic characterization of marine dissolved organic matter isolated using coupled reverse osmosis–electrodialysis, Geochem. Cosmochim. Acta 73 (14) (2009) 4215–4231.

G. Korshin, M. Benjamin, C.-W. Li, Use of differential spectroscopy to evaluate the structure and reactivity of humics, Water Sci. Technol. 40 (1999) 9–16.

G. Korshin, C. Chow, R. Fabris, M. Drikas, Absorbance spectroscopy-based examination of effects of coagulation on the reactivity of fractions of natural organic matter with varying apparent molecular weights, Water Res. 43 (2009) 1541–1548.

D.N. Kothawala, C.A. Stedmon, R.A. Müller, G.A. Weyhenmeyer, S.J. Köhler, L.J. Tranvik, Controls of dissolved organic matter quality: evidence from a large-scale boreal lake survey, Global Change Biol. 20 (2013) 1101–1114.

I. Kristiana, B. Allpike, C. Joll, A. Heitz, R. Trolio, Understanding the behaviour of molecular weight fractions of natural organic matter to improve water treatment processes, Water Sci. Technol. Water Supply 10 (2010) 59–68.

M. Kulovaara, N. Corin, P. Backlund, J. Tervo, Impact of UV254-radiation on aquatic humic substances, Chemosphere 33 (5) (1996) 783–790.

E. Kunenkov, A. Kononikhin, I. Perminova, N. Hertkorn, A. Gaspar, P. Scmitt-Kopplin, I. Popov, A. Garmash, E. Nikolaev, Total mass difference statistics algorithm: a new approach to identification of high-mass building blocks in electrospray ionization Fourier transform ion cyclotron mass spectrometry data of natural organic matter, Anal. Chem. 81 (2009) 10106–10115.

F. Laborda, S. Ruiz-Beguería, E. Bolea, J. Castillo, Functional speciation of metaldissolved organic matter complexes by size exclusion chromatography coupled to inductively coupled plasma mass spectrometry and deconvolution analysis, Spectrochim. Acta B 392 (2009) 392–398.

B. Lam, A.J. Simpson, Direct 1H NMR spectroscopy of dissolved organic matter in natural waters, Analyst 133 (2) (2008) 263–269.

U. Lankes, H.-D. Lüdemann, F. Frimmel, Search for basic relationships between "molecular size" and "chemical structure" of aquatic natural organic matter – answers from 13C and 15N CPMAS NMR spectroscopy, Water Res. 42 (2008) 1051–1060.

U. Lankes, M. Müller, M. Weber, F. Frimmel, Reconsidering the quantitative analysis of organic carbon concentrations in size exclusion chromatography, Water Res. 43 (2009) 915–924.

J.R. Lead, K.J. Wilkinson, E. Balnois, B.J. Cutak, C.K. Larive, S. Assemi, R. Beckett, Diffusion coefficients and polydispersities of the Suwannee river fulvic acid: comparison of fluorescence correlation spectroscopy, pulsed-field gradient nuclear magnetic resonance, and flow field-flow fractionation, Environ. Sci. Technol. 34 (2000) 3508–3513.

D.-J. Lee, Y.-L. Cheng, R.-J. Wong, Characteristics of natural organic matters in raw and treated drinking water: comparison study, J. Taiwan Inst. Chem. Eng. 44 (2013) 103–110.

S.T. Lee, B. Yang, J.Y. Kim, J.H. Park, M.H. Moon, Combining asymmetrical flow field-flow fractionation with on-and off-line fluorescence detection to examine biodegradation of riverine dissolved and particulate organic matter, J. Chromatogr. A 1409 (2015) 218–225.

J. Leenher, Comprehensive approach to preparative isolation and fractionation of dissolved organic carbon from natural waters and wastewaters, Environ. Sci. Technol. 15 (1981) 578–587.

J.A. Leenher, Fractionation techniques for aquatic humic substances, in: G. Aiken, D. McKnight, R. Wershaw, P. MacCarthy (Eds.), Humic Substances in Soil, Sediment and Water, John Wiley & Sons, 1985, pp. 409–429.

J. Leenher, J.-P. Croué, Characterizing dissolved aquatic organic matter, Environ. Sci. Technol. 37 (2003) 18A–26A.

A. Li, J. Hu, W. Li, W. Zhang, X. Wang, Polarity based fractionation of fulvic acids, Chemosphere 77 (2009) 1419–1426.

W.T. Li, Z.X. Xu, A.M. Li, W. Wu, Q. Zhou, J.N. Wang, HPLC/HPSEC-FLD with multi-excitation/emission scan for EEM interpretation and dissolved organic matter analysis, Water Res. 47 (3) (2013) 1246–1256.

C. Li, D. Wang, X. Xu, M. Xu, Z. Wang, R. Xiao, Tracking changes in composition and amount of dissolved organic matter throughout drinking water treatment plants by comprehensive two-dimensional gas chromatography–quadrupole mass spectrometry, Sci. Total Environ. 609 (2017) 123–131.

R. Liu, J. Lead, A. Baker, Fluorescence characterization of cross flow ultrafiltration derived freshwater colloidal and dissolved organic matter, Chemosphere 68 (2007) 1304–1311.

S. Liu, M. Lim, R. Fabris, C. Chow, K. Chiang, M. Drikas, R. Amal, Removal of humic acid using TiO2 photocatalytic process – fractionation and molecular weight characterization studies, Chemosphere 72 (2008) 263–271.

S. Liu, M. Lim, R. Fabris, C. Chow, M. Drikas, R. Amal, Comparison of photocatalytic degradation of natural organic matter in two Australian surface waters using multiple analytical techniques, Org. Geochem. 41 (2010) 124–129.

B. Lyvén, M. Hassellöv, D. Turner, C. Haraldsson, K. Andersson, Competition between iron- and carbon-based colloidal carriers for trace metals in a freshwater assessed using flow field-flow fractionation coupled to ICPMS, Geochimica et Cosmochimica Acta 67 (20) (2003), https://doi.org/10.1016/S0016-7037(03)00087-5.

H. Ma, H. Allen, Y. Yin, Characterization of isolated fractions of dissolved organic matter from natural waters and a wastewater effluent, Water Res. 35 (2001) 985–996.

R. Malcolm, P. MacCarthy, Quantitative evaluation of XAD-8 and XAD-4 resins used in tandem for removing organic solutes from water, Environ. Int. 18 (1992) 597–607.

J. Mao, R. Cory, D. McKnight, K. Schmidt-Rohr, Characterization of a nitrogenrich fulvic acid and its precursor algae from solid state NMR, Org. Geochem. 38 (2007a) 1277–1292.

J. Mao, L. Tremblay, J.-P. Gagné, S. Kohl, J. Rice, K. Schmidt-Rohr, Humic acids from particulate organic matter in the Saguenay Fjord and St. Lawrence Estuary investigated by advanced solid-state NMR, Geochem. Cosmochim. Acta 71 (2007b) 5483–5499.

J. Mao, X. Kong, K. Schmidt-Rohr, J.J. Pignatello, E.M. Perdue, Advanced solid-state NMR characterization of marine dissolved organic matter isolated using the coupled reverse osmosis/electrodialysis method, Environ. Sci. Technol. 46 (11) (2012) 5806–5814.

J. Mao, X. Cao, D.C. Olk, W. Chu, K. Schmidt-Rohr, Advanced solid-state NMR spectroscopy of natural organic matter, Prog. Nucl. Magn. Reson. Spectrosc. 100 (2017) 17–51.

T. Marhaba, Fluorescence technique for rapid identification of DOM fractions, J. Environ. Eng. Sci. 126 (2000) 145–152.

T.F. Marhaba, D. Van, R.L. Lippincott, Changes in NOM fractions through treatment: a comparison of ozonation and chlorination, Ozone Sci. Technol. 22 (2000) 249–266.

T. Marhaba, Y. Pu, K. Bengraine, Modified dissolved organic matter fractionation technique for natural water, J. Hazard Mater. 101 (2003) 43–53.

D. Markechova, M. Tomokova, J. Sadecka, Fluorescence excitation–emission matrix spectroscopy and parallel factor analysis in drinking water treatment: a review, Pol. J. Environ. Stud. 22 (2013) 1289–1295.

A. Matilainen, P. Iivari, J. Sallanko, E. Heiska, T. Tuhkanen, The role of ozonation and activated carbon filtration in the natural organic matter removal from drinking water, Environ. Technol. 27 (2006) 1171–1180.

A. Matilainen, E.T. Gjessing, T. Lahtinen, L. Hed, A. Bhatnagar, M. Sillanpää, An overview of the methods used in the characterization of natural organic matter (NOM) in relation to drinking water treatment, Chemosphere 83 (2011) 1431–1442.

D. Mawhinney, F. Rosario-Ortiz, S. Baik, B. Vanderford, S. Snyder, Characterization of fulvic acids by liquid chromatography-quadrupole time-of-flight mass spectrometry, J. Chromatogr. A 1216 (2009) 1319–1324.

F. Messaud, R. Sanderson, J. Runyon, T. Otte, H. Pasch, S. Williams, An overview on field-flow fractionation techniques and their applications in the separation and characterization of polymers, Prog. Polym. Sci. 34 (2009) 351–368.

E.C. Minor, M.M. Swenson, B.M. Mattson, A.R. Oyler, Structural characterization of dissolved organic matter: a review of current techniques for isolation and analysis, Environ. Sci.: Proc. Impacts 16 (9) (2014) 2064–2079.

C.S. Moody, A comparison of methods for the extraction of dissolved organic matter from freshwaters, Water Res. 184 (2020) 116114.

J. Moon, S. Kim, J. Cho, Characterizations of natural organic matter as nano particle using flow field-flow fractionation, Colloids Surf. A Physicochem. Eng. Aspects 287 (2006) 232–236.

K. Murphy, C. Stedmon, T. Waite, G. Ruiz, Distinguishing between terrestrial and autochthonous organic matter sources in marine environments using fluorescence spectroscopy, Mar. Chem. 108 (2008) 40–58.

E. Neubauer, F. vd Kammer, T. Hofmann, Using FLOWFFF and HPSEC to determine trace metal–colloid associations in wetland runoff, Water Res. 47 (8) (2013) 2757–2769.

R. Nimmagadda, C. McRae, Characterization of the backbone structures of several fulvic acids using a novel selective chemical reduction method, Org. Geochem. 38 (2007) 1061–1072.

U.G. Nwosu, R.L. Cook, 13C Nuclear magnetic resonance and electron paramagnetic spectroscopic comparison of hydrophobic acid, transphilic acid, and reverse osmosis May 2012 isolates of organic matter from the suwannee river, Environ. Eng. Sci. 32 (1) (2015) 14–22.

S. Odén, Die Huminsaure. Kolloidchem. Beih. 11 (1919) 75.

C.L. Osburn, D.P. Morris, K.A. Thorn, R.E. Moeller, Chemical and optical changes in freshwater dissolved organic matter exposed to solar radiation, Biogeochemistry 54 (3) (2001) 251–278.

E. O'Loughlin, Y.-P. Chin, Effect of detector wavelength on the determination of the molecular weight of humic substances by high-pressure size exclusion chromatography, Water Res. 35 (2001) 333–338.

Y. Pan, H. Li, X. Zhang, A. Li, Characterization of natural organic matter in drinking water: sample preparation and analytical approaches, Trends Environ. Anal. Chem. 12 (2016) 23–30.

J.-H. Park, Spectroscopic characterization of dissolved organic matter and its interactions with metals in surface waters using size exclusion chromatography, Chemosphere 77 (2009) 485–494.

Y. Park, S.A. MacIsaac, P. Kaur, M. Brophy, G.A. Gagnon, Monitoring the influence of wastewater effluent on a small drinking water system using EEM fluorescence spectroscopy coupled with a PARAFAC and PCA statistical approach, Environ. Sci. Process Impacts 23 (6) (2021) 880–889.

Z. Parsi, N. Hartog, T. Górecki, J. Poerschmann, Analytical pyrolysis as a tool for the characterization of natural organic matter — a comparison of different approaches, J. Anal. Appl. Pyrolysis 79 (2007) 9–15.

R.H. Peiris, C. Hallé, H. Budman, C. Moresoli, S. Peldszus, P.M. Huck, R.L. Legge, Identifying fouling events in a membrane-based drinking water treatment process using principal component analysis of fluorescence excitation-emission matrices, Water Res. 44 (2010) 185–194.

N.M. Peleato, R.C. Andrews, Comparison of three-dimensional fluorescence analysis methods for predicting formation of trihalomethanes and haloacetic acids, J. Environ. Sci. 27 (2015) 159–167.

N.M. Peleato, R.L. Legge, R.C. Andrews, Neural networks for dimensionality reduction of fluorescence spectra and prediction of drinking water disinfection by-products, Water Res. 136 (2018) 84–94.

Y. Penru, F. Simon, A. Guastalli, S. Esplugas, J. Llorens, S. Baig, Characterization of natural organic matter from Mediterranean coastal seawater, J. Water Supply Res. Technol. Aqua 62 (2013) 42–51.

J. Peuravuori, K. Pihlaja, N. Välimäki, Isolation and characterization of natural organic matter from lake water: two different adsorption chromatographic methods, Environ. Int. 23 (1997) 453–464.

J. Peuravuori, P. Ingman, K. Pihlaja, Critical comments on accuracy of quantitative determination of natural humic matter by solid state 13C NMR spectroscopy, Talanta 59 (2003) 177–189.

A.D. Pifer, J.L. Fairey, Improving on SUVA254 using fluorescence-PARAFAC analysis and asymmetric flow-field flow fractionation for assessing disinfection byproduct formation and control, Water Res. 46 (2012) 2927–2936.

M. Pivokonsky, J. Naceradska, T. Brabenec, K. Novotna, M. Baresova, V. Janda, The impact of interactions between algal organic matter and humic substances on coagulation, Water Res. 84 (2015) 278–285, https://doi.org/10.1016/j.watres.2015.07.047.

A. Posner, Importance of electrolyte in the determination of molecular weights by sephadex gel with specific reference to humic acids, Nature 198 (1963) 1161–1163.

T. Ratpukdi, J.A. Rice, G. Chilom, A. Bezbaruah, E. Khan, Rapid fractionation of natural organic matter in water using a novel solid-phase extraction technique, Water Environ. Res. 81 (11) (2009) 2299–2308.

P.S. Redwood, J.R. Lead, R.M. Harrison, I.P. Jones, S. Stoll, Characterization of humic substances by environmental scanning electron microscopy, Environ. Sci. Technol. 39 (7) (2005) 1962–1966.

T. Reemtsma, Determination of molecular formulas of natural organic matter molecules by (ultra-) high-resolution mass spectrometry – status and needs, J. Chromatogr. A 1216 (2009) 3687–3701.

T. Reemtsma, A. These, Comparative investigation of low-molecular-weight fulvic acids of different origin by SEC-Q-TOF-MS: new insights into structure and formation, Environ. Sci. Technol. 39 (2005) 3507–3512.

T. Reemtsma, A. These, A. Springer, M. Linscheid, Fulvic acids as transition state of organic matter: indications from high resolution mass spectrometry, Environ. Sci. Technol. 40 (2006) 5839–5845.

T. Reemtsma, A. These, A. Springer, M. Linscheid, Differences in the molecular composition of fulvic acid size fractions detected by size-exclusion chromatography-on line Fourier transform ion cyclotron resonance (FTICR) mass spectrometry, Water Res. 42 (2008) 63–72.

S. Richardson, M. Plewa, E. Wagner, R. Schoeny, D. DeMarini, Occurrence, genotoxicity, and carcinogenicity of regulated and emerging disinfection by-products in drinking water: a review and roadmap for research, Mutat. Res. Rev. Mutat. Res. 636 (2007) 178–242.

S. Richardson, F. Fasano, J. Ellington, F. Crumley, K. Buettner, J. Evans, B. Blount, L. Silva, T. Waite, G. Luther, A. McKague, R. Miltner, E. Wagner, M. Plewa, Occurrence and mammalian cell toxicity of iodinated disinfection by-products in drinking water, Environ. Sci. Technol. 42 (2008) 8330–8338.

J. Ritchie, E. Perdue, Analytical constraints on acidic functional groups in humic acids, Org. Geochem. 39 (2008) 783–799.

P. Roccaro, F. Vagliasindi, Differential vs. absolute UV absorbance approaches in studying NOM reactivity in DBPs formation: comparison and applicability, Water Res. 43 (2009) 744–750.

F.J. Rodríguez, P. Schlenger, M. García-Valverde, Monitoring changes in the structure and properties of humic substances following ozonation using UV–Vis, FTIR and 1H NMR techniques, Sci. Total Environ. 541 (2016) 623–637.

F. Rosario-Ortiz, S. Snyder, I. Suffet, Characterization of dissolved organic matter in drinking water sources impacted by multiple tributaries, Water Res. 41 (2007) 4115–4128.

F.L. Rosario-Ortiz, S. Snyder, I.H. Suffet, Characterization of the polarity of natural organic matter under ambient conditions by the polarity rapid assessment method (PRAM), Environ. Sci. Technol. 41 (14) (2007) 4895–4900.

N. Sanchez, A. Skeriotis, C. Miller, Assessment of dissolved organic matter fluorescence PARAFAC components before and after coagulation–filtration in a full scale water treatment plant, Water Res. 47 (2013) 1679–1690.

N.P. Sanchez, A.T. Skeriotis, C. Miller, A PARAFAC-based long-term assessment of DOM in a multi-coagulant drinking water treatment scheme, Environ. Sci. Technol. 48 (2014) 1582–1591.

L.M. Sanly, K. Chiang, R. Amal, R. Fabris, C. Chow, M. Drikas, A study on the removal of humic acid using advanced oxidation process, Separ. Sci. Technol. 42 (2007) 1391–1404.

S. Sarathy, M. Mohseni, The impact of UV/H2O2 advanced oxidation on molecular size distribution of chromophoric natural organic matter, Environ. Sci. Technol. 41 (2007) 8315–8320.
H. Schulten, Analytical pyrolysis of humic substances and soils: geochemical, agricultural and ecological consequences, J. Anal. Appl. Pyrolysis 25 (1993) 97–122.
B. Seredynska-Sobecka, A. Baker, A. Lead, Characterization of colloidal and particulate organic carbon in freshwaters by thermal fluorescence quenching, Water Res. 41 (2007) 3069–3076.
E. Sharp, P. Jarvis, S. Parsons, B. Jefferson, Impact of fractional character on the coagulation of NOM, Colloids Surf. A Physicochem. Eng. Aspects 286 (2006a) 104–111.
E. Sharp, S. Parsons, B. Jefferson, The impact of seasonal variations in DOC arising from a moorland peat catchment on coagulation with iron and aluminium salts, Environ. Pollut. 140 (2006b) 436–443.
Y. Shutova, A. Baker, J. Bridgeman, R. Henderson, Spectroscopic characterization of dissolved organic matter changes in drinking water treatment: from PARAFAC analysis to online monitoring wavelengths, Water Res. 54 (2014) 159–169.
F. Simon, Y. Penru, A. Guastalli, S. Esplugas, J. Llorens, S. Baig, NOM characterization by LC-OCD in a SWRO desalination line, Desalin. Water Treat. 51 (2013) 1776–1780.
A. Simpson, M. Simpson, W. Kingery, B. Lefebvre, A. Moser, A. Williams, M. Kvasha, B. Kvelleher, The application of 1H high-resolution magic-angle spinning NMR for the study of clay-organic associations in natural and synthetic complexes, Langmuir 22 (2006) 4498–4503.
Y. Soh, F. Roddick, J. van Leeuwen, The impact of alum coagulation on the character, biodegradability and disinfection by-product formation potential of reservoir natural organic matter (NOM) fractions, Water Sci. Technol. 58 (2008) 1173–1179.
N. Solihat, T. Acter, D. Kim, A. Plante, S. Kim, Analyzing solid-phase natural organic matter using laser desorption ionization ultrahigh resolution mass spectrometry, Anal. Chem. 91 (1) (2019) 951–957, https://doi.org/10.1021/acs.analchem.8b04032.
H. Song, O. Orr, Y. Hong, T. Karanfil, Isolation and fractionation of natural organic matter: evaluation of reverse osmosis performance and impact of fractionation parameters, Environ. Monit. Assess. 153 (2009) 307–321.
C. Specht, F. Frimmel, Specific interactions of organic substances in size-exclusion chromatography, Environ. Sci. Technol. 34 (2000) 2361–2366.
R. Spencer, L. Bolton, A. Baker, Freeze/thaw and pH effects on freshwater dissolved organic matter fluorescence and absorbance properties from a number of UK locations, Water Res. 41 (2007) 2941–2950.
C. Stedmon, R. Bro, Characterizing dissolved organic matter fluorescence with paral-lel factor analysis: a tutorial, Limnol. Oceanogr.: Methods 6 (2008) 572–579, https://doi.org/10.4319/lom.2008.6.572.
C. Stedmon, S. Markager, Resolving the variability in dissolved organic matter fluorescence in a temperate estuary and its catchment using PARAFAC analysis, Limnol. Oceanogr. 50 (2005) 686–697.
C. Stedmon, S. Markagera, R. Bro, Tracing dissolved organic matter in aquatic environments using a new approach to fluorescence spectroscopy, Mar. Chem. 82 (2003) 239–254.
A. Stenson, Reversed-phase chromatography fractionation tailored to mass spectral characterization of humic substances, Environ. Sci. Technol. 42 (2008) 2060–2065.
A.K. Stoddart, G.A. Gagnon, Application of photoelectrochemical chemical oxygen demand to drinking water, J. Am. Water Works Assoc. 106 (9) (2014) E383–E390.
B. Stolpe, L. Guo, A.M. Shiller, M. Hassellöv, Size and composition of colloidal organic matter and trace elements in the Mississippi River, Pearl River and the northern Gulf of

Mexico, as characterized by flow field-flow fractionation, Mar. Chem. 118 (2010) 119–128.
J. Swietlik, A. Dabrowska, U. Raczyk-Stanislawiak, J. Nawrocki, Reactivity of natural organic matter fractions with chlorine dioxide and ozone, Water Res. 38 (2004) 547–558.
J. Templier, S. Derenne, J.-P. Croué, C. Largeau, Comparative study of two fractions of riverine dissolved organic matter using various analytical pyrolytic methods and a 13C CP/MAS NMR approach, Org. Geochem. 36 (2005) 1418–1442.
L. Tercero Espinoza, E. ter Haseborg, M. Weber, F. Frimmel, Investigation of the photocatalytic degradation of brown water natural organic matter by size exclusion chromatography, Appl. Catal. B Environ. 87 (2009) 56–62.
K. Thorn, L. Cox, N-15 NMR spectra of naturally abundant nitrogen in soil and aquatic natural organic matter samples of the International Humic Substances Society, Org. Geochem. 40 (2009) 484–499.
E. Thurman, Organic Geochemistry of Natural Waters, Martinus Nijhoff/Dr W. Junk Publishers, Dordrecht, The Netherlands, 1985.
E. Thurman, R. Malcolm, Preparative isolation of aquatic humic substances, Environ. Sci. Technol. 15 (1981) 463–466.
B.F. Trueman, S.A. MacIsaac, A.K. Stoddart, G.A. Gagnon, Prediction of disinfection by-product formation in drinking water via fluorescence spectroscopy, Environ. Sci. J. Integr. Environ. Res.: Water Res. Technol. 2 (2) (2016) 383–389.
B.F. Trueman, T. Anaviapik-Soucie, V. L'Hérault, G.A. Gagnon, Characterizing colloidal metals in drinking water by field flow fractionation, Environ. Sci. J. Integr. Environ. Res.: Water Res. Technol. 5 (12) (2019) 2202–2209.
C. Uyguner, S. Suphandag, A. Kerc, M. Bekbolet, Evaluation of adsorption and coagulation characteristics of humic acids preceded by alternative advanced oxidation techniques, Desalination 210 (2007) 183–193.
C.S. Uyguner-Demirel, M. Bekbolet, Significance of analytical parameters for the understanding of natural organic matter in relation to photocatalytic oxidation, Chemosphere 84 (2011) 1009–1031.
S. Valencia, J. Marín, G. Restrepoa, F. Frimmel, Application of excitation–emission fluorescence matrices and UV/Vis absorption to monitoring the photocatalytic degradation of commercial humic acid, Sci. Total Environ. 442 (2013) 207–214.
S. Valencia, J. Marín, G. Restrepo, F.H. Frimmel, Evaluation of natural organic matter changes from Lake Hohloh by three-dimensional excitation–emission matrix fluorescence spectroscopy during TiO2/UV process, Water Res. 51 (2014) 124–133.
T. Virtanen, S.-P. Reinikainen, M. Kögler, M. Mänttäri, T. Viitala, M. Kallioinen, Real-time fouling monitoring with Raman spectroscopy, J. Membr. Sci. 525 (2017) 312–319, https://doi.org/10.1016/j.memsci.2016.12.005.
B. Warton, A. Heitz, B. Allpike, R. Kagi, Size-exclusion chromatography with organic carbon detection using a mass spectrometer, J. Chromatogr. A 1207 (2008) 186–189.
Q. Wei, C. Feng, D. Wang, B. Shi, L. Zhang, Q. Wei, H. Tang, Seasonal variations of chemical and physical characteristics of dissolved organic matter and trihalomethane precursors in a reservoir: a case study, J. Hazard Mater. 150 (2008) 257–264.
R.G. Wetzel, P.G. Hatcher, T.S. Bianchi, Natural photolysis by ultraviolet irradiance of recalcitrant dissolved organic matter to simple substrates for rapidbacterial metabolism, Limnol. Oceanogr. 40 (8) (1995) 1369–1380.
M.A. Wilson, A.H. Gillam, P.J. Collin, Analysis of the structure of dissolved marine humic substances and their phytoplanktonic precursors by 1H and 13C nuclear magnetic resonance, Chem. Geol. 40 (3–4) (1983) 187–201.
S. Wong, J. Hanna, S. King, T. Carroll, R. Eldridge, D. Dixon, B. Bolto, S. Hesse, G. Abbt-Braun, F.H. Frimmel, Fractionation of natural organic matter in drinking water and

characterization by 13C cross-polarization magic-angle spinning NMR spectroscopy and size exclusion chromatography, Environ. Sci. Technol. 36 (2002) 3497–3503.

F. Wu, R. Evans, P. Dillon, Separation and characterization of NOM by highperformance liquid chromatography and on-line three-dimensional excitation emission matrix fluorescence detection, Environ. Sci. Technol. 37 (2003) 3687–3693.

F. Wu, R. Evans, P. Dillon, Y. Cai, Rapid quantification of humic and fulvic acids by HPLC in natural waters, Appl. Geochem. 22 (2007a) 1598–1605.

F. Wu, D. Kothawala, R. Evans, P. Dillon, Y. Cai, Relationships between DOC concentration, molecular size and fluorescence properties of DOM in a stream, Appl. Geochem. 22 (2007b) 1659–1667.

Q.Y. Wu, T.H. Zhou, Y. Du, B. Ye, W.L. Wang, H.Y. Hu, Characterizing the molecular weight distribution of dissolved organic matter by measuring the contents of electron-donating moieties, UV absorbance, and fluorescence intensity, Environ. Int. 137 (2020) 105570.

U.J. Wünsch, K.R. Murphy, C.A. Stedmon, The one-sample PARAFAC approach reveals molecular size distributions of fluorescent components in dissolved organic matter, Environ. Sci. Technol. 51 (20) (2017) 11900–11908.

L. Xing, R. Fabris, C. Chow, J. van Leeuwen, M. Drikas, D. Wang, Prediction of DOM removal of low specific UV absorbance surface waters using HPSEC combined with peak fitting, J. Environ. Sci. 24 (2012) 1174–1180.

H. Xu, M. Yan, W. Li, H. Jiang, L. Guo, Dissolved organic matter binding with Pb (II) as characterized by differential spectra and 2D UV–FTIR heterospectral correlation analysis, Water Res. 144 (2018) 435–443.

H. Yamamura, K. Okimoto, K. Kimura, Y. Watanabe, Hydrophilic fraction of natural organic matter causing irreversible fouling of microfiltration and ultrafiltration membranes, Water Res. 54 (2014) 123–136.

Y. Yamashita, R. Jaffé, Characterizing the interactions between trace metals and dissolved organic matter using excitation–emission matrix and parallel factor analysis, Environ. Sci. Technol. 42 (2008) 7374–7379.

Y. Yamashita, T. Saito, Effects of weak organic acids on the size distribution and size-dependent metal binding of humic substances as studied by flow field-flow fractionation, J. Environ. Chem. Eng. 3 (4B) (2015) 3024–3029, https://doi.org/10.1016/j.jece.2015.03.026.

X. Yang, W. Guo, X. Zhang, F. Chen, T. Ye, W. Liu, Formation of disinfection by-products after pre-oxidation with chlorine dioxide or ferrate, Water Res. 47 (15) (2013) 5856–5864.

K. Yapsakli, F. Cecen, Use of an enriched inoculum for determination of biodegradable dissolved organic carbon (BDOC) in drinking water, Water Sci. Technol. 9 (2009) 149–157.

W. Yu, T. Liu, J. Crawshaw, T. Liu, N. Graham, Ultrafiltration and nanofiltration membrane fouling by natural organic matter: mechanisms and mitigation by pre-ozonation and pH, Water Res. 139 (2018) 353–362.

L. Zappia, M. Alessadrino, S. Garbin, P. Pringle, A. Heitz, C. Joll, D. Masters, B. Hiller, S. Capewell, P. Franzmann, J. Plumb, S. Von Eckstaedt, K. Cadee, Comparison of methods for the determination of biodegradable dissolved organic carbon in potable water supply: use of a novel biofilm, Water Sci. Technol. 8 (2008) 633–641.

M. Zark, T. Dittmar, Universal molecular structures in natural dissolved organic matter, Nat. Commun. 9 (1) (2018) 1–8.

T. Zhang, J. Lu, J. Ma, Z. Qiang, Fluorescence spectroscopic characterization of DOM fractions isolated from a filtered river water after ozonation and catalytic ozonation, Chemosphere 71 (2008) 911–921.

H. Zhang, Y. Zhang, Q. Shi, J. Hu, M. Chu, J. Yu, M. Yang, Study on transformation of natural organic matter in source water during chlorination and its chlorinated products using ultrahigh resolution mass spectrometry, Environ. Sci. Technol. 46 (2012a) 4396–4402.

H. Zhang, Y. Zhang, Q. Shi, S. Ren, J. Yu, F. Ji, W. Luo, M. Yang, Characterization of low molecular weight dissolved natural organic matter along the treatment trait of a waterworks using Fourier transform ion cyclotron resonance mass spectrometry, Water Res. 46 (2012b) 5197–5204.

X. Zhang, J. Han, X. Zhang, J. Shen, Z. Chen, W. Chu, Y. Zhou, Application of Fourier transform ion cyclotron resonance mass spectrometry to characterize natural organic matter, Chemosphere (2020) 127458.

Z.-Y. Zhao, D.-J. Gu, H.-B. Li, X.-Y. Li, K. Leung, Disinfection characteristics of the dissolved organic fractions at several stages of a conventional drinking water treatment plant in South China, J. Hazard Mater. 172 (2009) 1093–1099.

Z. Zhou, L. Guo, A critical evaluation of an asymmetrical flow field-flow fractionation system for colloidal size characterization of natural organic matter, J. Chromatogr. A 1399 (2015) 53–64.

J. Zhong, R. Sleighter, E. Salmon, G. McKee, P. Hatcher, Combining advanced NMR techniques with ultrahigh resolution mass spectrometry: A new strategy for molecular scale characterization of macromolecular components of soil and sedimentary organic matter, Org. Geochem. 42 (8) (2011) 903–916, https://doi.org/10.1016/j.orggeochem.2011.04.007.

Q. Zhou, S. Cabaniss, P. Maurice, Considerations in the use of high-pressure size exclusion chromatography (HPSEC) for determining molecular weights of aquatic chumic substances, Water Res. 34 (2000) 3505–3514.

CHAPTER 4

NOM removal by coagulation

Mika Sillanpää[1,2,3,4,5] and Anu Matilainen[6]
[1]Department of Chemical Engineering, School of Mining, Metallurgy and Chemical Engineering, University of Johannesburg, Doornfontein, South Africa; [2]Department of Applied Physics, Faculty of Science and Technology, Universiti Kebangsaan Malaysia, Bangi, Selangor, Malaysia; [3]Zhejiang Rongsheng Environmental Protection Paper Co. LTD, Pinghu Economic Development Zone, Zhejiang, P.R. China; [4]Department of Civil Engineering, University Centre for Research & Development, Chandigarh University, Gharuan, Mohali, Punjab, India; [5]International Research Centre of Nanotechnology for Himalayan Sustainability (IRCNHS), Shoolini University, Solan, Himachal Pradesh, India; [6]Finnish Safety and Chemicals Agency, Tampere, Finland

Abstract

The presence of natural organic matter (NOM) causes various problems in drinking water and drinking water treatment processes, including (1) negative effect on water quality by causing color, taste, and odor problems, (2) increased coagulant and disinfectant doses (which in turn results in increased sludge volumes and production of harmful disinfection by-products), (3) promoted biological growth in distribution system, and (4) increased levels of complexed heavy metals and adsorbed organic pollutants. NOM can be removed from drinking water through the use of several treatment options, of which the most common and economically feasible processes are considered to be coagulation and flocculation followed by sedimentation/flotation and sand filtration. Most of the NOM can be removed by coagulation, although the hydrophobic fraction and high molar mass compounds of NOM are removed more efficiently than hydrophilic fraction and the low molar mass compounds. Thus, enhanced and/or optimized coagulation, as well as new process alternatives for better removal of NOM by coagulation, have been suggested. In this chapter, we present an overview of the recent research dealing with coagulation and flocculation in the removal of NOM from drinking water.

Keywords: Coagulation; Drinking water; Natural organic matter; NOM; Water treatment.

Abbreviations

APAMs: Anionic polyacrylamides
BDOC: Biodegradable dissolved organic carbon
CNT: Carbon nanotube
CPAMs: Cationic polyacrylamides
DBP: Disinfection by-product
DOC: Dissolved organic carbon
FP: Formation potential
HAAs: Haloacetic acids

Natural Organic Matter in Water
ISBN 978-0-12-824274-2
https://doi.org/10.1016/B978-0-12-824274-2.00008-9

HFCVD:	Hot filament chemical vapor deposition
HMM:	High molar mass
HPAC:	Composite polyaluminum chloride
IPFs:	Inorganic polymer flocculants
LMM:	Low molar mass
NOM:	Natural organic matter
PAA:	Polyacrylamide
PACl:	Polyaluminum chloride
PAF-SiC:	Polyaluminum ferric silicate chloride
PAFC:	Polymer-polyaluminum ferric chloride
PAM:	Polyacrylamides
PAS:	Polyaluminum sulfate
PASiC:	Polyaluminum silicate chloride
PDADMAC:	Polydiallyldimethyl ammonium chloride
PFC:	Polyferric chloride
PFS:	Polyferric sulfate
PFSiS:	Polyferric silicate sulfate
PSF:	Polysilicate ferric
PTSS:	Polytitanium silicate sulfate
RSM:	Response surface method
SUVA:	Specific UV absorbance
THM:	Trihalomethane
TOC:	Total organic carbon

4.1 Introduction

Generally, the most common and economically feasible standard processes for the removal of natural organic matter (NOM) (and particulate matter) are considered to be coagulation and flocculation, followed by sedimentation/flotation and sand filtration (Jacangelo et al., 1995). Coagulation tends to balance cost and NOM removal efficiency well. Thus, enhanced and/or optimized coagulation appears to be the best option for better control of NOM and disinfection by-product (DBP) formation.

This chapter presents the options for NOM removal from drinking water using various coagulants and coagulation techniques. In the coagulation stage, solid particles are destabilized by a coagulant. This may include the formation of small aggregates by Brownian motion. The subsequent process by which large flocs are formed is known as *flocculation*. Chemical coagulation is achieved by the addition of inorganic coagulants (Duan and Gregory, 2003). Coagulants are most often iron or aluminum salts. When added to water, aqueous Al(III) and Fe(III) salts are dissociated to their respective trivalent ions, i.e., Al^{3+} and Fe^{3+}. They are then hydrolyzed and form several soluble complexes possessing high positive charges, thus adsorbing onto the surface of the negative colloids (Edzwald and Tobiason, 1999; Pham et al., 2006). According to the literature, at pH higher than the minimum solubility of the coagulants (e.g., 5.8 and 6.3 for ferric chloride and aluminum chloride, respectively), the hydrolysis products are high molar mass (HMM) polymers or colloidal/precipitated species, while at a pH slightly lower than the minimum solubility, the hydrolysis products are mainly medium polymers or monomers (Yan et al., 2008a). The degree of polymerization can be determined by ferron-timed spectroscopy. The different species can be classified into three categories: the mononuclear (Al_a and Fe_a), medium polynuclear (Al_b and Fe_b), and precipitated/colloidal (Al_c and Fe_c) species, in which the Al_b and Fe_b are considered to be the most efficient species for dissolved organic carbon (DOC) removal (Duan and Gregory, 2003; Zouboulis and Moussas, 2008; Yan et al., 2009a).

Coagulation is a physical–chemical process where the repulsive potential of electrical double layer of colloids is reduced in such a way that microparticles can be produced. These microparticles collide with each other to form larger structures, called flocs in the flocculation process. The use of coagulants for NOM removal was reviewed a few years ago (Matilainen et al., 2010) and this chapter is an update. Traditionally, coagulation has been employed in water treatment to decrease turbidity, resolve color problems, and remove pathogens. It must be emphasized, however, that optimal turbidity or color removal is not always achieved under the same conditions as NOM removal (Budd et al., 2004; Yu et al., 2007; Yan et al., 2008a; Vasyukova et al., 2012). For example, rapid changes in raw water composition can occur. Seasonal variations also play a key role (Vasyukova et al., 2012). In baseline coagulation, the coagulation conditions are optimized for turbidity removal, whereas optimized coagulation refers to dose and pH conditions optimized especially for organic matter reduction. Enhanced coagulation uses a higher amount of coagulant than would be needed for baseline coagulation, with accompanying changes in pH, order of chemical addition, or use of alternative coagulant chemicals. The goal is more efficient total organic carbon (TOC) and thus NOM removal (Budd et al., 2004; Bell-Ajy et al., 2000). Little work has been done on changes in water chemistry in NOM removal even by conventional ferric chloride and aluminum sulfate, but recently it has been discovered that if, during a treatment train, pH changes, removal of organic matter is affected (Slavik et al., 2012).

The aggregation mechanisms through which NOM is removed include a combination of charge neutralization, entrapment, adsorption, and complexation with coagulant metal ions into insoluble particulate aggregates (Fig. 4.1) (Jarvis et al., 2004). Because of the variable composition of NOM, the removal mechanism will be different for the specific types of NOM molecules in water (Sharp et al., 2006a; Jarvis et al., 2004); however, the charge-driven mechanisms are most evident when pH is optimized (Sharp et al., 2006a; Zhao et al., 2008; Bond et al., 2009). Moreover, the HMM polymer species (e.g., Al_c) efficiently remove particles through bridging or sweep flocculation, while medium polymer or monomer species appear to have a high ability to remove DOC by complexation,

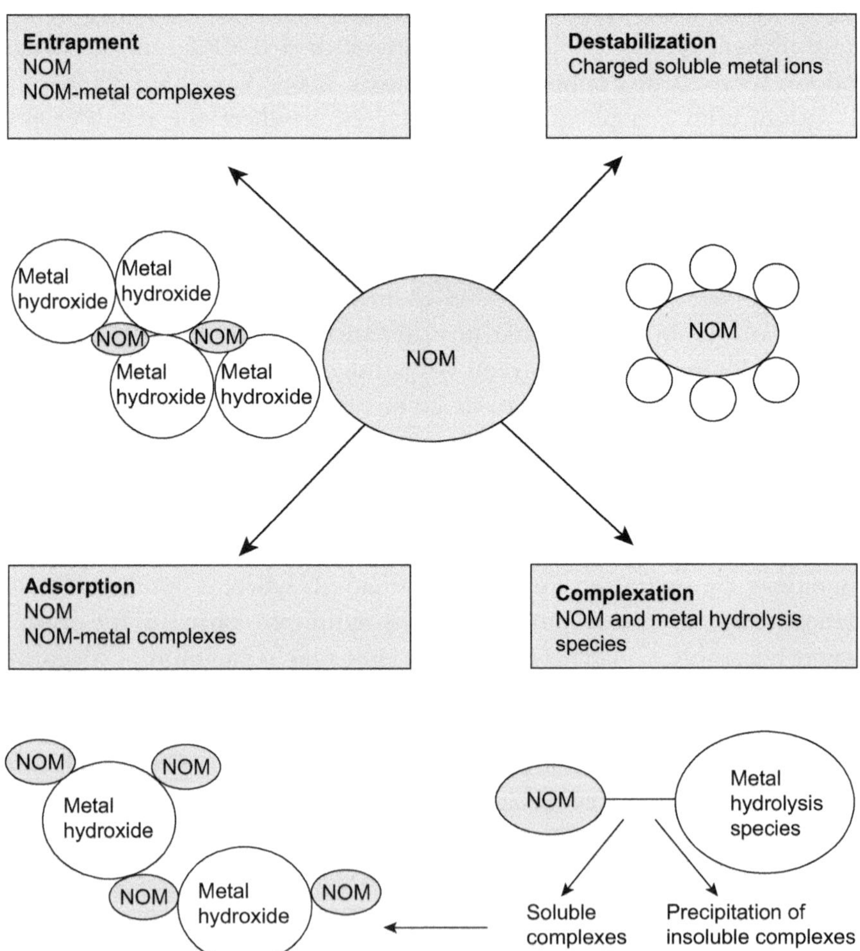

Figure 4.1 The possible removal mechanisms of natural organic matter (NOM) during coagulation. *(Adapted from Jarvis et al. (2004).)*

adsorption, charge neutralization, or coprecipitation (Yan et al., 2008a). The flocs formed during various coagulation mechanisms have different properties, including size, structure, and strength (Li et al., 2006).

The ability of coagulation to remove NOM and particles depends on several factors, including coagulant type and dosage, mixing conditions, pH, temperature, particle, and NOM properties (such as size, functionality, charge, and hydrophobicity) and the presence of divalent cations and concentrations of destabilizing anions including bicarbonate, chloride, and sulfate (Sharp et al., 2006a; Duan and Gregory, 2003; Kim et al., 2006; Sharp et al., 2006b; Qin et al., 2006). Ferric chloride, ferric sulfate, and alum are acidic, thus lowering the pH after application. Low-alkalinity waters therefore may require the addition of base, while high alkalinity may sometimes need to be lowered by adding acid to reach an efficient pH range (Jacangelo et al., 1995). The buffer capacity of high-alkalinity waters is higher than that of low-alkalinity waters against pH decrease (Budd et al., 2004).

The nature of NOM has a significant effect on the coagulant dose. The optimal coagulant dose for water with HMM NOM is expected to be low because the removal mechanism is mostly charge neutralization. If NOM consists of low molar mass (LMM) or nonhumic substances, the removal mechanism is expected to be adsorption onto metal hydroxide surfaces and the optimal coagulant dose is thus much larger. The hydrophobic fraction of NOM is generally removed more efficiently than the hydrophilic fraction by coagulation. Furthermore, HMM compounds are more readily removed than LMM compounds (Korshin et al., 2009; Sharp et al., 2006a; Bond et al., 2009; Soh et al., 2008; Chow et al., 2009; Szlachta and Adamski, 2009) most likely because HMM matter is quite hydrophobic in nature, consisting more of aromatic compounds (Swietlik et al., 2004). The ability of high specific UV absorbance (SUVA) values to reflect the better reactivity of organic compounds in water for removal by coagulation has generated a great deal of discussion (Edzwald and Tobiason, 1999; Archer and Singer, 2006; Bose and Reckhow, 2007). The suggested relationship between SUVA and DOC is illustrated in Table 4.1 (Edzwald and Tobiason, 1999). Compared to DOC, UV_{254} has been noted to reduce more, suggesting that aromatic materials are removed more effectively than other NOM fractions (Uyguner et al., 2007a; Uyak and Toroz, 2007; Uyak et al., 2007). Hydrophobic NOM has been observed to carry high levels of negative charge due to the presence of ionized groups such as carboxylic and phenolic groups. Thus, the hydrophobic fraction dominates the specific colloidal charge character of water, and fractions with higher charge are

Table 4.1 Relationship between specific UV absorbance (SUVA) and dissolved organic carbon (DOC) removal during coagulation.

SUVA	Composition	Coagulation	DOC removals
>4	Mostly aquatic humics, high hydrophobicity, high molar mass (MM) compounds	Natural organic matter (NOM) controls, good DOC removals	>50% for alum, little greater for ferric
2–4	Mixture of aquatic humics and other NOM, mixture of hydrophobic and hydrophilic NOM, mixture of MMs	NOM influences, DOC removals should be fair to good	25%–50% for alum, little greater for ferric
<2	Mostly nonhumics, low hydrophobicity, low MM compounds	NOM has little influence, poor DOC removals	<25% for alum, little greater for ferric

Adapted from Edzwald and Tobiason (1999).

more amenable to removal (Sharp et al., 2006c; Bond et al., 2009; Bose and Reckhow, 2007). NOM removal can be monitored by the removal of negative charge from the suspension. Maximum NOM removal has been achieved in a zeta potential window between −10 and + 5 mV (Sharp et al., 2006c; Bond et al., 2009; Henderson et al., 2006).

The most widely used coagulant in drinking water production has been aluminum sulfate (alum), but recently the use of ferric salts has become more popular (Duan and Gregory, 2003). This may be due to a suggestion that aluminum in drinking water may increase the risk of Alzheimer's disease (Flaten, 2001). Many investigations have been made to compare the effectiveness of one coagulant over another. According to these studies, ferric-based coagulants have been generally noted to remove NOM, measured as DOC and UV_{254}, more effectively than aluminum-based ones (Sharp et al., 2006c; Budd et al., 2004; Bell-Ajy et al., 2000; Uyguner et al., 2007a; Uyak and Toroz, 2007) and were found to be less sensitive to low temperatures than alum with regard to turbidity removal (Yu et al., 2007), although contradictory results have also been obtained in some cases (Uyguner et al., 2007b).

It has been proposed that the removal mechanisms of ferric and alum coagulants are different (Sharp et al., 2006c; Jarvis et al., 2005; Lei et al., 2009). Ferric-based coagulants have proved more effective, especially in removing NOM with intermediate MM (1000–4000 g mol^{-1}). This fraction may be largely composed of fulvic acids and more efficiently removed by ferric salts due to the higher charge density of ferric coagulants (Lindqvist et al., 2002, 2004). Moreover, the flocs formed during ferric coagulation are noted to be both larger and more numerous than those formed during aluminum coagulation, therefore increasing the likelihood of collisions and subsequent removal (Lindqvist et al., 2002, 2004; Ratnaweera et al., 1999). The choice of proper coagulant ultimately depends on the characteristics of the water to be treated. In addition, Cheng et al. (2010) discovered that intensive and prolonged rapid mixing enhances the NOM removal.

New inorganic polymer flocculants (IPFs) have been developed during the past few years based on inorganic salts. Prehydrolyzed coagulants are formed by the forced hydrolysis of a simple coagulant, such as alum. Concerning their coagulation–flocculation properties, IPFs are considered to be somewhere between traditional inorganic salts and organic flocculants (Moussas and Zouboulis, 2009). Besides aluminum and ferric-based polymeric flocculants, a combination of the two has also been tested: polyaluminum ferric silicate chloride (PAF-SiC) coagulant (Cheng et al., 2008). It was observed that flocs formed during coagulation were relatively large but nonsettleable, causing high turbidity when the dosage was insufficient, although turbidity was effective at high coagulant doses.

4.2 Aluminum-based coagulants

The most commonly used aluminum-based coagulant is alum ($Al_2(SO_4)_3$), with other coagulants including aluminum chloride ($AlCl_3$). Yet relatively high aluminum residuals may remain in the finished water, which can cause possible health hazards or other problems in the distribution system, e.g., spontaneous flocculation, especially during periods of cold temperatures or at low pH levels (Chow et al., 2009; Flaten, 2001; Matsukawa et al., 2006; Bérubê and Dorea, 2008). This can be avoided by pH control, but aluminum-based coagulants also affect turbidity and NOM removal (Matsukawa et al., 2006; Bérubê and Dorea, 2008). Table 4.2 presents pros and cons of aluminum-based coagulants.

Increasing the alum dose increases NOM removal up to a certain point. At very high doses (>100 mg L^{-1}), NOM removal is not significantly improved, suggesting the presence of components of NOM recalcitrant to removal by coagulation. Larger MM components are removed more efficiently, whereas LMM compounds are more resistant to removal

Table 4.2 Advantages and disadvantages of aluminum-based coagulants.

Coagulant type	Features	Advantages	Disadvantages	References
Alum, aluminum chloride	Trivalent aluminum ions are released into a solution from the respective salt. They are hydrolyzed and form soluble complexes possessing high positive charges. Coagulation efficiency depends on, e.g., coagulant dose, mixing, pH, temperature, particle and natural organic matter (NOM) properties. During coagulation, the most effective range of pH is 5–6.5.	Stable, easily handled, readily soluble. Better turbidity removal than with ferric salts in many cases. Can be more effective than ferric in low doses. Higher color removal efficiency.	Relatively high coagulant residuals in the finished water in some cases. Possibly increases risk of Alzheimer's disease. Ferric salts were found to be better than aluminum salts for NOM removal. High alkalinity consumption. Sulfate and/or chloride in finished water increases corrosivity. Net increase in the dissolved constituents.	Smith and Kamal (2009), Chow et al. (2008), Yu et al. (2007), Uyguner et al. (2007a), Qin et al. (2006), Soh et al. (2008), Chow et al. (2009), Uyak and Toroz (2007), Flaten (2001), Lindqvist et al. (2004), Matsukawa et al. (2006), Bérubê and Dorea (2008), Shin et al. (2008), Yan et al. (2008a), Rizzo et al. (2008), Yan et al. (2006)

(Chow et al., 2008, 2009; Soh et al., 2008). The hydrophilic neutral NOM fraction becomes difficult to remove with the addition of alum at any concentration. Furthermore, this fraction has been observed to have the highest biodegradable dissolved organic carbon (BDOC) contribution to the treated water, thus potentially supporting biofilm formation in the distribution system, and contributing to trihalomethane (THM) formation (Soh et al., 2008). The recalcitrant fraction of NOM is indicated to be polysaccharides and their derivatives (Chow et al., 2009). Coagulation pH has a significant effect on coagulation efficiency, and the observed optimal pH for alum coagulation is in the range of 5.0−6.5 (with alum doses between 5 and 100 mg Al L^{-1}), resulting in an average of 25%−67%, 44%−77%, 25%−66% and 97% of DOC, UV_{254}, THMFP and turbidity removal, respectively (Smith and Kamal, 2009; Yu et al., 2007; Uyguner et al., 2007a; Qin et al., 2006; Soh et al., 2008; Chow et al., 2009; Uyak and Toroz, 2007; Shin et al., 2008; Rigobello et al., 2011).

4.3 Ferric-based coagulants

Ferric salts commonly used in coagulation processes include ferric chloride ($FeCl_3$) and ferric sulfate ($Fe_2(SO_4)_3$) (Budd et al., 2004). The observed optimal pH level for ferric-based coagulation is pH 4 5−6, with NOM removal at about 29%−70% as DOC (Budd et al., 2004; Yu et al., 2007; Bell-Ajy et al., 2000; Bond et al., 2009; Uyak and Toroz, 2007; Abbaszadegan et al., 2007; Meyn et al., 2008; Park and Yoon, 2009; Rigobello et al., 2011). DOC removal and THMFP mitigation increase with increasing coagulant dose up to 100 mg L^{-1} (Uyak et al., 2007). At low coagulant doses, the anionic humic network reorganizes upon association with cationic coagulant species to yield more compact structures; when the coagulant dose is increased, the aggregation of humic acid with hydrolyzed-Fe species has been ascribed to competition between the humic network reconformation rate and collision rate of destabilized colloids (Siéliéchi et al., 2008). Table 4.3 presents pros and cons of iron-based coagulants.

4.4 Inorganic polymer flocculants

Prehydrolyzed aluminum coagulants such as polyaluminum chloride (PACl) and polyaluminum sulfate (PAS) have been developed in the past decade (Yan et al., 2008b, 2009a; Hu et al., 2006; Iriarte-Velasco et al., 2007; Ji et al., 2008; Li et al., 2008; Wang et al., 2009; Zhan et al., 2011; Trinh and Kang, 2011; Amin et al., 2012). PACl has been made by partially neutralizing $AlCl_3$ to different basicity ratios, and its use is becoming more widespread. Prehydrolyzing the $AlCl_3$ can enhance the amount of Al_{13} ($AlO_4Al_{12}(OH)_{24}^{+7}$) in the coagulation process, making it the most efficient known Al-species for contaminant removal (Hu et al., 2006; Dongsheng

Table 4.3 Advantages and disadvantages of iron-based coagulants.

Coagulant type	Features	Advantages	Disadvantages	References
Ferric chloride, ferric sulfate	Ferric salts hydrolyze similarly to aluminum salts when added to water and form different hydrolysis products. Coagulation effectiveness is dependent on the same factors as for alum salts. The most effective pH range is 4.5–6.	Many studies found ferric salts better than aluminum salts at removing natural organic matter (NOM), especially the middle size NOM fractions. Not as sensitive to temperature changes as alum. Larger flocs than those produced by Al coagulants.	Ferric-based coagulant acid strength and associated optimized flocculation pH ranges can produce purified water with less buffering capacity and require greater chemical addition for stabilization and corrosion control. High alkalinity consumption. Sulfate and/or chloride in finished water increases corrosion. Net increase in the dissolved constituents.	Sharp et al. (2006a), Budd et al. (2004), Yu et al. (2007), Bell-Ajy et al. (2000), Uyguner et al. (2007a), Uyak and Toroz (2007), Lindqvist et al. (2002), Lindqvist et al. (2004), Rizzo et al. (2008), Rigobello et al. (2011), Zhao et al. (2011)

et al., 2006; Yan et al., 2008c). These coagulant species (Al_{13} or Al_b) are considered to be the most efficient Al-species due to their larger size and higher positive charges. The properties of flocs formed under various aluminum coagulant species have been observed to differ (Hu et al., 2006; Li et al., 2008; Wang et al., 2009; Dongsheng et al., 2006; Shi et al., 2007), and the destabilization mechanisms of different aluminum species have also been investigated (Zhao et al., 2008; Yan et al., 2008c; Kazpard et al., 2006; Sarpola et al., 2007). Prehydrolyzed polymer coagulants have reported advantages over conventional coagulants, including reduced temperature or pH dependence, as well as lower alkalinity consumption (Iriarte-Velasco et al., 2007; Wang et al., 2009; Dongsheng et al., 2006; Shi et al., 2007; Yan et al., 2009b), but the characteristics of the water to be treated (e.g., alkalinity, pH, NOM content) play a major role in the choice of a proper coagulant (Hu et al., 2006; Yan et al., 2008b). Consequently, prehydrolyzed coagulants do not appear to enhance the removal efficiency of NOM in all cases (Hu et al., 2006; Shi et al., 2007). Lou et al. (2012) suggested that the drawback of using some polymeric salts is that the preformed species are stable and cannot be further hydrolyzed during coagulation. Thus, these coagulants may be relatively inefficient in removing a highly hydrophobic NOM (Lou et al., 2012).

Activated silica is an efficient coagulant aid and enhances the bridging ability of PACl (Yan et al., 2008a). Polymeric aluminum coagulant with supplemented activated SiO_2 has a larger molecular weight and smaller surface charge than PACl (Yan et al., 2008a; Cheng et al., 2008; Zouboulis and Tzoupanos, 2009). Tubić et al. (2013) studied the role of various NOM fractions in changes in THM and haloacetic acid formation during coagulation with $FeCl_3$ and a combination of PACl and $FeCl_3$. Based on their hydrophobicity, they fractionated the DOC in the raw water into four parts after coagulation and observed significant changes in the reactivities of the hydrophilic acidic and nonacidic fractions.

Polyaluminum silicate chloride (PASiC) can enhance the efficiency of NOM and turbidity removal, proving more efficient than PACl in the treatment of low turbidity waters (Cheng et al., 2008). Improved coagulation properties depend on both the basicity and silica content of a polymer and the preparation technique. PASiC is more resistant to pH variation (the effective pH range is 6.5–9) than PACl, and the flocs formed under optimal conditions are larger, contributing to greater removal efficiency (Zouboulis and Tzoupanos, 2009).

Instead of processing one factor at a time—which is time consuming and incapable of revealing the optimal combination of factors as it ignores the interrelation—Trinh and Kang (2011) applied the response surface method (RSM) to PACl coagulation. RSM includes the impacts of various individual factors and their interactions. The authors successfully optimized the coagulation conditions for turbidity and DOC removal.

More attention has recently been paid to inorganic polymeric iron coagulants, including polyferric sulfate (PFS), polymeric iron chloride (PICl), or polyferric chloride (PFC) (Jarvis et al., 2005; Zouboulis et al.,

2008; Liu and Chin, 2009; Cao et al., 2011; Wang et al., 2008; Lei et al., 2009; Zhan et al., 2010, 2011). These coagulants may have some advantages over monomeric forms of ferric salts, such as a wider working pH range, lower sensitivity to water temperature, reduced amounts of coagulant, and lower residual iron concentrations (Park and Yoon, 2009; Liu and Chin, 2009), although they are still in the development stage and the applications are limited. Prehydrolyzed coagulants are prepared by partial neutralization of metal salts, e.g., PFS from ferric sulfate. The prime advantage of these coagulants is that they produce only the most favorable hydrolysis and polymeric species for coagulation. The performance of the prehydrolyzed coagulants depends on the nature of the polymeric species formed in the process of preparation (Lei et al., 2009; Moussas and Zouboulis, 2009). Possible dominant removal mechanisms during PFC coagulation are adsorption, entrapment, and complexation rather than charge neutralization (Wei et al., 2009a). Zhan et al. (2011) reported the lack of research on coagulation performance and floc properties for NOM removal under various low-shear flows at the flocculation stage, especially at low NOM levels. They compared PFC and PACl as coagulants and observed that the G values in the slow-mixing phase had greater influence on the PFC flocs.

Polymeric coagulant efficiency can be improved by adding other components to the polymer such as polysilic acid, to produce new composite coagulants such as polysilicate ferric (PSF) and polyferric silicate sulfate (PFSiS) (Zouboulis and Moussas, 2008; Li et al., 2008; Fu and Yu, 2007; Fu et al., 2009; Moussas and Zouboulis, 2008; Xu et al., 2009). These coagulants work across a broad pH range and exhibit a markedly better coagulation performance than, e.g., ferric sulfate. Table 4.4 presents pros and cons of inorganic polymer flocculants.

4.5 Organic polyelectrolytes

A polymer is termed a *polyelectrolyte* when its monomeric unit contains ionizable groups (e.g., carboxyl, amino or sulfonic). Thus, polyelectrolytes are characterized by their charge, charge density, and molecular weight (MW). Consequently, polyelectrolytes can be cationic, anionic, or ampholytic, while polymers without ionizable groups are termed nonionic. Cationic polyelectrolytes include polydiallyldimethyl ammonium chloride (PDADMAC), cationic polyacrylamides (CPAMs), and natural cationic polymers (e.g., chitosan, starch), whereas anionic polyelectrolytes include

Table 4.4 Advantages and disadvantages of inorganic polymer flocculants.

Coagulant type	Features	Advantages	Disadvantages	References
PACl	Prepared by partially neutralized (prehydrolyzed) aluminum chloride. Enhanced amounts of high-charged, moderate MM hydrolysis species, e.g., Al_{13}.	Less temperature and pH dependent than alum salts. Lower alkalinity consumption. Better natural organic matter (NOM) removal capacity than alum in many cases. Lower dose requirement and less sludge produced. Lower residual aluminum in treated water.	Coagulant effectiveness is significantly affected by coagulant hydrolysis species speciation. Preformed Al species are stable and cannot be further hydrolyzed during coagulation. Might not be so efficient in removing HMM and highly hydrophobic NOM.	Yan et al. (2009a), Cheng et al. (2008), Yan et al. (2008b), Dongsheng et al. (2006), Yan et al. (2006, 2008a), Zhan et al. (2011), Trinh and Kang (2011), Amin et al. (2012)
PICl, PFC	Prepared by partially neutralized (prehydrolyzed) ferric chloride. Enhanced amounts of high-charged, moderate MM hydrolysis species.	Wider pH range, lower sensitivity to temperature, reduced amounts of coagulant, lower residual iron concentration. Less corrosive than ferric chloride. Lower sludge production.	Still in the development stage. Not so widely used.	Liu and Chin (2009), Wei et al. (2009a), Zhan et al. (2010, 2011), Cao et al. (2011)
PFS	Made by partially neutralized (prehydrolyzed) ferric	Broader pH range, lower sensitivity to temperature, reduced amounts of	Hydrolysis conditions have impact on speciation of hydrolysis/polymeric	Lei et al. (2009), Zouboulis et al. (2008)

Continued

Table 4.4 Advantages and disadvantages of inorganic polymer flocculants.—cont'd

Coagulant type	Features	Advantages	Disadvantages	References
	sulfate. Enhanced amounts of high-charged, moderate MM hydrolysis species.	coagulants, lower residual iron concentration. Less corrosive than ferric chloride.	species. Still in the development stage. Not so widely used.	
PASiC, PAF-SiC, PFSiS, PSF	Combination of aluminum and/or ferric-based polymeric flocculants with polysilicate.	Enhances bridging ability of coagulant. Relatively large flocs formed. More resistant to pH variations. Less corrosive. Wider pH range. Better performance. Produces little sludge. Effective in hydrophobic NOM removal. Only most favorable hydrolysis and polymeric species are produced in some cases.	Flocs' difficulty in settling creates high turbidity values at low dosages. Still in the development stage (especially ferric-based coagulants). High costs.	Yan et al. (2008b), Zouboulis and Moussas (2008), Cheng et al. (2008), Li et al. (2008), Zouboulis and Tzoupanos (2009), Fu and Yu (2007), Fu et al. (2009), Moussas and Zouboulis (2008), Xu et al. (2009), Lou et al. (2012)

HMM, high molar mass; *PACl*, polyaluminum chloride; *PAF-SiC*, polyaluminum ferric silicate chloride; *PASiC*, polyaluminum silicate chloride; *PFC*, polyferric chloride; *PFS*, polyferric sulfate; *PFSiS*, polyferric silicate sulfate; *PICl*, polymeric iron chloride; *PSF*, polysilicate ferric.

anionic PAMs or natural biopolymers. Nonionic polyelectrolytes include polyacrylamide (PAA) (Bolto and Gregory, 2007).

Synthetic or natural polymers are used as both primary coagulants and coagulant or flocculant aids. Cationic polyelectrolytes are most commonly used for NOM removal because NOM in water is likely to be negatively charged, favoring NOM removal by charge neutralization over bridging. The cationic polymer PDADMAC has been used as primary coagulant for NOM removal, since it is highly charged and has HMM, although its high cost has been a limiting factor (Bolto and Gregory, 2007; Gao et al., 2008). Polymer is used in combination with a conventional coagulant to lower the process costs by reducing coagulant demand, resulting in larger aggregates from already formed microflocs (Bolto and Gregory, 2007; Jarvis et al., 2006). Polyelectrolytes are more resistant to pH changes than conventional coagulants, and have achieved better floc characteristics in terms of size, strength, settleability, and filterability, thus producing a lower volume of sludge (Edzwald and Tobiason, 1999; Jarvis et al., 2004; Li et al., 2006; Bolto and Gregory, 2007; Barbot et al., 2008). Moreover, cationic polymers have proved efficient coagulation aids in improving PACl coagulation performance by enhancing the neutralization ability of PACl (Yan et al., 2008a) and improving the ability of PFS to remove DOC and turbidity (Wei et al., 2009a). In some cases, however, no improvement in floc characteristics or NOM removal efficiency was reported (Jarvis et al., 2006; Rizzo et al., 2008). The NOM removal capacity of different polyelectrolytes has varied from case to case (Yu et al., 2007; Yan et al., 2008a), so polyelectrolytes may not always be suitable coagulants, e.g., for turbidity removal (Yu et al., 2007). Cationic organic polyelectrolytes in particular are considered to be quite toxic to aquatic organisms and several countries have thus restricted their use in water purification (Bolto and Gregory, 2007; Rizzo et al., 2008). Table 4.5 presents the advantages and disadvantages of various organic polyelectrolytes, which were discussed above.

4.6 Composite coagulants

The combination of organic polymers or inorganics into IPFs to create a composite coagulant is an innovation in coagulation. These composites are synthesized by adding inorganic or organic modifiers to the respective coagulants (Zhu et al., 2012). The aim is to combine the advantages of both components and further enhance coagulation efficiency (Tzoupanos and Zouboulis, 2011). Several researchers have studied the composition of

Table 4.5 The advantages and disadvantages of organic polyelectrolytes.

Coagulant type	Features	Advantages	Disadvantages	References
Organic polyelectrolytes				
Cationic (e.g., PDADMAC, CPAMs, chitosan)	Cationic polyelectrolytes are the most commonly used polyelectrolytes due to their charge density. Charge neutralization is probably the main NOM removal mechanism.	High charge density polyelectrolytes remove hydrophobic NOM effectively. Produces smaller amounts of sludge, which is easier to dewater.	Higher the polymer dose required, higher the cost. Toxic effects. Formation of smaller flocs due to charge neutralization.	Yan et al. (2008b), Lindqvist et al. (2004), Wei et al. (2009a), Bolto and Gregory (2007), Gao et al. (2008), Rizzo et al. (2008)
Anionic (e.g., APAMs)	High MW carboxylic acid polymers. NOM removal mainly by bridging mechanism.	Improves low molecular weight (LMW) matter removal in some cases.	Not as effective as cationic polymers in every case.	Lindqvist et al. (2004); Bolto and Gregory (2007)
Inorganic salt + polyelectrolyte	Coagulation of NOM with inorganic salts using polyelectrolytes as a coagulant and/or flocculant aid, to maximize the benefits of both.	Larger and stronger flocs are formed than with either coagulant alone. Lower coagulant dose requirements. Smaller volume of sludge. Cost savings.	Higher the polymer dose required, higher the cost. Toxic effects.	Lindqvist et al. (2004); Bolto and Gregory (2007)

APAM, anionic polyacrylamides; *CPAM*, cationic polyacrylamides; *MW*, molecular weight; *NOM*, natural organic matter; *PDADMAC*, polydiallyldimethyl ammonium chloride.

PFC-PDADMAC (Wei et al., 2009a; Bolto and Gregory, 2007; Wang et al., 2008; Wei et al., 2009b,c), as well as PFS-PAA (Lindqvist et al., 2004). PFC-PDADMAC has efficiently removed NOM and created high levels of Fe_b species, the most active components in ferric salt coagulants (Gao et al., 2008; Wang et al., 2008; Wei et al., 2009b). Another novel composite coagulant is polytitanium silicate sulfate (PTSS) (Huang et al., 2014). Composites may be more suitable for surface water treatment than conventional metal coagulants or organic polymers, as they are efficient for DOC, SUVA, and turbidity removal through stronger charge neutralization (Wei et al., 2009a). PFS-PAA is particularly effective, generating flocs that are larger and have a broader size distribution range than those produced by a PSF + PAA mixture or either component alone (Moussas and Zouboulis, 2009).

A highly efficient composite polyaluminum chloride (HPAC) has been developed from PACl and other organic and inorganic additives such as active silicates and PDADMAC. HPAC has exhibited about 30% more efficiency on NOM removal than conventional coagulants or PACl (Yan et al., 2006, 2007; 2008a; 2009b), and has proved more efficient than conventional coagulants in basic conditions, thus being effective in waters with high pH or alkalinity (Yan et al., 2008a, 2009b). Polymer polyaluminum ferric chloride (PAFC) and composite $FeCl_3$/PAFC coagulate more efficiently than $FeCl_3$ (Lou et al., 2012). This composite could integrate the effects of adsorption, charge neutralization, entrapment, and complexation.

Bioflocculants are nontoxic and biodegradable (Yang et al., 2009) but have a relatively low flocculating capacity (Li et al., 2009). The compound bioflocculant-aluminum sulfate dual coagulant was found to exhibit a similar floc growth rate and strength to aluminum sulfate alone in the coagulation of kaolin-humic acid solution (Bo et al., 2012). Table 4.6 presents advantages and disadvantages of various composite coagulants, which have been discussed above.

4.7 Novel coagulants

Al and Fe coagulants meet the need for improved NOM removal. In addition to Fe and Al salts, the inorganic coagulant Zr may prove effective due to its increased positive charge (Jarvis et al., 2008, 2012). Zr provides a significantly lower DOC residual, lower THMs, and larger and stronger flocs than Fe. This was verified in pilot test runs in which flocs formed by the Zr coagulant were better removed by filtration and flotation processes. Yet, redestabilization occurred, so careful control of coagulation conditions

Table 4.6 Advantages and disadvantages of composite coagulants.

Coagulant type	Features	Advantages	Disadvantages	References
$FeCl_3$/PAFC inorganic-bioflocculant composites	Composite bioflocculant-aluminum sulfate	Nontoxic and biodegradable.	Relatively low flocculating capacity.	Bo et al. (2012)
HPAC	Composition of organic polymer (PDADMAC) and inorganic polymeric coagulant (PACl) with additives such as active silicates.	Effective in waters with high alkalinity or pH. Wider working pH range.	Flocs formed can be small or incompact and thus unfavorable to sedimentation, although flotation is much more efficient.	Yan et al. (2008b), Yan et al. (2009a), Yan et al. (2006), Yan et al. (2007)
PFC–PDADMAC	Composition of organic polymer (PDADMAC) and inorganic polymeric coagulant (PFC). High content of Fe_b hydrolysis species.	Stronger charge neutralization, hence efficient DOC, SUVA, and turbidity removal capacity.	Still in the development stage. Not so widely used.	Wei et al. (2009a), Gao et al. (2008), Wang et al. (2008), Wei et al. (2009b,c)

DOC, dissolved organic carbon; *HPAC*, composite polyaluminum chloride; *PAFC*, polymer-polyaluminum ferric chloride; *PDADMAC*, polydiallyldimethyl ammonium chloride; *PFC*, polyferric chloride; *SUVA*, specific UV absorbance.

is necessary (Jarvis et al., 2012). Zr was proposed as a good substitute when additional NOM removal is required, particularly due to its nontoxicity.

Aluminum formate was found to be as efficient in coagulation as conventional polyaluminum chloride and aluminum sulfate (Väänänen et al., 2012; Mikola et al., 2013). This coagulant is advantageous because it does not contribute harmful sulfate and chloride ions to waters. It was observed that aluminum formate functions selectively toward some NOM fractions. The composite coagulant polymeric aluminum ferric sulfate (PAFS) also proved effective (Zhu et al., 2012).

Carbon nanotubes (CNT) have demonstrated their efficiency as coagulants, although they lack dispersion and solubility properties (Simate et al., 2012a). Attempts have been made to overcome these challenges (Zhang et al., 2002, 2009; Pei et al., 2008). Due to their cost and potential toxicity, CNTs need to be separated by membrane filtration or ultracentrifugation and regenerated after use (Simate et al., 2012b). Another novel nanostructured coagulant that may prove useful for NOM removal is nanocellulose (Suopajärvi et al., 2013).

4.8 Conclusions

This chapter has reviewed recent research in the field of NOM removal from drinking water by coagulation. The character and concentration of NOM in raw water sources change both regionally and seasonally (Fabris et al., 2008; Sharp et al., 2006d). In addition, an overall increase in NOM content has been observed in various regions worldwide (Eikebrokk et al., 2004). Coagulation, particularly when enhanced or optimized, is a major treatment option to address raised NOM levels. Enhanced coagulation is an especially good option for high-alkalinity waters (Yan et al., 2008a).

Enhanced coagulation removed more NOM than baseline coagulation in the majority of the reported studies, yet some of the LMM compounds and nonhumic material in NOM still remained in the treated water. Applying increased coagulant dosages and controlling the raw water pH by the addition of acid or base are two common options for enhancing NOM removal by coagulation. Yet, these solutions increase the proportion of rapidly precipitated material as sludge or flocs, forming looser aggregates and thus exhibiting inferior dewatering properties (Verrelli et al., 2009, 2010). Sludge can be dewatered by thickening, centrifugation, or filtration. Fenton's reagent may also improve sludge dewaterability, and is considered a good option for conditioning an alum-based sludge (Tony et al., 2008).

Other ways of handling the sludge formed include atomization and use in the cement industry (Rodriguez et al., 2010). Enhanced coagulation also increases the iron corrosion of the finished water due to the consumption of alkalinity and introduction of sulfate, so water discoloration may result (Shi and Taylor, 2007). Instead of increasing coagulant doses, NOM removal may be better enhanced by recycling backwash streams that contain destabilized particles, especially in low-turbidity waters (Gottfried et al., 2008). Since enhanced coagulation aims to further NOM removal, the aesthetic quality of the water must be preserved, which includes efficient microbial and virus deterioration (Abbaszadegan et al., 2007).

Conventional coagulation has been performed with inorganic salts, including alum, ferric chloride, and ferric sulfate. A recent improvement is the development of prehydrolyzed ferric and alum coagulants called IPFs (e.g., PFC and PACl). Their advantages over conventional coagulants include a broader working pH range, lower sensitivity to water temperature, and lower residual coagulant concentrations (Dongsheng et al., 2006; Liu and Chin, 2009). Synthetic and natural organic polyelectrolytes have long been used in combination with metal coagulants to achieve better coagulation efficiency and floc characteristics (Bolto and Gregory, 2007). A new application combines IPFs and organic polymers, such as PFC-PDADMAC and PFS-PAA, with efficient NOM removal rates (Moussas and Zouboulis, 2009; Wei et al., 2009c). Other new coagulants include Mg/Al hydrotalcite (Santosa et al., 2008) and Zr^{4+}, which creates significantly larger flocs than Fe^{3+} coagulation, achieving a higher DOC and recalcitrant NOM removal efficiency; this technique is still in development (Jarvis et al., 2008). A highly promising new approach to removing organics from water is nanotechnology, which has introduced new types of material such as dendrimers, hyperbranched polymers, carbon nanotubes, cyclodextrins (Upadhyayula et al., 2009), and nanocellulose (Suopajärvi et al., 2013). These nanomaterials have exceptional adsorption properties and are able to entrap a variety of organic molecules. Nanomaterials are particularly effective in removing DBPs and other organic pollutants, and have also been observed to remove NOM (Lu and Su, 2007).

References

M. Abbaszadegan, B.K. Mayer, H. Ryu, N. Nwachuku, Efficacy of removal of CCL viruses under enhanced coagulation conditions, Environ. Sci. Technol. 41 (2007) 971–977.

M. Amin, M. Faraji, F. Momenbeik, A. Hasanzadeh, M. Sadani, Performance evaluation of an anaerobic migrating blanket reactor in the biodegradation of perchloroethylene from industrial wastewaters, Int. J. Environ. Health Eng. 1 (2012) 1–8.

A. Archer, P. Singer, An evaluation of the relationship between SUVA and NOM coagulation using the ICR database, J. Am. Water Works Assoc. 98 (2006) 110–123.
E. Barbot, S. Moustier, J. Bottero, P. Moulin, Coagulation and ultrafiltration: understanding of the key parameters of the hybrid process, J. Membr. Sci. 325 (2008) 520–527.
K. Bell-Ajy, M. Abbaszadegan, E. Ibrahim, D. Verges, M. LeChevallier, Conventional and optimized coagulation for NOM removal, J. Am. Water Works Assoc. 92 (2000) 44–58.
D. Bérubê, C. Dorea, Optimizing alum coagulation for turbidity, organics, and residual Al reductions, Water Sci. Technol. Water Supply 8 (2008) 505–511.
X. Bo, B. Gao, N. Peng, Y. Wang, Q. Yue, Y. Zhao, Effect of dosing sequence and solution pH on floc properties of the compound bioflocculant–aluminum sulfate dual-coagulant in kaolin–humic acid solution treatment, Bioresour. Technol. 113 (2012) 89–96.
B. Bolto, J. Gregory, Organic polyelectrolytes in water treatment, Water Res. 41 (2007) 2301–2324.
T. Bond, E.H. Goslan, S.A. Parsons, B. Jefferson, Disinfection by-product formation of natural organic matter surrogates and treatment by coagulation, MIEX and nanofiltration, Water Res. 44 (2009) 1645–1653.
P. Bose, D. Reckhow, The effect of ozonation on natural organic matter removal by alum coagulation, Water Res. 41 (2007) 1516–1524.
G. Budd, A. Hess, H. Shorney-Darby, J. Neemann, C. Spencer, J. Bellamy, P. Hargette, Coagulation applications for new treatment goals, J. Am. Water Works Assoc. 92 (2004) 102–113.
B. Cao, B. Gao, X. Liu, M. Wang, Z. Yang, Q. Yue, The impact of pH on floc structure characteristic of polyferric chloride in a low DOC and high alkalinity surface water treatment, Water Res. 45 (2011) 6181–6188.
W. Cheng, F. Chi, C. Li, R. Yu, A study on the removal of organic substances from low-turbidity and low-alkalinity water with metal-polysilicate coagulants, Colloids Surf. A Physicochem. Eng. Aspects 312 (2008) 238–244.
Y.-L. Cheng, R.-J. Wong, J. Lin, C. Huang, D.-J. Lee, A. Mujumdar, Water coagulation using electrostatic patch coagulation (EPC) mechanism, Dry. Technol. 28 (2010) 850–857.
C. Chow, R. Fabris, J. van Leeuwen, D. Wang, M. Drikas, Assessing natural organic matter treatability using high performance size exclusion chromatography, Environ. Sci. Technol. 42 (2008) 6683–6689.
C. Chow, J. van Leeuwen, R. Fabris, M. Drikas, Optimized coagulation using aluminium sulfate for the removal of dissolved organic carbon, Desalination 245 (2009) 120–134.
W. Dongsheng, L. Hong, L. Chunhua, T. Hongxiao, Removal of humic acid by coagulation with nano-Al_{13}, Water Sci. Technol. Water Supply 6 (2006) 59–67.
J. Duan, J. Gregory, Coagulation by hydrolysing metal salts, Adv. Colloid Interface Sci. 100–102 (2003) 475–502.
J. Edzwald, J. Tobiason, Enhanced coagulation: US requirements and a broader view, Water Sci. Technol. 40 (1999) 63–70.
B. Eikebrokk, R. Vogt, H. Liltved, NOM increase in Northern European source waters: discussion of possible causes and impacts on coagulation/contact filtration processes, Water Sci. Technol. Water Supply 4 (2004) 47–54.
R. Fabris, C. Chow, M. Drikas, B. Eikebrokk, Comparison of NOM character in selected Australian and Norwegian drinking waters, Water Res. 42 (2008) 4188–4196.
T. Flaten, Aluminium as a risk factor in Alzheimer's disease, with emphasis on drinking water, Brain Res. Bull. 55 (2001) 187–196.
Y. Fu, S. Yu, C. Han, Morphology and coagulation performance during preparation of poly-silicic-ferric (PSF) coagulant, Chem. Eng. J. 149 (2009) 1–10.

Y. Fu, S. Yu, Characterization and coagulation performance of solid poly-silicic—ferric (PSF) coagulant, J. Non-cryst. Solids 353 (2007) 2206—2213.

B.-Y. Gao, Y. Wang, Q.-Y. Yue, J.-C. Wei, Q. Li, The size and coagulation behaviour of a novel composite inorganic—organic coagulant, Separ. Purif. Technol. 62 (2008) 544—550.

A. Gottfried, A. Shepard, K. Hardiman, M. Walsh, Impact of recycling filter backwash water on organic removal in coagulation—sedimentation processes, Water Res. 42 (2008) 4683—4691.

R. Henderson, E. Sharp, P. Jarvis, S. Parsons, B. Jefferson, Identifying the linkage between particle characteristics and understanding coagulation performance, Water Sci. Technol. Water Supply 6 (2006) 31—38.

J. Hu, H. Liu, J. Qu, D. Wang, J. Ru, Coagulation behaviour of aluminum salts in eutrophic Water: significance of Al_{13} species and pH control, Environ. Sci. Technol. 40 (2006) 325—331.

X. Huang, B. Gao, Y. Wang, Q. Yue, Q. Li, Y. Zhang, Coagulation performance and flocs properties of a new composite coagulant: polytitanium-silicate-sulfate, Chem. Eng. J. 245 (2014) 173—179.

U. Iriarte-Velasco, J. Àlvarez-Uriarte, J. Gonzáles-Velasco, Enhanced coagulation under changing alkalinity-hardness conditions and its implications on trihalomethane precursors removal and relationship with UV absorbance, Separ. Purif. Technol. 55 (2007) 368—380.

J. Jacangelo, J. DeMarco, D. Owen, S. Randtke, Selected processes for removing NOM: an overview, J. Am. Water Works Assoc. 87 (1995) 64—77.

P. Jarvis, J. Banks, R. Molinder, T. Stephenson, S. Parsons, B. Jefferson, Processes for enhanced NOM removal: beyond Fe and Al coagulation, Water Sci. Technol. Water Supply 8 (2008) 709—716.

P. Jarvis, B. Jefferson, S. Parsons, Characterizing natural organic matter flocs, Water Sci. Technol. Water Supply 4 (2004) 79—87.

P. Jarvis, B. Jefferson, S. Parsons, Breakage, regrowth, and fractal nature of natural organic matter flocs, Environ. Sci. Technol. 39 (2005) 2307—2314.

P. Jarvis, B. Jefferson, S. Parsons, Floc structural characteristics using conventional coagulation for a high doc, low alkalinity surface water source, Water Res. 40 (2006) 2727—2737.

P. Jarvis, E. Sharp, M. Pidou, R. Molinder, S. Parsons, B. Jefferson, Comparison of coagulation performance and floc properties using a novel zirconium coagulant against traditional ferric and alum coagulants, Water Res. 46 (2012) 4179—4187.

Q. Ji, H. Liu, C. Hu, J. Qu, D. Wang, J. Li, Studies on the treatment of brilliant green solution by combination microwave induced oxidation with $CoFe_2SO_4$, Separ. Purif. Technol. 62 (2008) 464—469.

V. Kazpard, B. Lartiges, C. Frochot, J. d'Espinose de la Caillerie, M. Viriot, J. Portal, T. Görner, J. Bersillon, Fate of coagulant species and conformational effects during the aggregation of a model of a humic substance with Al_{13} polycations, Water Res. 40 (2006) 1965—1974.

H. Kim, S. Lee, S. Byun, M. Yu, Application of improved rapid mixing for enhanced removal of dissolved organic matter and DBPFP (disinfection by-product formation potential) control, Water Sci. Technol. Water Supply 6 (2006) 49—57.

G. Korshin, C. Chow, R. Fabris, M. Drikas, Absorbance spectroscopy-based examination of effects of coagulation on the reactivity of fractions of natural organic matter with varying apparent molecular weights, Water Res. 43 (2009) 1541—1548.

G. Lei, X. Ma, X. Guan, A. Song, Y. Cui, Effect of basicity on coagulation performance of polyferric chloride applied in eutrophicated raw water, Desalination 247 (2009) 518—529.

F. Li, Y. Akira, A. Yuka, Characterization of micro-flocs of NOM coagulated by PACl, alum and polysilicate-iron in terms of molecular weight and floc size, Water Sci. Technol. 57 (2008) 83–90.

T. Li, Z. Zhu, D. Wang, C. Yao, H. Tang, Characterization of floc size, strength and structure under various coagulation mechanisms, Powder Technol. 168 (2006) 104–110.

Z. Li, S. Zhong, H.-Y. Lei, R.-W. Chen, Q. Yu, H.-L. Li, Production of a novel bio-flocculant by *Bacillus licheniformis* X14 and its application to low temperature drinking water treatment, Bioresour. Technol. 100 (2009) 3650–3656.

N. Lindqvist, J. Jokela, T. Tuhkanen, in: In: Proceedings of the 11th International Gothenburg Symposium on Chemical Treatment of Water and Wastewater, November 8–11, Orlando, Florida, USA, 2004.

N. Lindqvist, S. Korhonen, J. Jokela, T. Tuhkanen, in: H. Hahn, E. Hofmann, H. Odegaard (Eds.), Chemical Water and Wastewater Treatment VIII, IWA Publishing, London, 2002, p. 133.

T.-K. Liu, C. Chin, Improved coagulation performance using preformed polymeric iron chloride (PICl), Coll. Surf. A Physicochem. Eng. Aspects 339 (2009) 192–198.

I. Lou, S. Gong, X. Huang, Y. Liu, Coagulation optimization for low temperature and low turbidity source water using combined coagulants: a case study, Desalination Water Treat. 46 (2012) 107–114.

C. Lu, F. Su, Adsorption of natural organic matter by carbon nanotubes, Separ. Purif. Technol. 58 (2007) 113–121.

M. Mikola, J. Rämö, A. Sarpola, J. Tanskanen, Removal of different NOM fractions from surface water with aluminium formate, Separ. Purif. Technol. 118 (2013) 842–846.

A. Matilainen, M. Vepsäläinen, M. Sillanpää, Natural organic matter removal by coagulation during drinking water treatment: a review, Adv. Colloid Interface Sci. 159 (2010) 189–197.

S. Matsukawa, S. Itoho, S. Habutnu, T. Aizawa, Water Sci. Technol. Water Supply 6 (2006) 67.

T. Meyn, A. Bahn, T. Leiknes, Significance of flocculation for NOM removal by coagulation–ceramic membrane microfiltration, Water Sci. Technol. Water Supply 8 (2008) 691–700.

P. Moussas, A. Zouboulis, A study on the properties and coagulation behaviour of modified inorganic polymeric coagulant—polyferric silicate sulphate (PFSiS), Separ. Purif. Technol. 63 (2008) 475–483.

P. Moussas, A. Zouboulis, A new inorganic–organic composite coagulant, consisting of Polyferric Sulphate (PFS) and Polyacrylamide (PAA), Water Res. 43 (2009) 3511–3524.

S. Park, T. Yoon, Effects of iron species and inert minerals on coagulation and direct filtration for humic acid removal, Desalination 239 (2009) 146–158.

X. Pei, L. Hu, W. Liu, J. Hao, Synthesis of water-soluble carbon nanotubes via surface initiated redox polymerization and their tribological properties as water-based lubricant additive, Eur. Polym. J. 44 (2008) 2458–2464.

A. Pham, A. Rose, A. Feitz, T. Waite, Kinetics of Fe(III) precipitation in aqueous solutions at pH 6.0–9.5 and 25°C, Geochem. Cosmochim. Acta 70 (2006) 640–650.

J.-J. Qin, M. Oo, K. Kekre, F. Knops, P. Miller, Impact of coagulation pH on enhanced removal of natural organic matter in treatment of reservoir water, Separ. Purif. Technol. 49 (2006) 295–298.

H. Ratnaweera, E. Gjessing, E. Oug, Influence of physical–chemical characteristics of natural organic matter (NOM) on coagulation properties: an analysis of eight Norwegian water sources, Water Sci. Technol. 40 (1999) 89–95.

E. Rigobello, A. Dantas, L. Di Bernardo, E. Vieira, Influence of the apparent molecular size of aquatic humic substances on colour removal by coagulation and filtration, Environ. Technol. 32 (2011) 1767–1777.

L. Rizzo, A. Di Gennaro, M. Gallo, V. Belgiorno, Coagulation/chlorination of surface water: a comparison between chitosan and metal salts, Separ. Purif. Technol. 62 (2008) 79–85.

N. Rodriguez, S. Ramirez, M. Blanco Varela, M. Guillem, J. Puig, Re-use of drinking water treatment plant (DWTP) sludge: characterization and technological behaviour of cement mortars with atomized sludge additions, Cement Concr. Res. 40 (2010) 778–786.

S.J. Santosa, E.S. Kunarti, Karmanto, Synthesis and utilization of Mg/Al hydrotalcite for removing dissolved humic acid, Appl. Surf. Sci. 254 (2008) 7612–7617.

A. Sarpola, H. Hellman, V. Hietapelto, J. Jalonen, J. Jokela, J. Rämö, J. Saukkoriipi, Hydrolysis products of water treatment chemical aluminium sulfate octadecahydrate by electrospray ionization mass spectrometry, Polyhedron 26 (2007) 2851–2858.

E. Sharp, P. Jarvis, S. Parsons, B. Jefferson, Impact of fractional character on the coagulation of NOM, Coll. Surf. A Physicochem. Eng. Aspects 286 (2006a) 104–111.

E. Sharp, P. Jarvis, S. Parsons, B. Jefferson, The impact of zeta potential on the physical properties of ferric–NOM flocs, Environ. Sci. Technol. 40 (2006b) 3934–3940.

E. Sharp, S. Parsons, B. Jefferson, The impact of seasonal variations in DOC arising from a moorland peat catchment on coagulation with iron and aluminium salts, Environ. Pollut. 140 (2006c) 436–443.

E. Sharp, S. Parsons, B. Jefferson, Seasonal variations in natural organic matter and its impact on coagulation in water treatment, Sci. Total Environ. 363 (2006d) 183–194.

B. Shi, J. Taylor, Potential impact of enhanced coagulation on corrosion by-product release in a distribution system, Desalination 208 (2007) 260–268.

B. Shi, Q. Wei, D. Wang, Z. Zhu, H. Tang, Coagulation of humic acid: the performance of preformed and non-preformed Al species, Coll. Surf. A Physicochem. Eng. Aspect. 296 (2007) 141–148.

J. Shin, R. Spinette, C. O'Melia, Stoichiometry of coagulation revisited, Environ. Sci. Technol. 42 (2008) 2582–2589.

J.-M. Siéliéchi, B. Lartiges, G. Kayem, S. Hupont, C. Frochot, J. Thieme, J. Ghanbaja, J. d'Espinose de la Caillerie, O. Barrés, R. Kamga, P. Levitz, L. Michot, Changes in humic acid conformation during coagulation with ferric chloride: implications for drinking water treatment, Water Res. 42 (2008) 2111–2123.

G. Simate, S. Iyuke, S. Ndlovu, M. Heydenrych, L. Walubita, Human health effects of residual carbon nanotubes and traditional water treatment chemicals in drinking water, Environ. Int. 39 (2012a) 38–49.

G. Simate, S. Iyuke, S. Ndlovu, M. Heydenrych, The heterogeneous coagulation and flocculation of brewery wastewater using carbon nanotubes, Water Res. 46 (2012b) 1185–1197.

I. Slavik, S. Müller, R. Mokosch, J. Azongbilla, W. Uhl, Impact of shear stress and pH changes on floc size and removal of dissolved organic matter (DOM), Water Res. 46 (2012) 6543–6553.

E. Smith, Y. Kamal, Optimizing treatment for reduction of disinfection by-product (DBP) formation, Water Sci. Technol. Water Supply 9 (2009) 191–198.

Y. Soh, F. Roddick, J. van Leeuwen, The impact of alum coagulation on the character, biodegradability and disinfection by-product formation potential of reservoir natural organic matter (NOM) fractions, Water Sci. Technol. 58 (2008) 1173–1179.

T. Suopajärvi, H. Liimatainen, O. Hormi, J. Niinimäki, Coagulation–flocculation treatment of municipal wastewater based on anionized nanocelluloses, Chem. Eng. J. 231 (2013) 59–67.

J. Swietlik, A. Dabrowska, U. Raczyk-Stanislawiak, J. Nawrocki, Reactivity of natural organic matter fractions with chlorine dioxide and ozone, Water Res. 38 (2004) 547–558.
M. Szlachta, W. Adamski, Effects of natural organic matter removal by integrated processes: alum coagulation and PAC-adsorption, Water Sci. Technol. 59 (2009) 1951–1957.
M. Tony, Y. Zhao, J. Fu, A. Tayeb, Conditioning of aluminium-based water treatment sludge with Fenton's reagent: effectiveness and optimizing study to improve dewaterability, Chemosphere 72 (2008) 673–677.
T. Trinh, L. Kang, Response surface methodological approach to optimize the coagulation–flocculation process in drinking water treatment, Chem. Eng. Res. Des. 89 (2011) 1126–1135.
A. Tubić, J. Agbaba, B. Dalmacija, J. Molnar, S. Maletić, M. Watson, S.U. Perović, Insight into changes during coagulation in NOM reactivity for trihalomethanes and haloacetic acids formation, J. Environ. Manage. 118 (2013) 153–160.
N. Tzoupanos, A. Zouboulis, Preparation, characterization and application of novel composite coagulants for surface water treatment, Water Res. 45 (2011) 3614–3626.
V. Upadhyayula, S. Deng, M. Mitchell, G. Smith, Application of carbon nanotube technology for removal of contaminants in drinking water: a review, Sci. Total Environ. 408 (2009) 1–13.
V. Uyak, I. Toroz, Disinfection by-product precursors reduction by various coagulation techniques in Istanbul water supplies, J. Hazard Mater. 141 (2007) 320–328.
V. Uyak, S. Yavuz, I. Toroz, S. Ozaydin, E. Genceli, Disinfection by-products precursors removal by enhanced coagulation and PAC adsorption, Desalination 216 (2007) 334–344.
C. Uyguner, M. Bekbolet, H. Selcuk, A comparative approach to the application of a physico-chemical and advanced oxidation combined system to natural water samples, Separ. Sci. Technol. 42 (2007a) 1405–1419.
C. Uyguner, S. Suphandag, A. Kerc, M. Bekbolet, Evaluation of adsorption and coagulation characteristics of humic acids preceded by alternative advanced oxidation techniques, Desalination 210 (2007b) 183–193.
E. Vasyukova, W. Uhl, F. Braga, C. Simões, T. Baylão, K. Neder, Drinking water production from surface water sources in the tropics: Brasília DF, Brazil, Environ. Earth Sci. 65 (2012) 1587–1599.
D. Verrelli, D. Dixon, P. Scales, Effect of coagulation conditions on the dewatering properties of sludges produced in drinking water treatment, Colloids Surf. A Physicochem. Eng. Aspects 348 (2009) 14–23.
D. Verrelli, D. Dixon, P. Scales, Assessing dewatering performance of drinking water treatment sludges, Water Res. 44 (2010) 1542–1552.
M. Väänänen, L. Kupiainen, J. Rämö, A. Sarpola, J. Tanskanen, Speciation and coagulation performance of novel coagulant – aluminium formate, Separ. Purif. Technol. 86 (2012) 242–247.
Y. Wang, B.-Y. Gao, X.-M. Xu, W.-Y. Xu, G.-Y. Xu, Characterization of floc size, strength and structure in various aluminum coagulants treatment, J. Colloid Interface Sci. 332 (2009) 354–359.
Y. Wang, B. Gao, Q. Yue, J. Wei, Q. Li, The characterization and flocculation efficiency of composite flocculant iron salts–polydimethyldiallylammonium chloride, Chem. Eng. J. 142 (2008) 175–181.
J. Wei, B. Gao, Q. Yue, Y. Wang, W. Li, X. Zhu, Comparison of coagulation behavior and floc structure characteristic of different polyferric-cationic polymer dual-coagulants in humic acid solution, Water Res. 43 (2009a) 724–732.
J. Wei, B. Gao, Q. Yue, Y. Wang, L. Lu, Performance and mechanism of polyferric-quaternary ammonium salt composite flocculants in treating high organic matter and high alkalinity surface water, J. Hazard Mater. 165 (2009b) 789–795.

J. Wei, B. Gao, Q. Yue, Y. Wang, Strength and regrowth properties of polyferric-polymer dual-coagulant flocs in surface water treatment, J. Hazard Mater. 175 (2009c) 949–954.

X. Xu, S.-I. Yu, W. Shi, Z. Jiang, C. Wu, Effect of acid medium on the coagulation efficiency of polysilicate-ferric (PSF)—a new kind of inorganic polymer coagulant, Separ. Purif. Technol. 66 (2009) 486–491.

M. Yan, H. Liu, D. Wang, J. Ni, J. Qu, Natural organic matter removal by coagulation: effect of kinetics and hydraulic power, Water Sci. Technol. Water Supply 9 (2009a) 21–30.

M. Yan, D. Wang, J. Ni, J. Qu, C.W.K. Chow, H. Liu, Mechanism of natural organic matter removal by polyaluminum chloride: effect of coagulant particle size and hydrolysis kinetics, Water Res. 42 (2008a) 3361–3370.

M. Yan, D. Wang, J. Ni, J. Qu, W. Ni, J. Van Leeuwen, Natural organic matter (NOM) removal in a typical North-China water plant by enhanced coagulation: targets and techniques, Separ. Purif. Technol. 68 (2009b) 320–327.

M. Yan, D. Wang, J. Qu, J. Ni, C.W.K. Chow, Enhanced coagulation for high alkalinity and micro-polluted water: the third way through coagulant optimization, Water Res. 42 (2008b) 2278–2286.

M. Yan, D. Wang, B. Shi, M. Wang, Y. Yan, Effect of pre-ozonation on optimized coagulation of a typical North-China source water, Chemosphere 69 (2007) 1695–1702.

M. Yan, D. Wang, S. You, J. Qu, H. Tang, Enhanced coagulation in a typical North-China water treatment plant, Water Res. 40 (2006) 3621–3627.

M. Yan, D. Wang, J. Yu, J. Ni, M. Edwards, J. Qu, Enhanced coagulation with polyaluminum chlorides: role of pH/alkalinity and speciation, Chemosphere 71 (2008c) 1665–1673.

Z.-H. Yang, J. Huang, G.-M. Zeng, M. Ruan, C.-S. Zhou, L. Li, Optimization of flocculation conditions for kaolin suspension using the composite flocculant of MBFGA1 and PAC by response surface methodology, Bio Technol. 100 (2009) 4233–4239.

J. Yu, D. Wang, M. Yan, C. Ye, M. Yang, X. Ge, Optimized coagulation of high alkalinity, low temperature and particle water: pH adjustment and polyelectrolytes as coagulant aids, Environ. Monit. Assess. 131 (2007) 377–386.

X. Zhan, B. Gao, Q. Yue, Y. Wang, B. Gao, Coagulation behavior of polyferric chloride for removing NOM from surface water with low concentration of organic matter and its effect on chlorine decay model, Separ. Purif. Technol. 75 (2010) 61–68.

X. Zhan, B. Gao, Y. Wang, Q. Yue, Influence of velocity gradient on aluminum and iron floc property for NOM removal from low organic matter surfacewater by coagulation, Chem. Eng. J. 166 (2011) 116–121.

N. Zhang, J. Xie, V. Varadan, Functionalization of carbon nanotubes by potassium permanganate assisted with phase transfer catalyst, Smart Mater. Struct. 11 (2002) 962–965.

L. Zhang, Q.-Q. Ni, Y. Fu, T. Natsuki, One-step preparation of water-soluble single-walled carbon nanotubes, Appl. Surf. Sci. 255 (2009) 7095–7099.

Y. Zhao, B. Gao, H. Shon, B. Cao, J.-H. Kim, Coagulation characteristics of titanium (Ti) salt coagulant compared with aluminum (Al) and iron (Fe) salts, J. Hazard Mater. 185 (2011) 1536–1542.

H. Zhao, C. Hu, H. Liu, X. Zhao, J. Qu, Role of aluminum speciation in the removal of disinfection byproduct precursors by a coagulation process, Environ. Sci. Technol. 42 (2008) 5752–5758.

G. Zhu, H. Zheng, W. Chen, W. Fan, P. Zhang, T. Tshukudu, Preparation of a composite coagulant: polymeric aluminum ferric sulfate (PAFS) for wastewater treatment, Desalination 285 (2012) 315–323.

A. Zouboulis, P. Moussas, F. Vasilakou, Polyferric sulphate: preparation, characterization and application in coagulation experiments, J. Hazard Mater. 155 (2008) 459–468.

A. Zouboulis, P. Moussas, Polyferric silicate sulphate (PFSiS): preparation, characterization and coagulation behaviour, Desalination 224 (2008) 307–316.
A. Zouboulis, N. Tzoupanos, Polyaluminium silicate chloride—a systematic study for the preparation and application of an efficient coagulant for water or wastewater treatment, J. Hazard Mater. 162 (2009) 1379–1389.

CHAPTER 5

NOM removal by electrochemical methods

Mika Sillanpää[1,2,3,4,5], Heikki Särkkä[6] and Mikko Vepsäläinen[7]
[1]Department of Built Environment, Aalto University, Espoo, Finland; [2]Department of Applied Physics, Faculty of Science and Technology, Universiti Kebangsaan Malaysia, Bangi, Selangor, Malaysia; [3]Zhejiang Rongsheng Environmental Protection Paper Co. LTD, Pinghu Economic Development Zone, Zhejiang, P.R. China; [4]Department of Civil Engineering, University Centre for Research & Development, Chandigarh University, Gharuan, Mohali, Punjab, India; [5]International Research Centre of Nanotechnology for Himalayan Sustainability (IRCNHS), Shoolini University, Solan, Himachal Pradesh, India; [6]Department of Built Environment, Aalto University, Espoo, Finland; [7]CSIRO Mineral Resources Flagship, Clayton, VIC, Australia

Abstract

Electrochemical techniques such as electrocoagulation (EC) and electrooxidation (EO) have proved their efficiency in the removal of humic acid (HA), coliform, and algae from surface waters. Many investigations have also been conducted with synthetic wastewaters. EC combined with membrane filtration hybrid systems can increase natural organic matter (NOM) removal rates remarkably. In EO technology, electrolysis efficiency is strongly linked to electrode composition. Efficiency could be increased by changing the reactor design, using commercial electrodes, and exploring the semiconducting properties of oxide mixtures. Electrochemical methods may present an attractive alternative to other NOM removal techniques, such as conventional coagulation and chemical oxidation methods, for natural waters. Surface water treatment with EC can produce high-quality water for either potable or industrial use. This technology appears to remove some toxic pollutants from wastewater and could be used as a pretreatment in combination with some other purification technology. Boron-doped diamond (BDD) anodes have proved effective in HA removal from aqueous solutions and potentially their total mineralization.

Keywords: Electrochemical methods; Electrocoagulation; Electrooxidation; Natural organic matter (NOM); Water treatment.

Abbreviations

AOPs:	Advanced oxidation processes
BDD:	Boron-doped diamond
BIEC:	Doubled EC
COD:	Chemical oxygen demand
DBP:	Disinfection by-product
DOC:	Dissolved organic carbon
DSA:	Dimensional stable anode

Natural Organic Matter in Water
ISBN 978-0-12-824274-2
https://doi.org/10.1016/B978-0-12-824274-2.00007-7

EC:	Electrocoagulation
EEM:	Excitation-emission matrix
EF:	Electroflotation
eMBR:	Electro-MBR
EMT-BIEC:	Electromagnetic treatment-doubled EC
EO:	Electrooxidation
EPS:	Extracellular polymeric substance
HA:	Humic acids
HFCVD:	Hot filament chemical vapor deposition
NF:	Nanofiltration
NOM:	Natural organic matter
ORP:	Oxidation reduction potential
ROS:	Reactive oxygen species
SHE:	Standard hydrogen electrode
SPEF:	Solar photoelectro-Fenton
TOC:	Total organic carbon

5.1 Introduction

Electrochemical technologies for water purification and disinfection have been the subject of growing interest in recent years (Chen, 2004). Techniques such as electrocoagulation (EC), also known as electroflocculation, electroflotation (EF), and electrooxidation (EO) can easily be distributed (Mollah et al., 2004). Conventional water purification techniques such as chemical coagulation, biological treatment, or UV oxidation are not effective against some toxic and refractory organic pollutants, and electrochemical techniques can provide a more efficient means of treating these. Electrochemical methods are innovative, inexpensive, and effective in purifying industrial wastewaters before discharge into water systems or circulation back into processes (Mollah et al., 2001, 2004; Holt et al., 2005). They could also be regarded as "green technologies" because little or no chemicals are needed to facilitate water treatment.

EC has a long history: the first plant was built in London in 1889 for the treatment of sewage (Moreno et al., 2009; Vik et al., 1984). Despite some promising results, this technology has enjoyed limited success. Scientific, economic, and environmental interest in this technology has revived in recent years, however, due the demand for alternative water treatment technologies. EC understandably has several similarities to chemical coagulation but also significant differences, such as side reactions.

A relatively new technology, used especially for the disinfection of drinking water or treatment of wastewaters, is the electrochemical mineralization of organic compounds (Patermarakis and Fountoukidis, 1990; Comninellis et al., 2008; Chen, 2004). An electrochemical oxidation mechanism involves the production of OH-radicals at the active sites of the anode. The activity of these electrogenerated OH radicals is strongly linked to their interaction with the anode surface (Comninellis et al., 2008). Electrochemical oxidation over anodes made of graphite, Pt, TiO_2, IrO_2, PbO_2, several Ti-based alloys and more recently, boron-doped diamond (BDD) electrodes in the presence of a suitable electrolyte (typically NaCl) has been employed for the decontamination of various inorganic and organic pollutants (Comninellis et al., 2008; Klavarioti et al., 2009; Särkkä et al., 2009), as well as microbial inactivation (Bergmann et al., 2009; Jeong et al., 2009).

5.2 Principles of EC and EO technologies

5.2.1 Theory of EC

In the EC system, multiple electrochemical reactions occur simultaneously at the anodes and cathodes. These can be divided into the main mechanisms that cause the destabilization of pollutants, and side reactions, such as hydrogen formation. Electrodes that produce coagulants in water are made from either iron or aluminum. In addition, there can be inert electrodes, typically cathodes, that are sometimes used in the system as counter-electrodes. Iron and aluminum cations dissolve from the anodes according to Eqs. (5.1) and (5.2):

$$Fe(s) \rightarrow Fe^{n+}(aq) + ne^- \quad (5.1)$$

$$Al(s) \rightarrow Al^{3+}(aq) + 3e^- \quad (5.2)$$

In typical aqueous environments and conditions of the EC process, iron can dissolve in divalent Fe(II) and trivalent Fe(III) forms, whereas aluminum dissolves only in trivalent form Al(III). Fe(II) can further oxidize to Fe(III) if oxidation reduction potential and pH conditions are suitable. Oxygen has to be present and pH has to be neutral or alkaline to achieve a reasonable reaction rate (Morgan and Lahav, 2007).

Moreno et al. (2009) studied the electrochemical reactions of iron electrodes in an EC system. According to their results and the thermodynamic data they presented, the potential of the iron anode in an EC system is in the region where Fe(III) iron is produced. Some authors suggest, however, that the potential of the cathode is in the region where Fe(III) is reduced to Fe(II) form and that, therefore, both forms exist in the EC system (Parga et al., 2005; Gomes et al., 2007). Iron is produced on the anodes mainly in the form of Fe(II) (Ben-Sasson et al., 2009). At low pH, the chemical dissolution of iron can be significant and total iron concentration can be higher than would theoretically be expected. Ben-Sasson et al. (2009) also studied the oxidation of produced Fe(II). According to their results, no significant oxidation occurred at pH 5, the oxidation rate was moderate at pH 6, and very rapid oxidation occurred at pH 7–9.

The amount of metal cations dissolved during the reactions at the anode can be calculated according to Faraday's law (Eq. 5.3):

$$m = \frac{It\text{Mw}}{zF} \tag{5.3}$$

where I is the current, t is the operation time (s), Mw is molecular weight of the substance (g mol$-^{1}$), F is the Faraday constant (96,485 C mol$-^{1}$), z is the number of electrons involved in the reaction (2 for Fe^{2+} and 3 for Fe^{3+} and Al^{3+}), and m is the quantity of metal dissolved (g). Several studies have reported current yields of over 100% for the dissolution of aluminum electrodes (Picard et al., 2000). It seems that aluminum also dissolves from the cathodes. This occurs when the pH on the surface of the cathode decreases due to the formation of HO^- (Eq. 5.4) or by the consumption of hydronium ions/protons (Eq. 5.5):

$$2H_2O + 2e^- \rightarrow H_2(g) + 2HO^-(aq) \tag{5.4}$$

$$2H^+\ (aq) + 2e^- \rightarrow H_2(g) \tag{5.5}$$

At high pH, aluminum dissolves as aluminate (Eq. 5.6):

$$2Al(s) + 6H_2O + 2HO^-(aq) \rightarrow 2\left[2Al(OH)_4\right]^-(aq) + 3H_2(g) \tag{5.6}$$

Picard et al. (2000) studied cathodic dissolution during the EC process. They compared the hydrogen evolution from stainless-steel cathodes and aluminum cathodes. Hydrogen produced at the stainless-steel cathodes followed Faraday's law. In tests with the aluminum cathodes, hydrogen production was higher due to the chemical dissolution of aluminum according to Eq. (5.6). Hydrogen formation and aluminum dissolution from the cathodes increased exponentially with current intensity.

In contrast to chemical coagulation, EC treatment raises the pH of the solution when it is acidic, neutral, or slightly alkaline, and lowers the pH when the solution is highly alkaline. This change of pH during EC treatment affects the speciation of aluminum and iron hydroxides. At highly acidic pH (pH 2) the alkalinity produced during EC is not sufficient to increase the pH of the solution, whereas at initial pH 3 and above, pH rises during the treatment (Mouedhen et al., 2008). This is easily understandable, as the concentration of hydronium ions increases by a factor of 10 when pH decreases from pH 3 to pH 2. When initial pH is significantly alkaline ($pH > 9$), pH probably decreases due to the formation of aluminate $[Al(OH)_4]$, which is an alkalinity consumer (Chen et al., 2000; Kobya et al., 2006). It seems that the pH change rate and final steady-state pH depend on the concentration of anions in the solution. According to Trompette and Vergnes (2009), pH increases more in sulfate solution than in chloride solution. Sulfates can replace hydroxyl ions in the hydroxide precipitates, and therefore fewer hydroxyl ions are bound to hydroxides.

5.2.2 Treatment parameters in EC

Various parameters affect EC efficiency in removing the pollutants from water. Parameters that are known to have an effect are as follows:

- Electrode materials can be iron, aluminum and/or inert material (typically cathodes). Iron and aluminum ions and hydroxides have different chemistries and applications.
- Solution pH has an effect on the speciation of metal hydroxides in the solution, the ζ-potential of the colloidal particles, and the dissolution of aluminum cathodes.
- Current density is proportional to the amount of electrochemical reactions taking place on the electrode surface.
- Treatment time or electric charge added per volume is proportional to the amount of coagulants produced in the EC system and other reactions taking place in the system.
- Electrode potential defines which reactions occur on the electrode surface.
- Pollutant concentration affects removal efficiency because coagulation does not follow zero-order reaction kinetics but rather pseudo second- or first-order kinetics.
- The concentration of anions, such as sulfate or fluoride, affects the composition of hydroxides, because they can replace hydroxide ions in the precipitates.
- Temperature affects flocs formation, reaction rates, and conductivity. Depending on the pollutant, increasing temperature can have a negative or a positive effect on removal efficiency.

Other parameters, such as hydrodynamic conditions and interelectrode distance, may affect treatment efficiency and electricity consumption. These parameters either have not been widely studied or are found to have a negligible effect on overall removal efficiency.

5.2.3 Reactor design in EC

EC systems are typically constructed of plate electrodes, and water flows through the space between the electrodes (Chen, 2004). Electrodes can be arranged in the EC system in either a monopolar or bipolar way. In the monopolar systems (Fig. 5.1), all anodes are connected to each other and similarly all cathodes are also connected to each other. In the bipolar systems (Fig. 5.1), the outermost electrodes are connected to a power source and current passes through the other electrodes, thus polarizing them. In the bipolar systems, the side of the electrode facing the anode is negatively polarized and vice versa on the other side facing the cathode. Water flow between the electrodes can follow a vertical or horizontal direction.

5.2.4 Theory of EO

Thermodynamically, the electrochemical degradation of any soluble organic compound in water should be achieved at low potentials, before the thermodynamic potential of water oxidation to molecular oxygen (1.23 V per standard hydrogen electrode (SHE) under standard conditions), as indicated in Eq. (5.7) (Kapalka et al., 2010):

$$2H_2O \rightarrow O_2 + 4H^+ + 4e^- \tag{5.7}$$

In acidic media, water can be discharged on the electrode, producing highly oxidative absorbed hydroxyl radicals (Eq. 5.8):

$$H_2O + M \rightarrow M(^*OH) + H^+ + e^- \tag{5.8}$$

where M is the electrode surface.

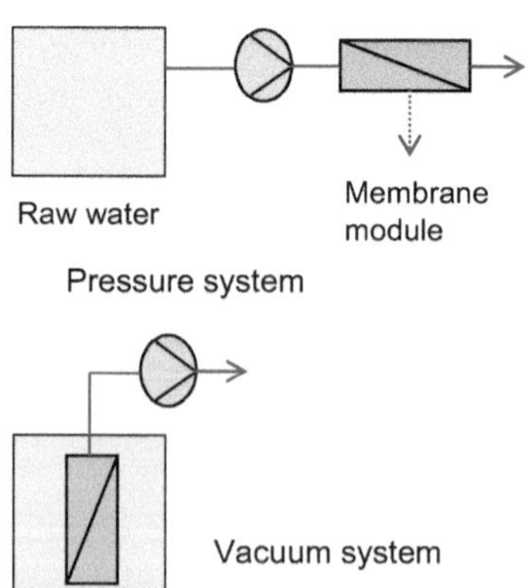

Figure 5.1 (A) Bipolar and (B) monopolar electrode arrangements. *(Adapted from Chen (2004).)*

These radicals are physisorbed on the anode surface where the organic pollutant R can be oxidized as follows (Eq. 5.9):

$$R_{(aq)} + M(^{*}OH)_{n/2} \rightarrow M + \text{Oxidation products} + \frac{n}{2}H^{+} + \frac{n}{2}e^{-} \quad (5.9)$$

where n is the number of electrons involved in the oxidation reaction of R. The reaction of organics with electrogenerated hydroxyl radicals (Eq. 5.9) is in competition with the side reaction of the anodic discharge of these radicals to oxygen (Eq. 5.10):

$$M(^{*}OH) \rightarrow M + \frac{1}{2}O_2 + H^{+} + e^{-} \quad (5.10)$$

Anodic activity by electrodes depends on their overpotential for oxygen evolution (Chen, 2004). Platinum electrodes have a much lower potential value for oxygen evolution reactions (1.3 V vs. SHE) than, for example, SnO_2 electrodes (1.9 V vs. SHE) or BDD electrodes (Ti/BDD, 2.7 V vs. SHE). This signifies that anodic oxidation by hydroxyl radicals can take place on a Ti/BDD electrode surface at a significantly higher current density with a minimal oxygen evolution side reaction.

Electrochemical oxidation can also occur indirectly through the formation of oxidants such as chlorine, hypochlorous acid, and hypochlorite (Rajkumar and Kim, 2006; Rajkumar et al., 2007; Kraft et al., 1999; Bergmann and Koparal, 2005; Chatzisymeon et al., 2006; Sakalis et al., 2006; Gotsi et al., 2005) or hydrogen peroxide/ozone (Polcaro et al., 2007; Dhaouadi et al., 2009; Chu et al., 2012; Guinea et al., 2010) at the electrodes by the following Eqs. (5.11)–(5.17):

$$2Cl^{-} \rightarrow Cl_2 + 2e^{-} \quad (5.11)$$

$$Cl_2 + H_2O \rightarrow HOCl + H^{+} + Cl^{-} \quad (5.12)$$

$$HOCl \rightarrow H^{+} + OCl^{-} \quad (5.13)$$

$$H_2O \rightarrow {}^{*}OH + H^{+} + e^{-} \quad (5.14)$$

$$2^{*}OH \rightarrow H_2O_2 \quad (5.15)$$

$$H_2O_2 \rightarrow O_2 + 2H^{++2}e^{-¿} \quad (5.16)$$

$$O_2 + {}^{*}O \rightarrow O_3 \quad (5.17)$$

At acidic pH, chlorine is present in the solution in the form of hypochlorous acid, which has a higher oxidation potential (1.49 V) than hypochlorite (0.94 V). Under alkaline conditions, hypochlorite is the dominant species (Rajkumar et al., 2007). Under higher pH conditions, more chlorate or perchlorate is formed than chlorine/hypochlorite, which decreases oxidation efficiency. Higher initial chloride concentration in the electrolyte solution naturally encourages greater chlorine/hypochlorite production (Rajkumar and Kim, 2006; Rajkumar et al., 2007; Kraft et al., 1999). A similar effect can be observed when the temperature of the electrolyte solution is lowered, but this depends on the electrode material used (Kraft et al., 1999). Higher current density also increases chlorine production (Rajkumar et al., 2007; Kraft et al., 1999).

Electrochemical production of reactive oxygen species, such as ozone and hydrogen peroxide, has proved effective in water disinfection without the mediation of active chlorine products (Polcaro et al., 2007). In an electro-Fenton oxidation process, H_2O_2 is continuously generated in acidic solutions by the reduction of O_2 at the graphite cathodes or carbon felt (Dhaouadi et al., 2009). When Fe^{2+} is introduced as the catalyst, a Fenton reaction takes place in the solution, generating hydroxyl radicals to decompose pollutants or disinfect bacteria. Chu et al. (2012) effectively degraded 4-nitrophenol by employing a dual-cathode system to generate H_2O_2 and Fe^{2+} simultaneously at two cathodes to encourage an electro-Fenton reaction in the solution together with anodic oxidation by hydroxyl radicals at the anode. Guinea et al. (2010) achieved almost total mineralization of enrofloxacin solutions by solar photoelectro-Fenton treatment using an O_2-diffusion cathode.

Efficient organic pollutant removal can also be achieved through generating Fe^{2+} in situ by an EC process in which an iron anode is dissolved into a solution, causing a Fenton reaction when hydrogen peroxide is added to it (Martins et al., 2006; Zhao et al., 2012). Martins et al. (2006) were able to remove 95% of nonylphenol polyethoxylate in 5 min from aqueous solution and in 10 min from wastewater, respectively.

5.2.5 Electrodes and reactors used in EO

Several electrodes have been used for water and wastewater treatment by EO. Anodes used include lead and lead dioxide (El-Ashtoukhy et al., 2009; Hamza et al., 2011; Awad and Abo Galwa, 2005; Flox et al., 2009), dimensional stable anode (DSA) electrodes (Tavares et al., 2012; Wang et al., 2007b; Qu et al., 2012; Miwa et al., 2006), graphite (Patel and

Suresh, 2008; Wang et al., 2007a; Ma et al., 2007), and BDD electrodes (Flox et al., 2009; Montilla et al., 2002; Kraft et al., 2003; Rodrigo et al., 2010; Panizza and Cerisola, 2005; Chen and Chen, 2006, 2011; Faouzi et al., 2007; Migliorini et al., 2011).

Lead and lead dioxide are effective anodes due to their stability, low cost, and high oxygen evolution potential, which delays O_2 evolution in favor of Cl_2 evolution (El-Ashtoukhy et al., 2009). Hamza et al. (2011) completely mineralized 1,3,5-trimethoxybenzene in acid media at a Ta/PbO_2 anode. They discovered that carboxylic acids finally oxidized all oxidation products to CO_2. Awad and Abo Galwa (2005) found that the electrocatalytic activity of a lead dioxide electrode depends on the conductive electrolyte. They concluded that electrode poisoning occurred in the presence of an H_2SO_4 electrolyte, since an adherent film was formed on the anode surface. The dissolution of toxic Pb^{2+} ions also hinders the use of lead and lead dioxide as anodes (Chen, 2004).

DSA electrodes are catalytic oxide electrodes that can effectively generate active hydroxyl radicals and active chloride species (Wang et al., 2007b). They also have a relatively high overpotential for oxygen evolution. Efficient degradation of paper mill wastewater has been achieved using 3D electrodes ($Ti/Co/SnO_2{-}Sb_2O_5$) combined with activated carbon treatment (Wang et al., 2007a). This result was mainly due to the fact that their large specific surface area in comparison to conventional 2D electrodes can substantially increase the conversion rate within an electrochemical reactor. "Nonactive" electrodes, such as SnO_2, form hydroxyl radicals on their surface more easily, which can result in the complete oxidation of the organic molecules to CO_2 (Miwa et al., 2006). With "active electrodes" such as RuO_2 and IrO_2, only selective oxidation of the organic species in the solution occurs.

Effective removal of chemical oxygen demand (COD) (>96%) was achieved when the electrochemical degradation process was catalyzed by transition metals (Co and Cu; Wang et al., 2007a,b) or molybdenum and phosphate (Mo−P) modified kaolin with graphite as the anode and cathode (Ma et al., 2007). Pollutants were adsorbed on the surface of the kaolin, where they were oxidized by hydroxyl radicals produced at the graphite cathode by the reaction of hydrogen peroxide and transition metals (Wang et al., 2007a,b). This process is similar to the electro-Fenton process.

The potential of conducting diamond films for water treatment has recently been recognized. These films have an inert surface with low adsorption properties, remarkable corrosion stability even in strong acidic media, and an extremely wide potential window in aqueous and nonaqueous media (Kraft et al., 2003; Panizza and Cerisola, 2005). They also have the highest oxygen evolution overpotential value (Chen, 2004; Kraft et al., 2003), meaning that more hydroxyl radicals are formed on the anode surface during treatment. BDD electrodes can also degrade refractory organic pollutants completely, and the nature of the pollutant does not significantly affect the efficiency of the process (Rodrigo et al., 2010). Furthermore, besides hydroxyl radical formation on the electrode surface, diamond electrodes also increase mediated oxidation by other electrochemically formed compounds such as persulfate, perphosphate, percarbonate, or hypochlorite, depending on the electrolyte used.

The low pressure conversion of carbon to diamond crystals has made it possible to grow a thin layer of diamond film on suitable substrates such as silicon, niobium, tungsten, molybdenum, and titanium (Chen and Chen, 2011). The hot filament chemical vapor deposition (HFCVD) technique has been applied primarily to fabricate active and stable BDD electrodes, mainly using titanium as a substrate material (Chen and Chen, 2006, 2011; Migliorini et al., 2011). Migliorini et al. (2011) used an additional H_2 gas flux passing through a bubbler containing a solution of B_2O_3 dissolved in CH_3OH with a B/C ratio of 30,000 ppm during the HFCVD coating of diamond films. Two different fluxes were used to produce heavily BDD films.

The main advantages and disadvantages of different electrodes in EO treatment are presented in Table 5.1.

The simplest EO reactor design is the bipolar cell. Besides plane electrodes, the cylindrical electrodes can also be employed (Chen, 2004). Typical reactor design for EO treatment is presented in Fig. 5.2. It includes a tank, a pump, a DC power supply, and an electrochemical reactor with electrodes (Zhu et al., 2011).

5.3 EC and EO technologies in NOM removal

5.3.1 EC in NOM removal

EC has been proposed as a promising alternative to chemical coagulation for removing various pollutants from freshwater and wastewater (Kobya et al., 2011; Meas et al., 2010; Kabdaslı et al., 2009). When iron or aluminum is used as a sacrificial anode, it dissolves and produces $Al^{3\pm}$, $Fe^{2\pm}$, and $Fe^{3\pm}$ ions in the solution. These ions form metal hydroxides, which dissolve less readily, such as $Al(OH)_3$, $Fe(OH)_2$, and $Fe(OH)_3$, by linking

Table 5.1 Comparison of electrodes in EO treatment.

Electrode	Advantages	Disadvantages	Compared to other electrodes
Ti	Stable	Passive, expensive	
Pt	Inert, low oxygen. evolution overpotential	Expensive	Poor efficiency in anodic oxidation of organic compounds
PbO_2	Good current efficiency, cheap, effective in oxidizing pollutants, high oxygen evolution overpotential, easy to prepare	Corrosive, toxic Pb^{2+}-ions could be released	
Dimensional stable anode electrodes	Supports indirect oxidation, good current efficiency, high oxygen evolution overpotential, lower cost, higher availability	Short lifespan, lack of electrochemical stability	
Boron-doped diamond	Inert under tough conditions, high oxygen evolution overpotential and electrochemical stability, good current efficiency, high corrosion stability, good conductivity	Very expensive	Higher activity

hydroxides ions in water. Metal hydroxide species that adsorb readily are effective in the destabilization of suspended pollutants. They can remove pollutants from water either by complexation or by electrostatic attraction, followed by coagulation, sedimentation, or flotation.

Several researchers have considered EC for surface water treatment and natural organic matter (NOM) removal from waters (Vepsäläinen et al., 2009; Yildiz et al., 2007; Dubrawski and Mohseni, 2013a). Vepsäläinen et al. (2009) studied the effect of initial pH and temperature on EC treatment and found that high NOM removal was obtained using an Al

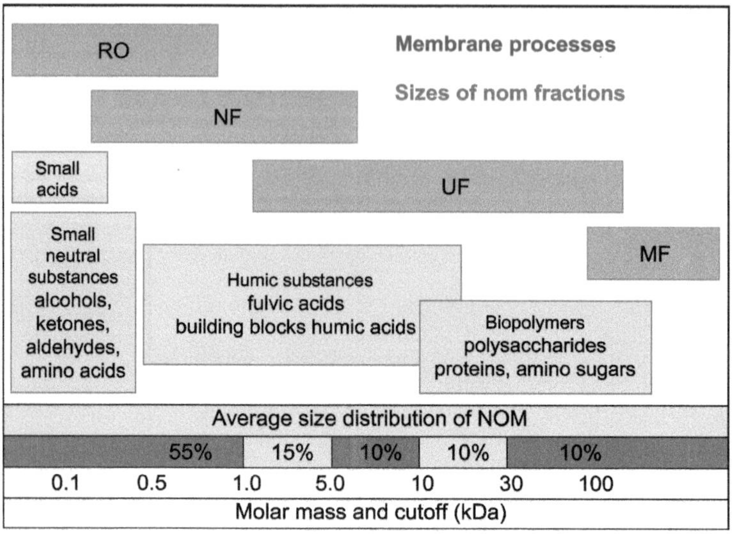

Figure 5.2 Typical electrooxidation setup. *(Adapted from Zhu et al. (2011).)*

electrode at a low pH of 4.3 and a temperature of 22.25 °C. The dissolving rate of the electrodes in different water temperatures is shown in Fig. 5.3. Temperature had significant effect on the dissolving speed of the aluminum, but the impact of temperature on NOM removal was minor

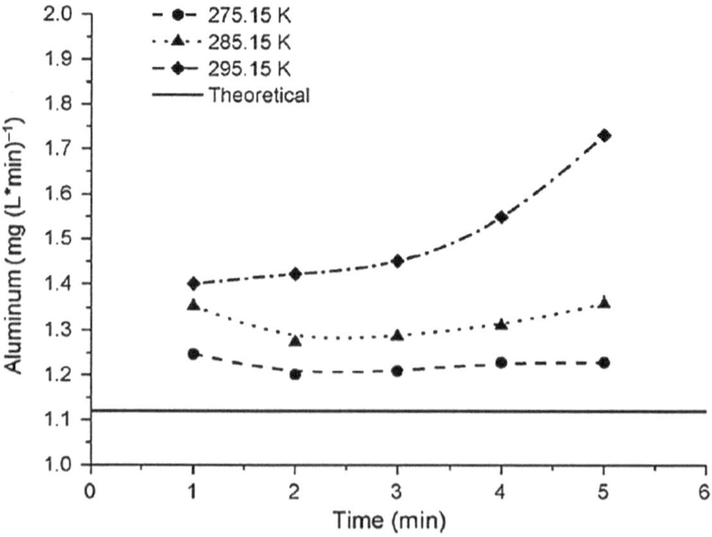

Figure 5.3 Dissolving rate of the aluminum (Al) electrodes in different water temperatures. Theoretical values are calculated according to Faraday's law for anodic dissolving reaction (Vepsäläinen et al., 2009).

compared to that of initial pH (Vepsäläinen et al., 2009). Yildiz et al. (2007) studied the influence of different supporting electrolyte solutions on NOM removal, and Na_2SO_4 was observed to be the most favorable electrolyte type. Dubrawski and Mohseni (2013a) applied an iron EC reactor to drinking water treatment. Significant EC design variables were examined: current density, coagulant, and charge-loading rate. They observed a correlation between increased current density and decreased current efficiency, and optimal NOM removal was found at a current density of 10 mA cm^{-2} (Dubrawski and Mohseni, 2013a). Dubrawski et al. (2013) also compared zinc, iron, and aluminum electrodes for dissolved organic carbon (DOC) removal, finding that zinc electrodes could achieve better results than aluminum electrodes.

Mohora et al. (2012) studied the performance of a laboratory-scale EC/flotation reactor in removing NOM and arsenic from groundwater. Removal rates were rather high, satisfying the drinking water standards in both cases. Electrode passivation was reduced by adding NaCl to the electrolyte solution, although this did not improve the NOM removal rate significantly (Mohora et al., 2012). Ghernaout et al. (2009a) investigated the reduction of humic acids (HAs) by continuous electromagnetic (EM) treatment followed by EC. EM treatment seemed to increase the reactivity of HA macromolecules before their removal by EC, but alone in continuous mode, EM did not have any positive effect on the HA system. In another study by Ghernaout et al. (2010), EC in series using two electrocoagulators (BIEC or doubled EC) and BIEC assisted by an electromagnetic field (EMT-BIEC) proved efficient for HA removal. Ghernaout et al. (2008) also investigated EC for *Escherichia coli* removal in surface water treatment, rapidly achieving total removal of coliforms and algae. Aluminum electrodes were found to be slightly more efficient than ordinary steel or stainless-steel electrodes.

Kourdali et al. (2014) conducted a series of experimental assays to determine the optimal operating conditions for HA removal from surface waters. Under optimal conditions, up to 72% of HA could be removed. EC treatment mechanisms were proposed for HA removal from various waters (Ghernaout et al., 2009b, 2014). Based on current intensity as a function of applied voltage variation and the relevant literature, three mechanisms were proposed for acid, neutral, and alkaline pH (Ghernaout et al., 2009b).

Koparal et al. (2008) investigated the influence of initial pH on the removal of a high concentration of humic substances by EC. They observed the two possible effects of pH on EC systems: transformation of humic substances is related to the initial pH and gel layer formation on the surface of the electrode starts at higher humic substance concentrations. The authors prevented gel layer formation by adjusting the initial pH value of the wastewater to 5.0. Vepsäläinen et al. (2012) analyzed the effect of EC cell construction on NOM removal. Three types of cells were investigated: one that only had aluminum electrodes, one that had aluminum anodes and inert cathodes, and one that had inert anodes and aluminum cathodes. The reactions between NOM and aluminum were found to occur in the bulk water and not on the electrode surface. NOM destabilization mechanisms were similar to those for chemical coagulation; at low pH, double layer compression dominated, in contrast to adsorption and bridging at higher pH (Vepsäläinen et al., 2012).

El-Masry et al. (2004) successfully tested the ability of EC to purify raw surface water. EC efficiently reduced bacterial load, water turbidity, and chemical contaminants such as sulfates, nitrates, and phosphates. EC also removed 97.8% of HA from groundwater when the electrode interval, current density, and the initial pH value were optimized (Feng et al., 2007).

NOM was found to influence arsenic removal by the continuous-flow EC treatment of weakly mineralized waters (Pallier et al., 2011). Flocs generated during electroflocculation treatment were smaller than those generated during chemical coagulation/flocculation. The presence of HAs negatively influenced floc size because the more negative zeta potential involved influenced coagulation mechanisms. The As(V) removal rate was lower than that of chemical coagulation/flocculation, but during the electroflocculation process As(III) was oxidized to As(V), considerably improving removal rates regardless of the matrix conditions (Pallier et al., 2011).

Combined EC−EF treatment successfully removed suspended and organic matter from surface waters (Ricordel et al., 2010). In this process, coagulant ions Al^{3+} or Fe^{2+} are produced at the anode while the cathode reaction creates hydrogen bubbles that contribute to a flotation process. In contrast to chemical coagulation, less coagulant is needed; consequently, less sludge is formed using this technique.

Studies in which EC technology has been used for NOM removal are presented in Table 5.2.

The potential benefits of combining electric treatment with membrane technologies have been recognized in the last few years (Liu et al., 2012a,b,c). The fouling of membranes in a membrane bioreactor (MBR) can be reduced by EC to form bigger particles; an electric field can also successfully reduce fouling. Liu et al. (2012a) applied an electric field to a

Table 5.2 NOM removal by EC technology.

Matrix	Pollutant	Electrodes	Removal efficiency	Current densities used (mA cm−^{2})	References
Synthetic water	Humic substances	Iron	92.69% for Na_2SO_4 electrolyte	30, 40 and 50 V	Yildiz et al. (2007)
Synthetic water	Humic substance	Aluminum	96.84% for Na_2SO_4 electrolyte	20–50 V	Yildiz et al. (2008)
Synthetic water	NOM isolate (DOC)	Iron	73%	2.43–26.8	Dubrawski and Mohseni (2013a)
Natural water	DOC	Iron	71%	1–25	Dubrawski and Mohseni (2013b)
Synthetic and natural waters	DOC	Iron, aluminum, and zinc	c. 70% for all electrodes by 1 min of electrolysis	2.43	Dubrawski et al. (2013)
Raw surface water	TOC	Aluminum	89%	10–30 V	Mahvi et al. (2011)
Synthetic water	NOM	Iron and aluminum	80%	0.3 A	Ben-Sasson et al. (2013)
Groundwater	DOC	Aluminum	71%	5.78	Mohora et al. (2012)
Natural water (lake water)	MS2 bacteriophage	Iron	1.0–1.5 log reduction	—	Tanneru and Chellam (2012)
Surface water	Algae	Aluminum, stainless steel, and ordinary steel	100%	0.25 A	Ghernaout et al. (2008)

Continued

Table 5.2 NOM removal by EC technology.—cont'd

Matrix	Pollutant	Electrodes	Removal efficiency	Current densities used (mA cm$-^{2}$)	References
Synthetic water	Humic acid (HA)	Aluminum	96%	3.3	Ghernaout et al. (2009a)
Synthetic water	Humic substance	Aluminum	Almost 100%	50 V	Koparal et al. (2008)
Surface water	Color	Aluminum and inert	>90%	0.48	Vepsäläinen et al. (2012)
Surface water	Turbidity	Aluminum	90%	0.6325	El-Masry et al. (2004)
Groundwater	HA	Aluminum	97.8%	4.76	Feng et al. (2007)
Surface water	Permanganate index	Aluminum	47%	1 A	Ricordel et al. (2010)
Paper mill wastewater	Color	Iron	75%	0.4	Kliaugaite et al. (2013)
Paper mill wastewater	Color	Aluminum	91%	0.36−1.79	Vepsäläinen et al. (2007)
Paper mill wastewater	COD	Aluminum	>60%	10	Mansouri et al. (2012)

COD, chemical oxygen demand; *DOC*, dissolved organic carbon; *NOM*, natural organic matter; *TOC*, total organic carbon.

conductive membrane cathode and an iron anode effectively to reduce membrane fouling in an electro-MBR (eMBR). Electric currents of 1–5 mA could extend the membrane filtration cycle and reduce cleaning frequency. The other positive effects of EC included an increase in sludge activity and a decrease in both the extracellular polymeric substance (EPS) and the specific filtration resistances (Liu et al., 2012a).

Novel configurations of electrodes and membrane modules made it possible to achieve significant flux enhancement and fouling reduction using low voltage intensity (Liu et al., 2012b). The electrophoresis and/or electrostatic repulsion of sludge/EPS in an electric field decreased fouling. Liu et al. (2012c) also deposited electrically conductive polypyrrole membranes by coating them in the vapor phase. These membranes were used as cathodes to induce an electric field in a laboratory-scale MBR. EC in an eMBR further improved performance by reducing COD and total phosphorus in the effluent.

EC-membrane filtration (EC-MF) hybrid systems can drastically cut filtration times. EC pretreatment mitigated fouling by reducing the secondary effluent filtration period up to 36-fold (Ben-Sasson et al., 2011). Filtration energy was minimized by 90% when electroflocculation was applied in combination with microfiltration (Ben-Sasson and Adin, 2010a). The extent to which fouling was mitigated depended on fouling intensity, the dominant fouling mechanism, suspension pH, and electroflocculation operation time. Ben-Sasson and Adin (2010b) also investigated energy minimization and fouling mitigation by iron-based electroflocculation in dead-end microfiltration. They concluded that when internal fouling was the dominant mechanism, the iron hydroxide solids formed a layer that filtered out the primary particles. When external fouling was more significant, iron hydroxide solids also reduced the hydraulic resistance of the cake layer.

Using a novel anode material, titanium, Chen and Deng (2012) combined titanium-based EC with an ultrafiltration membrane process. This combination increased HA rejection, reduced membrane fouling, and decreased trans-membrane pressure. Hydrous TiO_2 flocs were formed, which had a large specific surface area and demonstrated high photocatalytic activity.

Ayu Sari and Chellam (2013) assessed the capability of four different advanced pretreatment processes to reduce the fouling of a thin-film composite nanofiltration (NF) membrane during surface water treatment. They observed that EC-MF was inefficient in removing the hydrophilic fraction of NOM, making this the major NF foulant. EC pretreatment significantly increased permeate fluxes during dead-end surface-water microfiltration

(Gamage and Chellam, 2011; Gamage et al., 2012) by reducing both the cake mass and the cumulative hydraulic resistance. Also, as more aluminum was added, the floc zeta potential became less negative, indicating charge neutralization in destabilization (Gamage and Chellam, 2011).

Harif and Adin (2011) examined the first time size and structural evolution of flocs formed in during EC, using a static light scattering technique to elucidate kinetic aspects of the process. EC was found to be capable of generating a range of flocs, and their growth rates and structural characteristics were highly dependent on the operating conditions.

Membrane electrolysis and EC were used for the removal and recovery of humiclike substances from industrial secondary effluent (Kliaugaite et al., 2013). Membrane electrolysis was suitable for the decolorization of colored wastewater but was less efficient than EC treatment in reducing COD. The operational costs for EC were also lower than for membrane electrolyses. Thus, electrochemical methods proved an attractive alternative to other techniques, such as conventional coagulation and chemical oxidation (Kliaugaite et al., 2013).

5.3.2 Comparison of EC and chemical coagulation

Published comparative studies of chemical coagulation and EC are surprisingly few in number. Since these can be considered competing technologies, comparisons of treatment efficiencies are needed, but these studies are difficult to conduct due to the dynamic nature of the EC process.

Change of pH during EC does not significantly affect the aluminum species formed (Cañizares et al., 2009). The formation of monomeric and polymeric aluminum hydroxides was compared when aluminum was added as $AlCl_3$ or by EC. The results showed no significant differences in the speciation of aluminum obtained by these two methods. Under the same dynamic conditions, doses of aluminum, and steady-state pH, removal of wastewater pollutants was similar in both processes.

EC and chemical coagulation have been compared for the treatment of ultrafine quartz suspensions (Kiliç et al., 2009), the removal of Cr(VI) (Golder et al., 2007), and as a pretreatment before microfiltration membranes (Bagga et al., 2008; Zhu et al., 2005). Kiliç et al. (2009) used aluminum anodes and stainless-steel cathodes to compare chemical coagulation and EC and the effect of initial pH and aluminum dosage on the removal efficiency of a model pollutant (ultrafine quartz particles). The optimal pH for particle removal was 9 for EC and 6–9 for chemical coagulation. The broader pH range of the latter was probably a result of higher sulfate concentration because aluminum sulfate was used as a chemical coagulant. The effect of aluminum dosage on the destabilization of pollutants and optimal aluminum dosages was similar with both technologies. Both technologies removed turbidity equally well and the researchers suggested, based on the ζ-potential measurements, that destabilization mechanisms are similar with both technologies.

EC was more effective (42%) than chemical coagulation (approximately 11.5%) at removing Cr(VI) from aqueous solution (Golder et al., 2007). Some Cr(VI) was removed by deposition on the cathode. Alum and aluminum sulfate showed poor Cr(VI) removal, which was probably due to competitive adsorption of sulfate ions on the sludge matrix. Zhu et al. (2005) compared EC and chemical coagulation pretreatment for enhanced virus removal using microfiltration membranes. Both technologies improved virus removal, but EC outperformed chemical coagulation in removal efficiency, which was attributed to electrophoretic movement of viruses on the electrodes, although this has not been reliably verified by experimental studies. Rather different results were obtained from a comparison of chemical coagulation and EC as pretreatment for surface water microfiltration (Bagga et al., 2008). Chemical coagulation with ferric chloride was superior to EC with an iron anode and a stainless-steel cathode. Iron dissolved in Fe(II) form, which had higher solubility and poorer coagulation performance than Fe(III) in ferric chloride. NOM can prevent the oxidation of Fe(II) to the more effective Fe(III) form (Theis and Singer, 1974). Thus, the better performance of EC in research by Zhu et al. (2005) can be explained by the fact that they used synthetic freshwaters that did not include NOM.

These comparative studies seem to indicate that EC and chemical coagulation have similar destabilization mechanisms. Yet, a chemical coagulant is added with an anion, such as sulfate or chloride, which may have some influence on the efficiency of the process. In some cases side reactions, such as reduction on the cathode surface, may play an important role in the destabilization of pollutants. Iron is dissolved from iron electrodes in Fe(II) form, hence oxidation to the more effective Fe(III) form is a significant factor in high removal efficiency. Therefore, it would be more descriptive to compare Fe(II) coagulants, such as ferrous sulfate, with EC using iron electrodes.

5.3.3 EO in NOM removal

Some work has been done on the removal of small model organic compounds (Scialdone et al., 2008, 2009a,b), although not many apply this specifically to NOM removal in drinking water treatment (Motheo and

Pinhedo, 2000). One example is the removal of NOM by photoelectrocatalysis, where TiO_2 coated surfaces were employed in an electrochemical process (Selcuk et al., 2003, 2006; Fu et al., 2007).

Photoelectrocatalytic degradation of a model HA was compared to the photocatalytic destruction process (Selcuk et al., 2003). After 2 h of treatment, 96% of the color, 98% of the HA, and 85% of the TOC were removed in the photoanode cell, which was a TiO_2-coated electrode. Photoelectrocatalysis was much more effective than the photocatalytic method, especially at high pH values (Selcuk et al., 2003). Photoelectrocatalytic treatment can form bromate as a side reaction, however, when bromine is produced by this technique (Selcuk et al., 2006). Using response surface methodology, Fu et al. (2007) optimized the operating parameters (such as initial pH, potassium peroxodisulfate [$K_2S_2O_8$] and bias potential) of fulvic acid photoelectrocatalytic degradation. Regression analysis with an R_2 value of 0.9754 showed that the experimental results closely correlated with the model predictions.

EO was effective in removing HA and algae from pond water. Liao et al. (2008) observed that the degradation followed first-order kinetics. Motheo and Pinhedo (2000) investigated EO of natural peat HA by DSA electrodes, demonstrating that electrolysis efficiency was highly dependent on electrode composition. Efficiency could be increased by changing the reactor design, using commercial electrodes, and exploring the semiconducting properties of these oxide mixtures (Motheo and Pinhedo, 2000).

Electrochemical oxidation of microcystin-LR in the presence of HA in a parallelepipedic electrolytic cell was studied by Tran and Drogui (2013). The pollutant was decomposed by an electrophilic attack by the oxidizing species, including hydroxyl radicals. Thus, the degradation was mainly attributed to direct anodic oxidation, although electrolyte composition influenced the treatment. The highest rates of color removal were recorded using sodium chloride, due to the generation of active chlorine, which is a powerful oxidizing agent.

Anglada et al. (2009) used BDD anodes to analyze the EO process of landfill leachate on a pilot-plant scale. They observed that organic matter and ammonia oxidation were highly influenced by the current density value, and the oxidation mechanism was changed when high current densities were applied. A later analysis (Anglada et al., 2011) showed that organic matter oxidation led to the formation of different disinfection byproducts such as chloroform, dichloroacetonitrile, 1,2-dichloroethane, and 1,1-dichloroacetone. COD and ammonia removals of 51% and 32%, respectively, could be achieved.

Reverse osmosis concentrate streams contain NOM, a major fraction of which are soluble humic substances. EO has been applied also in the treatment of these waters (Bagastyo et al., 2011; Radjenovic et al., 2011). Considerable differences between electrode materials were observed, and the main removal mechanism was recognized as indirect oxidation. Hazardous halogenated oxidation by-products were formed during the treatments, however, which may represent a significant barrier to the environmental application of EO treatment in NOM removal; thus, more detailed investigations are needed.

Studies of NOM removal by EO technology are shown in Table 5.3.

Table 5.3 Removal of NOM by EO technology.

Matrix	Pollutant	Electrodes	Removal efficiency	Current densities used (mA cm^{-2})	References
Landfill leachate	Color	BDD	84%	18 A	Anglada et al. (2011)
Reverse osmosis concentrate	DOC	$Ti/Ru_{0.7}Ir_{0.3}O_2$	25%	0.1–25	Radjenovic et al. (2011)
Reverse osmosis concentrate	DOC	Titanium coated with $IrO_2-Ta_2O_5$, RuO_2-IrO_2, $Pt-IrO_2$, PbO_2, SnO_2-Sb	<16%	10	Bagastyo et al. (2011)
Synthetic water	Color	Ti/BDD	81%	38	Tran and Drogui (2013)
Humic acid (HA) extracted from peat area	HA	DSA electrodes	100%	40	Motheo and Pinhedo (2000)
Pond water	HA	BDD	70%	17.5	Liao et al. (2008)
Synthetic water	TOC	DSA electrode	25%	20	Pinhedo et al. (2005)

BDD, boron-doped diamond; *DOC*, dissolved organic carbon; *DSA*, dimensional stable anode.

5.4 Conclusions

The advantages of EC technology over conventional coagulation include a reduction in the number of different chemicals needed and sludge formed, alkalinity consumption mitigation, and ease of adaptation to existing treatment units (Bagga et al., 2008). EC has removed not only hydrophobic, but also hydrophilic colloids (Vepsäläinen et al., 2009), which are not easily eliminated by conventional coagulation processes. Thus, EC is demonstrably effective for NOM removal, but the process conditions (e.g., pH, electric charge, and supporting electrolyte) must be determined carefully. Moreover, the EC mechanism is highly dependent on the chemistry of the aqueous medium, especially its conductivity (Yildiz et al., 2007, 2008; Vepsäläinen et al., 2009). Hence, it remains to be determined whether the benefits of this method are great enough for feasible application (Bagga et al., 2008).

EC has several advantages over other water treatment technologies (Mollah et al., 2001, 2004). Some of the important benefits for wastewater treatment are as follows:

- EC produces low amounts of sludge, and the sludge is dewatered easily.
- Flocs formed in EC are larger and more stable than those formed in chemical coagulation.
- Due to electrophoretic movement, the removal of DOC from water is better with EC than with chemical coagulation.
- Competing reactions, such as hydrogen and hypochlorite formation, can be beneficial in water treatment.
- EC can be run by green processes, such as solar or wind power and fuel cells.
- Adsorption of hydroxide on mineral surfaces is 100 times greater than precipitated hydroxides.
- No harmful anions, such as sulfate and chloride, are added to water in EC.

The disadvantages of EC in NOM removal during water treatment are the passivation of electrodes due to the presence of oxides and precipitation layers on the electrode surfaces, the dissolution of the electrodes, and the low conductivity of the surface waters (Mollah et al., 2001). High voltages can be avoided by reducing the distance between electrodes. Electrodes need to be replaced regularly as the electrode material dissolves, raising maintenance costs. Trihalomethane formation can occur in the treatment of waters with a high humic and fulvic acid content (Mollah et al., 2004).

Optimal DOC removal and low residual aluminum can be obtained simultaneously with EC, which may be a significant advantage of this method over chemical coagulation in surface water treatment. EC can produce high-quality water that is potable or suitable as fresh water for industrial applications. This technology seems capable of removing some toxic pollutants from wastewater and could be used as pretreatment prior to combination with some other purification technology. A thorough economic and ecological comparison of chemical coagulation and EC is recommended, however, because these methods seem to have similar pollutant destabilization mechanisms, metal consumption, and removal efficiencies in most applications. This investigation should measure operating and investment costs, including the coagulation, mixing, and separation phases of the treatment. Side reactions of EC, such as flotation due to hydrogen bubble production, should also be considered. These calculations would show what applications of EC are economically viable. If EC is more expensive than chemical coagulation, it may still have small-scale applications, for example in emergency water treatment systems or small communities, as it could be simpler to operate than chemical coagulation, especially if EF is used for particle separation. In addition to economic and ecological research, new technological innovations and investigations are also required to prevent electrode passivation and fouling, and to further optimize treatment parameters. Full-scale systems are likely to operate in continuous-flow mode, so it is proposed that bench-scale continuous-flow systems are used in future studies.

The electrochemical mineralization of organic compounds has interested several researchers. Nevertheless, BDD anodes have been found to be particularly effective for HA removal from aqueous solutions (Liao et al., 2008). Due to the complexity of humic substances, however, the formation of intermediates causes degradation. The observed disadvantages of using BDD anodes are the formation of disinfection by-products (DBPs), mainly perchlorate (Bergmann et al., 2009), and their high manufacturing costs.

Some traditional techniques such as chemical coagulation, adsorption, and advanced oxidation processes are still often more economical than EO for NOM removal. EO can provide almost complete mineralization of organic compounds, however, which makes it an attractive process in some applications. EO can be an effective pretreatment or posttreatment to guarantee the necessary water quality.

The efficiency of the EO process largely depends on cell configuration, electrode material, electrolyte composition, and other experimental parameters, such as current density or the temperature of the treated water. In future NOM treatment studies, it would be beneficial to measure different radical reactions at the surface of the anode and by-products formed after the treatment.

The discovery of an effective and stable, yet economical, electrode material would speed up the use of EO techniques for NOM removal. This would be further assisted by the development of some novel electric power source, such as solar energy. Combining treatment with current competitive techniques would enhance purification results. The economic viability of EO in the treatment of NOM-containing waters also needs to be calculated.

References

A. Anglada, A. Urtiaga, I. Ortiz, Pilot scale performance of the electro-oxidation of landfill leachate at boron-doped diamond anodes, Environ. Sci. Technol. 43 (2009) 2035–2040.

A. Anglada, A. Urtiaga, I. Ortiz, D. Mantzavinos, E. Diamadopoulos, Boron-doped diamond anodic treatment of landfill leachate: evaluation of operating variables and formation of oxidation by-products, Water Res. 45 (2011) 828–838.

H.S. Awad, N. Abo Galwa, Electrochemical degradation of Acid Blue and Basic Brown dyes on Pb/PbO2 electrode in the presence of different conductive electrolyte and effect of various operating factors, Chemosphere 61 (2005) 1327–1335.

M. Ayu Sari, S. Chellam, Surface water nanofiltration incorporating (electro) coagulation-microfiltration pretreatment: fouling control and membrane characterization, J. Membr. Sci. 437 (2013) 249–256.

A.Y. Bagastyo, J. Radjenovic, Y. Mu, R.A. Rozendal, D.J. Batstone, K. Rabaey, Electrochemical oxidation of reverse osmosis concentrate on mixed metal oxide (MMO) titanium coated electrodes, Water Res. 45 (2011) 4951–4959.

A. Bagga, S. Chellam, D.A. Clifford, Evaluation of iron chemical coagulation and electrocoagulation pretreatment for surface water microfiltration, J. Membr. Sci. 309 (2008) 82–93.

M. Ben-Sasson, A. Adin, Fouling mechanisms and energy appraisal in microfiltration pretreated by aluminum-based electroflocculation, J. Membr. Sci. 352 (2010) 86–94.

M. Ben-Sasson, A. Adin, Fouling mitigation by iron-based electroflocculation in microfiltration: mechanisms and energy minimization, Water Res. 44 (2010) 3973–3981.

M. Ben-Sasson, Y.M. Lin, A. Adin, Electrocoagulation-membrane filtration hybrid system for colloidal fouling mitigation of secondary-effluent, Separ. Purif. Technol. 82 (2011) 63–70.

M. Ben-Sasson, W. Calmano, A. Adin, Iron-oxidation processes in an electroflocculation (electrocoagulation) cell, J. Hazard Mater. 171 (2009) 704–709.

M. Ben-Sasson, Y. Zidon, R. Calvo, A. Adin, Enhanced removal of natural organic matter by hybrid process of electrocoagulation and dead-end microfiltration, Chem. Eng. J. 232 (2013) 338–345.

M.E.H. Bergmann, A.S. Koparal, Studies on electrochemical disinfectant production using anodes containing RuO2, J. Appl. Electrochem. 35 (2005) 1321–1329.

M.E.H. Bergmann, J. Rollin, T. Iourtchouk, The occurrence of perchlorate during drinking water electrolysis using BDD anodes, Electrochim. Acta 54 (2009) 2102–2107.

P. Cañizares, C. Jiménez, F. Martínez, M.A. Rodrigo, C. Sáez, The pH as a key parameter in the choice between coagulation and electrocoagulation for the treatment of wastewaters, J. Hazard Mater. 163 (2009) 158–164.

E. Chatzisymeon, N.P. Xekoukoulotakis, A. Coz, N. Kalogerakis, D. Mantzavinos, Electrochemical treatment of textile dyes and dyehouse effluents, J. Hazard Mater. B137 (2006) 998−1007.

G. Chen, Electrochemical technologies in wastewater treatment, Separ. Purif. Technol. 38 (2004) 11−41.

X. Chen, G. Chen, Anodic oxidation of Orange II on Ti/BDD electrode: variable effects, Separ. Purif. Technol. 48 (2006) 45−49.

X. Chen, G. Chen, Fabrication and application of Ti/BDD for wastewater treatment, in: E. Brillas, C.A. Martinez-Huitle (Eds.), Synthetic Diamond Films: Preparation, Electrochemistry, Characterization, and Applications, John Wiley & Sons, New Jersey, USA, 2011, pp. 353−371.

X. Chen, H. Deng, Removal of humic acids from water by hybrid titanium-based electrocoagulation with ultrafiltration membrane processes, Desalination 300 (2012) 51−57.

X. Chen, G. Chen, P.L. Yue, Separation of pollutants from restaurant wastewater by electrocoagulation, Separ. Purif. Technol. 19 (2000) 65−76.

Y.Y. Chu, Y. Qian, W.J. Wang, X.L. Deng, A dual-cathode electro-Fenton oxidation coupled with anodic oxidation system used for 4-nitrophenol degradation, J. Hazard Mater. (2012) 199−200, 179−185.

C. Comninellis, A. Kapalka, S. Malato, S.A. Parsons, I. Poulios, D. Mantzavinos, Advanced oxidation processes for water treatment: advances and trends for R&D, J. Chem. Technol. Biotechnol. 83 (2008) 769−776.

A. Dhaouadi, L. Monser, N. Adhoum, Anodic oxidation and electro-Fenton treatment of rotenone, Electrochim. Acta 54 (2009) 4473−4480.

K.L. Dubrawski, M. Mohseni, Standardizing electrocoagulation reactor design: iron electrodes for NOM removal, Chemosphere 91 (2013) 55−60.

K.L. Dubrawski, M. Mohseni, In-situ identification of iron electrocoagulation speciation and application for natural organic matter (NOM) removal, Water Res. 47 (2013) 5371−5380.

K.L. Dubrawski, M. Fauvel, M. Mohseni, Metal type and natural organic matter source for direct filtration electrocoagulation of drinking water, J. Hazard Mater. 244−245 (2013) 135−141.

E.-S.Z. El-Ashtoukhy, N.K. Amin, O. Abdelwahab, Treatment of paper mill effluents in a batch-stirred electrochemical tank reactor, Chem. Eng. J. 146 (2009) 205−210.

M.H. El-Masry, O.M. Sadek, W.K. Mekhemer, Purification of raw surface water using electro-coagulation method, Water Air Soil Pollut. 158 (2004) 373−385.

A.M. Faouzi, B. Nasr, G. Abdellatif, Electrochemical degradation of anthraquinone dye Alizarin Red S by anodic oxidation on boron-doped diamond, Dyes Pigments 73 (2007) 86−89.

Q. Feng, X. Li, Y. Cheng, L. Meng, Q. Meng, Removal of humic acid from groundwater by electrocoagulation, J. China Univ. Min. Technol. 17 (2007) 513−520.

C. Flox, C. Arias, E. Brillas, A. Savall, K. Groenen-Serrano, Electrochemical incineration of cresols: a comparative study between PbO_2 and boron-doped diamond anodes, Chemosphere 74 (2009) 1340−1347.

J. Fu, Y. Zhao, Q. Wu, Optimizing photoelectrocatalytic oxidation of fulvic acid response surface methodology, J. Hazard Mater. 144 (2007) 499−505.

N.P. Gamage, S. Chellam, Aluminum electrocoagulation pretreatment reduces fouling during surface water microfiltration, J. Membr. Sci. 379 (2011) 97−105.

N.P. Gamage, J.D. Rimer, S. Chellam, Improvements in permeate flux by aluminum electroflotation pretreatment during microfiltration of surface water, J. Membr. Sci. 411−412 (2012) 45−53.

D. Ghernaout, A. Badis, A. Kellil, B. Ghernaout, Application of electrocoagulation in *Escherichia coli* culture and two surface waters, Desalination 219 (2008) 118−125.

D. Ghernaout, B. Ghernaout, A. Saiba, A. Boucherit, A. Kellil, Removal of humic acids by continuous electromagnetic treatment followed by electrocoagulation in batch using aluminium electrodes, Desalination 239 (2009) 295–308.

D. Ghernaout, S. Irki, A. Boucherit, Removal of Cu^{2+} and Cd^{2+}, and humic acid and phenol by electrocoagulation using iron electrodes, Desalination Water Treat. 52 (2014) 3256–3270.

D. Ghernaout, B. Ghernaout, A. Boucherit, M.W. Naceur, A. Khelifa, A. Kellil, Study on mechanism of electrocoagulation with iron electrodes in idealized conditions and electrocoagulation of humic acids solution in batch using aluminium electrodes, Desalination Water Treat. 8 (2009) 91–99.

D. Ghernaout, A. Mariche, B. Ghernaout, A. Kellil, Electromagnetic treatment doubled electrocoagulation of humic acid in continuous mode using response surface method for its optimization and application on two surface waters, Desalination Water Treat. 22 (2010) 311–329.

A. Golder, A. Chanda, A. Samanta, S. Ray, Removal of Cr(VI) from aqueous solution: electrocoagulation vs chemical coagulation, Separ. Sci. Technol. 42 (2007) 2177–2193.

J.A.G. Gomes, P. Daida, M. Kesmez, M. Weir, H. Moreno, J.R. Parga, G. Irwin, H. McWhinney, T. Grady, E. Peterson, D.L. Cocke, Arsenic removal by electrocoagulation using combined Al–Fe electrode system and characterization of products, J. Hazard Mater. 139 (2007) 220–231.

M. Gotsi, N. Kalogerakis, E. Psillakis, P. Samaras, D. Mantzavinos, Electrochemical oxidation of olive oil mill wastewaters, Water Res. 39 (2005) 4177–4187.

E. Guinea, J.A. Garrido, R.M. Rodriguez, P.-L. Cabot, C. Arias, F. Centellas, E. Brillas, Degradation of the fluoroquinolone enrofloxacin by electrochemical advanced oxidation processes based on hydrogen peroxide electrogeneration, Electrochim. Acta 55 (2010) 2101–2115.

M. Hamza, S. Ammar, R. Abdelhedi, Electrochemical oxidation of 1,3,5-trimethoxybenzene in aqueous solutions at gold oxide and lead dioxide electrodes, Electrochim. Acta 56 (2011) 3785–3789.

T. Harif, A. Adin, Size and structure evolution of kaolin-Al(OH)3 flocs in the electroflocculation process: a study using static light scattering, Water Res. 45 (2011) 6195–6206.

P.K. Holt, G.W. Barton, C.A. Mitchell, The future for electrocoagulation as a localized water treatment technology, Chemosphere 59 (2005) 355–367.

J. Jeong, C. Kim, J. Yoon, The effect of electrode material on the generation of oxidants and microbial inactivation in the electrochemical disinfection processes, Water Res. 43 (2009) 895–901.

I. Kabdaslı, B. Vardar, I. Arslan-Alaton, O. Tunay, Effect of dye auxiliaries on color and COD removal from simulated reactive dye bath effluent by electrocoagulation, Chem. Eng. J. 148 (2009) 89.

A. Kapalka, G. Foti, C. Comninellis, Basic principles of the electrochemical mineralization of organic pollutants for wastewater treatment, in: C. Comninellis, G. Chen (Eds.), Electrochemistry for the Environment, Springer, New York, USA, 2010, pp. 1–23.

M.G. Kiliç, Ç. Hosten, S. Demirci, A parametric comparative study of electrocoagulation and coagulation using ultrafine quartz suspensions, J. Hazard Mater. 171 (2009) 247–252.

M. Klavarioti, D. Mantzavinos, D. Kassinos, Removal of residual pharmaceuticals from aqueous systems by advanced oxidation processes, Environ. Int. 35 (2009) 402–417.

D. Kliaugaite, K. Yasadi, G.-J. Euverink, M.F.M. Bijmans, V. Racys, Electrochemical removal and recovery of humic-like substances from wastewater, Separ. Purif. Technol. 108 (2013) 37–44.

M. Kobya, U. Gebologlu, F. Ulu, S. Oncel, E. Demirbas, Removal of arsenic from drinking water by the electrocoagulation using Fe and Al electrodes, Electrochim. Acta 56 (2011) 5060−5070.

M. Kobya, H. Hiz, E. Senturk, C. Aydiner, E. Demirbas, Treatment of potato chips manufacturing wastewater by electrocoagulation, Desalination 190 (2006) 2201−2211.

A.S. Koparal, Y.S. Yildiz, B. Keskinler, N. Demircioglu, Effect of initial pH on the removal of humic substances from wastewater by electrocoagulation, Separ. Purif. Technol. 59 (2008) 175−182.

S. Kourdali, A. Badis, A. Saiba, A. Boucherit, H. Boutoumi, Humic acid removal by electrocoagulation using aluminium sacrificial anode under influencing operational parameters, Desalination Water Treat. 52 (2014) 5442−5453.

A. Kraft, M. Stadelmann, M. Blaschke, Anodic oxidation with doped diamond electrodes: a new advanced oxidation process, J. Hazard. Mater B103 (2003) 247−261.

A. Kraft, M. Stadelmann, M. Blaschke, D. Kreysig, B. Sandt, F. Schröder, J. Rennau, Electrochemical water disinfection part I: hypochlorite production from very dilute chloride solutions, J. Appl. Electrochem. 29 (1999) 861−868.

A.A. Liao, M. Spitzer, A.J. Motheo, R. Bertazzoli, Electrocombustion of humic acid and removal of algae from aqueous solutions, J. Appl. Electrochem. 38 (2008) 721−727.

J. Liu, L. Liu, B. Gao, F. Yang, Cathode membrane fouling reduction and sludge property in membrane bioreactor integrating electrocoagulation and electrostatic repulsion, Separ. Purif. Technol. 100 (2012) 44−50.

L. Liu, J. Liu, B. Gao, F. Yang, Minute electric field reduced membrane fouling and improved performance of membrane bioreactor, Separ. Purif. Technol. 86 (2012) 106−112.

L. Liu, J. Liu, B. Gao, F. Yang, S. Chellam, Fouling reductions in a membrane bioreactor using an intermittent electric field and cathodic membrane modified by vapor phase polymerized pyrrole, J. Membr. Sci. (2012) 394−395, 202−208.

H. Ma, B. Wang, Y. Wang, Application of molybdenum and phosphate modified kaolin in electrochemical treatment of paper mill wastewater, J. Hazard Mater. 145 (2007) 417−423.

A.H. Mahvi, M. Malakootian, M.R. Heidari, Comparison of polyaluminum silicate chloride and electrocoagulation process, in natural organic matter removal from surface water in Ghochan, Iran, J. Water Chem. Technol. 33 (2011) 377−385.

K. Mansouri, A. Hannachi, A. Abdel-Wahab, N. Bensalah, Electrochemically dissolved aluminum coagulants for the removal of natural organic matter from synthetic and real industrial wastewaters, Ind. Eng. Chem. Res. 51 (2012) 2428−2437.

A.F. Martins, M.L. Wilde, T.G. Vasconcelos, D.M. Henriques, Nonylphenol polyethoxylate degradation by means of electrocoagulation and electrochemical Fenton, Separ. Purif. Technol. 50 (2006) 249−255.

Y. Meas, J.A. Ramirez, M.A. Villalon, T.W. Chapman, Industrial wastewaters treated by electrocoagulation, Electrochim. Acta 55 (2010) 8165−8171.

F.L. Migliorini, N.A. Braga, S.A. Alves, M.R.V. Lanza, M.R. Baldan, N.G. Ferreira, Anodic oxidation of wastewater containing the Reactive Orange 16 Dye using heavily boron-doped diamond electrodes, J. Hazard Mater. 192 (2011) 1683−1689.

D.W. Miwa, G.R.P. Malpass, S.A.S. Machado, A.J. Motheo, Electrochemical degradation of carbaryl on oxide electrodes, Water Res. 40 (2006) 3281−3289.

E. Mohora, S. Roncevic, B. Dalmacija, J. Agbaba, M. Watson, E. Karlovic, M. Dalmacija, Removal of natural organic matter and arsenic from water by electrocoagulation/flotation continuous flow reactor, J. Hazard Mater. (2012) 235−236, 257−264.

M.Y.A. Mollah, P. Morkovsky, J.A.G. Gomez, M. Kesmez, J. Parga, D.L. Cocke, Fundamentals, present and future perspectives of electrocoagulation, J. Hazard Mater. B114 (2004) 199−210.

M.Y.A. Mollah, R. Schennach, J.R. Parga, D.L. Cocke, Electrocoagulation (EC) – science and applications, J. Hazard Mater. B84 (2001) 29–41.
F. Montilla, P.A. Michaud, E. Morallon, J.L. Vazquez, C. Comninellis, Electrochemical oxidation of benzoic acid at boron-doped diamond electrodes, Electrochim. Acta 47 (2002) 3509–3513.
H. Moreno, D. Cocke, J. Gomes, P. Morkovsky, J. Parga, E. Peterson, C. Garcia, Electrochemical reactions for electrocoagulation using iron electrodes, Ind. Eng. Chem. Res. 48 (2009) 2275–2282.
B. Morgan, O. Lahav, The effect of pH on the kinetics of spontaneous Fe(II) oxidation by O_2 in aqueous solution – basic principles and a simple heuristic description, Chemosphere 68 (2007) 2080–2084.
A.J. Motheo, L. Pinhedo, Electrochemical degradation of humic acid, Sci. Total Environ. 256 (2000) 67–76.
G. Mouedhen, M. Feki, M.D.P. Wery, H.F. Ayedi, Behavior of aluminum electrodes in electrocoagulation process, J. Hazard Mater. 150 (2008) 124–135.
V. Pallier, G. Feuillade-Cathalifaud, B. Serpaud, Influence of organic matter on arsenic removal by continuous flow electrocoagulation treatment of weakly mineralized waters, Chemosphere 83 (2011) 21–28.
M. Panizza, G. Cerisola, Application of diamond electrodes to electrochemical processes, Electrochim. Acta 51 (2005) 191–199.
J. Parga, D. Cocke, V. Valverde, J. Gomes, M. Kesmez, H. Moreno, M. Weir, D. Mencer, Characterization of electrocoagulation for removal of chromium and arsenic, Chem. Eng. Technol. 28 (2005) 605–612.
U.D. Patel, S. Suresh, Electrochemical treatment of pentachlorophenol in water and pulp bleaching effluent, Separ. Purif. Technol. 61 (2008) 115–122.
G. Patermarakis, E. Fountoukidis, Disinfection of water by electrochemical treatment, Water Res. 24 (1990) 1491–1496.
T. Picard, G. Cathalifaud-Feuillade, M. Mazet, C. Vandensteendam, Cathodic dissolution in the electrocoagulation process using aluminium electrodes, J. Environ. Monit. 2 (2000) 77–80.
L. Pinhedo, R. Pelegrini, R. Bertazzoli, A.J. Motheo, Photoelectrochemical degradation of humic acid on a $(TiO_2)0.7(RuO_2)0.3$ dimensionally stable anode, Appl. Catal. B Environ. 57 (2005) 75–81.
A.M. Polcaro, A. Vacca, M. Mascia, S. Palmas, R. Pompei, S. Laconi, Characterization of a stirred tank electrochemical cell for water disinfection processes, Electrochim. Acta 52 (2007) 2595–2602.
X. Qu, W.J. Gao, M.N. Han, A. Chen, B.Q. Liao, Integrated thermophilic submerged aerobic membrane bioreactor and electrochemical oxidation for pulp and paper effluent treatment – towards system closure, Bioresour. Technol. 116 (2012) 1–8.
J. Radjenovic, A. Bagastyo, R.A. Rozendal, Y. Mu, J. Keller, K. Rabaey, Electrochemical oxidation of trace organic contaminants in reverse osmosis concentrate using RuO_2/IrO_2-coated titanium anodes, Water Res. 45 (2011) 1579–1586.
D. Rajkumar, J.G. Kim, Oxidation of various reactive dyes with in situ electro-generated active chlorine for textile dyeing industry wastewater treatment, J. Hazard Mater. B136 (2006) 203–212.
D. Rajkumar, B.J. Song, J.G. Kim, Electrochemical degradation of Reactive Blue 19 in chloride medium for the treatment of textile dyeing wastewater with identification of intermediate compounds, Dyes Pigments 72 (2007) 1–7.
C. Ricordel, A. Darchen, D. Hadjiev, Electrocoagulation–electroflotation as a surface water treatment for industrial uses, Separ. Purif. Technol. 74 (2010) 342–347.
M.A. Rodrigo, P. Cañizares, A. Sanchez-Carretero, C. Saez, Use of conductive-diamond electrochemical oxidation for wastewater treatment, Catal. Today 151 (2010) 173–177.

A. Sakalis, K. Fytianos, U. Nickel, A. Voulgaropoulos, A comparative study of platinized titanium and niobe/synthetic diamond as anodes in the electrochemical treatment of textile wastewater, Chem. Eng. J. 119 (2006) 127–133.
O. Scialdone, A. Galia, C. Guarisco, S. Randazzo, G. Filardo, Electrochemical incineration of oxalic acid at boron doped diamond anodes: the role of operative parameters, Electrochim. Acta 53 (2008) 2095–2108.
O. Scialdone, S. Randazzo, A. Galia, G. Filardo, Electrochemical oxidation of organics at metal oxide electrodes: the incineration of oxalic acid at IrO_2-TaO_2O_5, Electrochim. Acta 54 (2009) 1210–1217.
O. Scialdone, S. Randazzo, A. Galia, G. Silvestri, Electrochemical oxidation of organics in water: role of operative parameters in the absence and in the presence of NaCl, Water Res. 43 (2009) 2260–2272.
H. Selcuk, H.Z. Sarikaya, M. Bekbolet, M.A. Anderson, Bromate formation on the non-porous TiO_2 photoanode in the photoelectrocatalytic system, Chemosphere 62 (2006) 715–721.
H. Selcuk, J.J. Sene, M.A. Anderson, Photoelectrocatalytic humic acid degradation kinetics and effect of pH, applied potential and inorganic ions, J. Chem. Technol. Biotechnol. 78 (2003) 979–984.
H. Särkkä, K. Kuhmonen, M. Vepsäläinen, M. Pulliainen, J. Selin, P. Rantala, E. Kukkamäki, M. Sillanpää, Electrochemical oxidation of sulphides in paper mill wastewater by using mixed oxide anodes, Environ. Technol. 30 (2009) 885–892.
C.T. Tanneru, S. Chellam, Mechanisms of virus control during iron electrocoagulation – microfiltration of surface water, Water Res. 46 (2012) 2111–2120.
M.G. Tavares, L.V.A. da Silva, A.M. Sales Solano, J. Tonholo, C.A. Martinez-Huitle, C.L.P.S. Zanta, Electrochemical oxidation of Methyl Red using Ti/Ru0.3Ti0.7O2 and Ti/Pt anodes, Chem. Eng. J. (2012) 204–206, 141–150.
T.L. Theis, P.C. Singer, Complexation of iron(II) by organic matter and its effect on iron(II) oxygenation, Environ. Sci. Technol. 8 (1974) 569–573.
N. Tran, P. Drogui, Electrochemical removal of microcystin-LR from aqueous solution in the presence of natural organic pollutants, J. Environ. Manag. 114 (2013) 253–260.
J.L. Trompette, H. Vergnes, On the crucial influence of some supporting electrolytes during electrocoagulation in the presence of aluminum electrodes, J. Hazard Mater. 163 (2009) 1282–1288.
M. Vepsäläinen, M. Ghiasvand, J. Selin, J. Pienimaa, E. Repo, M. Pulliainen, M. Sillanpää, Investigations of the effects of temperature and initial sample pH on natural organic matter (NOM) removal with electrocoagulation using response surface method (RSM), Separ. Purif. Technol. 69 (2009) 255–261.
M. Vepsäläinen, M. Pulliainen, M. Sillanpää, Effect of electrochemical cell structure on natural organic matter (NOM) removal from surface through electrocoagulation (EC), Separ. Purif. Technol. 99 (2012) 20–27.
M. Vepsäläinen, J. Selin, M. Pulliainen, M. Sillanpää, Combined electrocoagulation and chemical coagulation of paper mill mechanically cleaned water, J. Pulp Pap. Sci. 33 (2007) 233–239.
E.A. Vik, D.A. Carlson, A.S. Eikum, E.T. Gjessing, Electrocoagulation of potable water, Water Res. 18 (1984) 1355–1360.
B. Wang, L. Gu, H. Ma, Electrochemical oxidation of pulp and paper making wastewater assisted by transition metal modified kaolin, J. Hazard Mater. 143 (2007) 198–205.
B. Wang, W. Kong, H. Ma, Electrochemical treatment of paper mill wastewater using three-dimensional electrodes with Ti/Co/SnO_2-Sb_2O_5 anode, J. Hazard Mater. 146 (2007) 295–301.
Y.S. Yildiz, A.S. Koparal, B. Keskinler, Effect of initial pH and supporting electrolyte on the treatment of water containing high concentration of humic substances by electrocoagulation, Chem. Eng. J. 138 (2008) 63–72.

Y.S. Yildiz, A.S. Koparal, S. İrdemez, B. Keskinler, Electrocoagulation of synthetically prepared waters containing high concentration of NOM using iron cast electrodes, J. Hazard Mater. B139 (2007) 373–380.

X. Zhao, B. Zhang, H. Liu, F. Chen, A. Li, J. Qu, Transformation characteristics of refractory pollutants in plugboard wastewater by an optimal electrocoagulation and electro-Fenton process, Chemosphere 87 (2012) 631–636.

B. Zhu, D.A. Clifford, S. Chellam, Comparison of electrocoagulation and chemical coagulation pretreatment for enhanced virus removal using microfiltration membranes, Water Res. 39 (2005) 3098–3108.

X. Zhu, J. Ni, J. Wei, P. Chen, Scale-up of B-doped diamond anode system for electrochemical oxidation of phenol simulated wastewater in batch mode, Electrochim. Acta 56 (2011) 9439–9447.

CHAPTER 6

Membranes

Mika Sillanpaa[1,2,3,4,5], Sari Metsämuuronen[6] and Mika Manttari[7]
[1]Department of Chemical Engineering, School of Mining, Metallurgy and Chemical Engineering, University of Johannesburg, Doornfontein, South Africa; [2]Department of Applied Physics, Faculty of Science and Technology, Universiti Kebangsaan Malaysia, Bangi, Selangor, Malaysia; [3]Zhejiang Rongsheng Environmental Protection Paper Co. LTD, Pinghu Economic Development Zone, Zhejiang, P.R. China; [4]Department of Civil Engineering, University Centre for Research & Development, Chandigarh University, Gharuan, Mohali, Punjab, India; [5]International Research Centre of Nanotechnology for Himalayan Sustainability (IRCNHS), Shoolini University, Solan, Himachal Pradesh, India; [6]Department of Chemical Technology, LUT University, Lappeenranta, Finland; [7]Department of Separation Science, LUT University, Lappeenranta, Finland

Abstract

Aquatic natural organic matter (NOM) is a heterogeneous mixture of biopolymers and their degradation products that cause harmful by-products during drinking water production. The great variability in NOM composition makes it difficult to remove from drinking water completely by conventional coagulation or by any single technique. This chapter reviews NOM removal by membranes—namely, micro-, ultra-, and nano-filtration, and also reverse osmosis.

Keywords: Fouling; Membrane technology; Microfiltration; Nanofiltration; Natural organic matter (NOM); Reverse osmosis; Ultrafiltration; Water treatment.

Abbreviations

AOC	Assimilable organic carbon
BDOC	Biodegradable dissolved organic carbon
DBP	Disinfection by-product
DOC	Dissolved organic carbon
HAAs	Haloacetic acids
MF	Microfiltration
MM	Molar mass
NF	Nanofiltration
NOM	Natural organic matter
PES	Polyethersulfone
PVDF	Polyvinylidene fluoride
RO	Reverse osmosis
SUVA	Specific UV absorbance
THM	Trihalomethane
THMFP	Trihalomethane formation potential
TOC	Total organic carbon
UF	Ultrafiltration

Natural Organic Matter in Water
ISBN 978-0-12-824274-2
https://doi.org/10.1016/B978-0-12-824274-2.00006-5

6.1 Introduction

Since the late 1980s, membrane processes have increasingly been considered an effective means of removing turbidity, organics, microorganisms, and disinfection by-product (DBP) precursors from drinking water in compliance with the more stringent regulations. This is the result of significant improvements in membrane technology, design, and materials during the past three decades, making these drinking water production processes more environmentally friendly.

It is estimated that, today, the overall plant capacity for microfiltration (MF) and ultrafiltration (UF) is around 20 million m^3 day^{-1}, with 60% for drinking water production (Schrotter and Bozkaya-Schrotter, 2010), and UF applications represent about 74% of the total capacity (Fane et al., 2008).

Membrane processes use a semipermeable film called a *membrane* to separate particles and solutes from water. Most membrane processes use pressure differences across the membrane as a driving force. These can be divided into pressure and vacuum systems (Fig. 6.1). In pressure systems, membranes are installed in modules, which are pressurized from the retentate side of the membrane. In a dead-end filtration feed, raw water permeates through the membrane, and solutes larger than the membrane pore size are retained, whereas in a cross-flow configuration, the majority of the feed flows tangentially across the membrane surface. In vacuum systems, membrane elements are submerged in open tanks.

The membranes reject uncharged macromolecules mainly by size exclusion, and thus the rejection is determined by the membrane pore size and pore size distribution compared with the size, size distribution, and

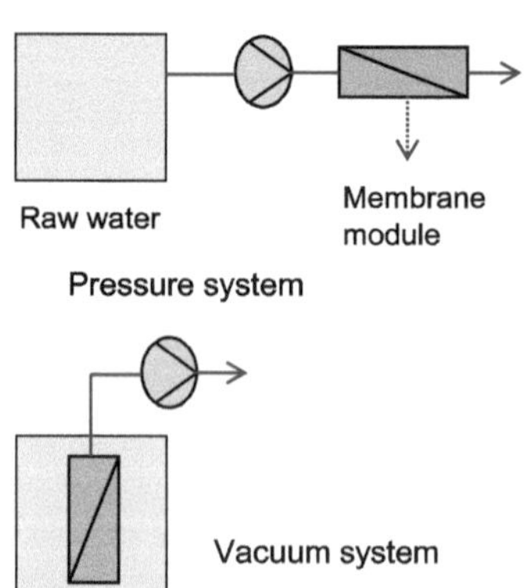

Figure 6.1 Schematic representation of pressure and vacuum membrane configurations.

shape of the natural organic matter (NOM) molecules. The molar mass (MM) of NOM is in the colloidal size range, from a few 100 Da (Daltons equals grams per mole) to more than 100 kDa (Ericsson and Tragardh, 1997). The size distribution of NOM varies between different water sources, but generally a very high proportion of small molecules, less than 1 kDa, have been observed (Fan et al., 2001; Kim andYu, 2005; Zularisam et al., 2007b; Zularisam et al., 2009) (Fig. 6.2). Thus, the average MM is quite low. Weight-averaged MM values of 833–1031 Da (Spain (de la Rubia et al., 2008)), 1254–1493 Da (Australia (Fabris et al., 2008)), 1442–1812 Da (Norway (Fabris et al., 2008)), and 2190 Da (Georgia, USA (Chin et al., 1994)) have been reported for surface waters. The smallest NOM fractions consist of hydrophilic compounds, organic acids, and building blocks, which are degradation products of humic substances (Huber et al., 2011). Polysaccharides, proteins, and amino sugars are the highest MM components in NOM, whereas the most aromatic compounds (i.e., lignin and tannin derivatives) are abundant in the intermediate-high MM fractions of NOM (Lankes et al., 2008).

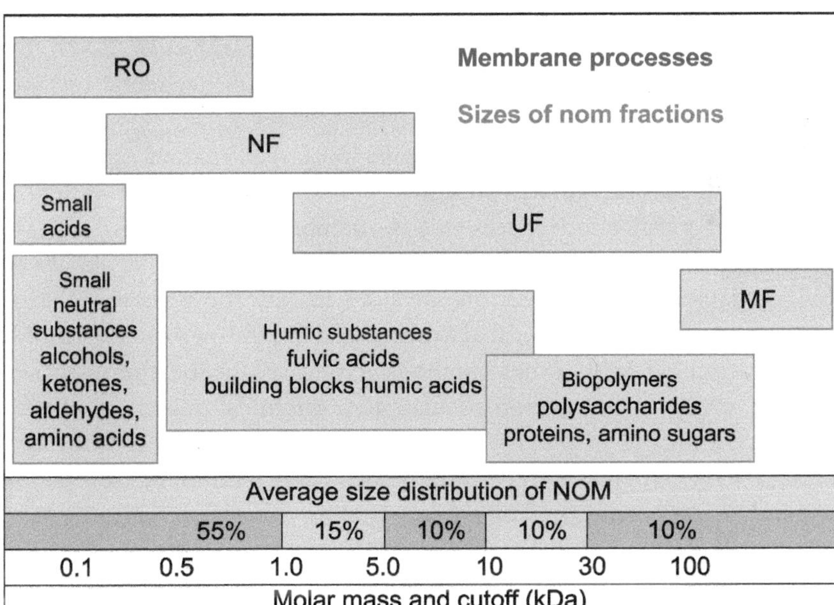

Figure 6.2 Cutoffs of different membrane processes and the sizes of natural organic matter (NOM) fractions. *MF*, microfiltration; *NF*, nanofiltration; *RO*, reverse osmosis; *UF*, ultrafiltration. *(Data from Fan et al. (2001), Zularisam et al. (2007a), Huber et al. (2011), Thorsen (2004), Kennedy et al. (2005), Kaewsuk and Seo (2011), and Jeong et al. (2013).)*

In addition to size exclusion, membrane rejection depends on solute-solute and solute-membrane interactions, and hydrodynamic conditions.

Rejection can be greater than what is expected according to the nominal cutoff value of the membrane if there is a repulsive electrostatic force between the membrane and solutes. Therefore, the properties of both the membrane and NOM are important factors in NOM removal, including acidic group content and hydrophobic/hydrophilic character. Other physicochemical effects that affect rejection when charge repulsion is used are pH, ionic strength, and divalent cation—especially Ca^{2+}—content. This chapter reviews NOM removal by membrane technologies and is an update of an earlier publication (Metsamuuronen et al., 2014).

6.2 Microfiltration

MF membranes were first used for color, turbidity, and pathogen removal in municipal drinking water plants during the late 1980s (Ericsson et al., 1997; Jacangelo et al., 1997; Thorsen, 1998; Laine et al., 2000). The pore sizes of MF membranes (0.1−5 μm or >100 kDa) are much larger than the size of most NOM fractions; thus, MF alone is not effective in the removal of NOM substances other than polysaccharides (Kennedy et al., 2005). MF membranes are able to remove less than 10% of NOM, whereas turbidity removal is generally more than 97%.

The advantage of MF is that its large pore size enables operation at relatively low transmembrane pressures. MF is used widely in variety of surface water purification processes and membrane bioreactors, and as a pretreatment method in seawater desalination plants (Hilal et al., 2006). Submerged membrane bioreactors are used mainly for wastewater treatment, but also increasingly for freshwater polishing (Sutherland, 2012). MF is usually coupled with other methods, either at the beginning of the treatment chain to remove particular and microbial materials or after coagulation (Sanchez et al., 2013; Schrotter and Bozkaya-Schrotter, 2010). Precoagulation can enhance NOM removal of MF membranes several fold and simultaneously mitigates membrane fouling (Li et al., 2011; Liu et al., 2011; Ben-Sasson et al., 2013).

Flat plate, hollow fiber, tubular, and spiral wound membranes are used (Sutherland, 2012), but hollow fiber is the most used MF and UF configuration in drinking water plants (Schrotter and Bozkaya-Schrotter, 2010; Rach-wal and Judd, 2006; Kajitvichyanukul et al., 2008; Fane et al., 2008). Many industrial plants use inside-out hollow fibers (Remize et al.,

2010). The MF and UF membranes are generally made of cellulose acetate, polysulfone, polyethersulfone (PES), and polyvinylidene fluoride (PVDF) polymers (Schrotter and Bozkaya-Schrotter, 2010). The hydrophilic regenerated cellulose and cellulose acetate membranes are only slightly charged, so the primary mechanism of NOM removal with these membranes is size exclusion. The other membrane materials exhibit greater surface charge densities and hydrophobicities based on their composition. Membrane surface charge is caused also by the adsorption of anions from water. Hence, the membrane surface charge depends on the concentration of ions in the solution (Bodzek et al., 2011).

6.3 Ultrafiltration

The pore size of UF membranes is 10–100 nm and their cutoff value is 1–100 kDa, which matches the size of the higher MM humic substances and biopolymers. In general, NOM removal increases as the membrane pore size decreases (Amy and Cho, 1999; Cho et al., 2000a; Schafer et al., 2000; Syafei et al., 2008; Ates et al., 2009). When considering surface water treatment by UF membranes, loose UF membranes (cutoff, >60 kDa) have shown dissolved organic carbon (DOC) removal ranging from 14% to 49% and medium membranes (cutoff, 10–60 kDa) ranging from 20% to 75% (Table 6.1). The pore sizes of tight UF membranes (cutoff, <10 kDa) are in the range of the larger NOM molecules, so their DOC removal is already quite high (38%–87%).

Yet, it is not straightforward to predict NOM removal on the basis of the membrane cutoff value. Apart from size exclusion, electrostatic repulsion is also an important separation mechanism in UF. The lowest NOM removal rates have been obtained with uncharged membranes, whereas with charged UF membranes, much greater removal rates have been observed than would be expected on the basis of the membrane cutoff value and the size of the surface water NOM (Cho et al., 1999, 2000a,b; Yoon et al., 2005; Lee et al., 2005).

Lee et al. (2005) filtered surface waters with a negatively charged 8-kDa cutoff membrane and noticed that the MM of NOM reduced from feed water values of 1210 and 2360 Da to 620 and 900 Da in the permeates, respectively. In another study, an 8-kDa cutoff membrane was observed to remove almost completely the MM fraction larger than 2 kDa (Cho et al., 2000b).

Hydrophobic interactions and electrostatic exclusion between the hydrophobic and charged membrane surface and NOM molecules can

Table 6.1 Studies concerning NOM removal by MF and UF membranes from natural waters and surrogate waters containing humic substances.

Solution	Membrane, material, cutoff/pore diameter, module type, TMP	Variable	DOC removal (%)	Quality of permeate; content and/or removal (%)	References
6.3% NOM, Australia	Different membranes, stirred batch cell, lab scale, 100–300 kPa	GVWP, mod; PVDF, 220 nm	17		Schäfer et al. (2000)
		PLHK, RC, 100 kDa	9		
		PLTK, 30 kDa	10		
		PLGC, RC, 10 kDa	51		
		PLCC, 5 kDa	74		
		PLBC, 3 kDa	77		
		PLAC, 1 kDa	86		
Humic acid, 15 mg L^{-1}	MF, polycarbonate track etched, 0.1 μm	pH 6		TOC, 20%	Ben-Sasson et al. (2013)
		pH 7		TOC, 33%	
		pH 8		TOC, 14%	
	UF, UE50, PES, 100 kDa	pH 6		TOC, 59%	
		pH 7		TOC, 49%	
		pH 8		TOC, 56%	
Surface water; DOC, 1.3 mg L^{-1}	PVDF, 100 nm, hollow fiber	Submerged	35		Chae et al. (2009)
		Pressurized	35		

Surface water; DOC, 2.2 mg L^{-1}; SUVA, 2.9; THMFP, 75 μg L^{-1}; HAAFP, 80 μg L^{-1}	VS-70-10, PE textile, 100 nm, cross-flow		18	DOC, 1.8 mg L^{-1}; THMFP, 55 μg L^{-1}, 8%; HAAFP, 48 μg L^{-1}, 4%	Siddiqui et al. (2000)
Surface water; DOC, 5.3 mg L^{-1}; SUVA, 1.8	ZeeWeed 1000, 20 nm, full scale		66	DOC, 1.9 mg L^{-1}; THMFP, 54%; HAAFP, 77%	Lamsal et al. (2012)
Different NOM fractions; DOC, 12.5 mg L^{-1}	PLHK, RC, 100 kDa/18.2 nm, stirred batch cell, lab scale, 100 kPa	IHSS humic acid IHSS fulvic acid 6.3% NOM, Australia NOM–humic acid, Australia NOM–fulvic acid, Australia NOM–hydrolysate, Australia	~45 ~10 ~10 10 <5 >10		Aoustin et al. (2001)
	PLGC, RC, 10 kDa/5.2 nm, stirred batch cell, lab scale, 300 kPa	IHSS humic acid IHSS fulvic acid 6.3% NOM,	75 45 ~50 51		

Continued

Table 6.1 Studies concerning NOM removal by MF and UF membranes from natural waters and surrogate waters containing humic substances.—cont'd

Solution	Membrane, material, cutoff/pore diameter, module type, TMP	Variable	DOC removal (%)	Quality of permeate; content and/or removal (%)	References
		Australia NOM —humic acid, Australia NOM—fulvic acid, Australia NOM —hydrolysate, Australia	41 19		
Moorland water; TOC, 9.8 mg L^{-1}	PSu, 100 kDa, flat sheet, bench scale, 100 kPa		18	UV254, 22%	Bessiere et al. (2009)

Solution	Membrane, material, cutoff/pore diameter, module type, TMP	Variable	DOC removal (%)	Quality of permeate; content and/or removal (%)	References
Humic acid (Aldrich), 2 mg L^{-1}; DOC, 8.7 mg L^{-1}; different pHs	Different flat sheet membranes or charges, stirred cell, lab scale, 69 kPa	RC, 100 kDa; neutral; pH 3.5 RC, 100 kDa; neutral; pH 7.5 RC, 100 kDa;		UV_{254}, 55% UV_{254}, 59–97% UV_{254}, 79% UV_{254}, 92–98% UV_{254}, 66%	Shao et al. (2011)

		charged; pH 3.5			
		RC, 100 kDa; charged; pH 7.5			
		PES, 100 kDa; zeta; −12.3 mV			
Humic acid (Aldrich); DOC, 5 mg L^{-1}; different size fractions	PSu, 100 kDa, hydrophobic, hollow fiber, cross-flow	MM, 6.5 −22.6 kDa; DOC, 10.9 mg L^{-1}	96	THMFP, <20 μg L^{-1}, 95%	Lin et al. (1999)
		MM, 2.2 −6.5 kDa; DOC, 7.0 mg L^{-1}	74	THMFP, ~50 μg L^{-1}	
		MM, 0.65 −2.2 kDa; DOC, 7.5 mg L^{-1}	64	THMFP, ~60 μg L^{-1}	
		MM, 180 −650 Da; DOC, 5.8 mg L^{-1}	14	THMFP, ~480 μg L^{-1}	
		PSu, 100 kDa, hydrophobic,	60 ~25		Lin et al. (2001)

Continued

Table 6.1 Studies concerning NOM removal by MF and UF membranes from natural waters and surrogate waters containing humic substances.—cont'd

Solution	Membrane, material, cutoff/pore diameter, module type, TMP	Variable	DOC removal (%)	Quality of permeate; content and/or removal (%)	References
Humic acid (Aldrich); DOC, 5 mg L^{-1}; different pH_s	Different membranes, hollow fiber, cross-flow, 150−170 kPa	pH 4 PSu, 100 kDa, hydrophobic, pH 7 PSu, 10 kDa, hydrophobic, pH 7 RCA, 10 kDa, hydrophilic, pH 7	~80 ~80		
6.3% NOM, Australia	Different membranes, stirred batch cell, lab scale, 100−300 kPa	PLHK, RC, 100 kDa PLTK, 30 kDa PLGC, RC, 10 kDa PLCC, 5 kDa PLBC, 3 kDa PLAC, 1 kDa	9 10 51 74 77 86		Schäfer et al. (2000)
Surface (SW) and reservoir (RW) waters	PSu, 68 kDa, hydrophobic, submerged hollow fiber	SW; DOC, 6.7 mg L^{-1}; SUVA, 1.3 RW; DOC, 8.0 mg L^{-1}; SUVA, 2.8	~17 ~38	UV_{254}, ~26% UV_{254}, 44%−50%	Zularisam et al. (2007a)

Surface water; DOC, 6.8 mg L^{-1}; SUVA, 1.3; different fractions	MRUTM55, PSu, 68 kDa, hollow fiber MRUTM66, CA, 50 kDa, hollow fiber	Hydrophobic fraction Transphilic fraction Hydrophilic fraction Hydrophobic fraction Transphilic fraction Hydrophilic fraction	49 31 14 34 27 22		Zularisam et al. (2007a)

Solution	Membrane, material, cutoff/pore diameter, module type, TMP	Variable	DOC removal (%)	Quality of permeate; content and/or removal (%)	References
Surface water; DOC, 2.2 mg L^{-1}; SUVA, 2.9; THMFP, 75 μg L^{-1}; HAAFP, 80 μg L^{-1}	192-PT3, TFC, 60 kDa, spiral wound		27	DOC, 1.6 mg L^{-1}; THMFP, 32 μg L^{-1}, 47%; HAAFP, 36 μg L^{-1}, 28%	Siddiqui et al. (2000)
Humic acid (Aldrich); DOC, 8.7–9.4 mg L^{-1}; pH 7	Different pore size ceramic membranes, flat sheet, cross-flow	50 kDa 15 kDa 1 kDa	~61 ~83 ~84		Syafei et al. (2008)
Surface water; DOC, 11.7 mg L^{-1}; SUVA, 2.0	YMJWSP3001, PVDF, 30 kDa, negative charge	11–17			Byun et al. (2011)

Continued

Table 6.1 Studies concerning NOM removal by MF and UF membranes from natural waters and surrogate waters containing humic substances.—cont'd

Solution	Membrane, material, cutoff/pore diameter, module type, TMP	Variable	DOC removal (%)	Quality of permeate; content and/or removal (%)	References
Humic acid (Aldrich); DOC, 10 mg L^{-1}	KERMBMU1, ceramic, 15 kDa/ 3.54 nm, single tube, bench scale	pH 2.4; zeta, −2.9 mV; pI, 1 mmol pH 7.9; zeta, −15.6 mV; pI, 1 mmol pI, 1 $mmolL^{-1}$; pH 7 pI, 152 $mmolL^{-1}$; pH	59 99 98 74	UV254 85%	de la Rubia et al. (2006)
Surface water; DOC, 3.1 mg L^{-1}; SUVA, 2.9	Different flat-sheet membranes dead-end stirred cell, bench scale	NTR7410, S-PES, 20 kDa, hydrophobic PM10, PES, 10 kDa, hydrophobic 10PESUF40, PES, 10 kDa hydrophobic GM, S-PES, 8 kDa, hydrophilic	~75 ~70 ~75 ~86		Yoon et al. (2005)
Surface water; TOC, 2.3 mg L^{-1}; SUVA, 1.7; THMFP, 70 μg L^{-1}	Different flat-sheet membranes, stirred cell, lab scale	NTR-7410, S-PSu, 20 kDa GR90, PSu, 10 kDa ETNA01A, HPC, 1 kDa HEKLA01A, amine + DIC, 1 kDa	20 40 44 48	UV254, 25% UV254, 50% UV254, 47% UV254, 53%	Lee and Lee (2007)
Surface water; DOC, 4.8 ± 2.2 mg L^{-1}	PW-4040F, 10 kDa, spiral wound, pilot plant		29	DOC, 3.4 ± 1.9 mg L^{-1}; UV254, 40%	Mierzwa et al. (2012)

Solution	Membrane, material, cutoff/pore diameter, module type, TMP	Variable	DOC removal (%)	Quality of permeate; content and/or removal (%)	References
Surface water DOC 2.0 mg L^{-1}; SUVA, 2.4	Different flat-sheet membranes, cross-flow, bench scale	PM10, PES, 10 kDa	0		Cho et al. (1999)
		GM, PA TFC, 8 kDa, negative charge	38		
		YM3, RC, 3 kDa	0		
Agricultural drain; DOC, 47.8 mg L^{-1}; SUVA, 3.7		PM10, PES, 10 kDa	37		Cho et al. (1999)
		GM, PA TFC, 8 kDa, negative charge	56	UV_{254}, 60%	
		YM3, RC, 3 kDa	51		
Surface water; DOC, 3.1 mg L^{-1}; SUVA, 3.0		NTR-7410, S-PES, 20 kDa; zeta, −22.6 mV	32	UV_{254}, 49%	Amy and Cho (1999) and Cho et al. (2000a)
		10PESUF40, PES, 10 kDa; zeta, −12.8 mV	32	UV_{254}, 49%	
		GM, PA TFC, 8 kDa; zeta, −17.0 mV	57	UV_{254}, 68%	
Groundwater; DOC, 6.8 mg L^{-1}; SUVA, 5.7		NTR7410, S-PES, 20 kDa; zeta, −22.6 mV	74	UV254, 86%	Amy and Cho (1999) and Cho et al. (2000a)
		10PESUF40, PES, 10 kDa; zeta, −12.8 mV	85	UV_{254}, 88%	
		GM, PA TFC, 8 kDa; zeta, −17.0 mV	87	UV_{254}, 93%	

Continued

Table 6.1 Studies concerning NOM removal by MF and UF membranes from natural waters and surrogate waters containing humic substances.—cont'd

Solution	Membrane, material, cutoff/pore diameter, module type, TMP	Variable	DOC removal (%)	Quality of permeate; content and/or removal (%)	References
Surface water; DOC, 3.07 mg L^{-1}; SUVA, 2.7	GM, PA TFC, 8 kDa; zeta, −20.0 mV; cross-flow; bench scale	Colloidal fraction; SUVA, 1.8; pH 4	60		Kim et al. (2011)
		Colloidal fraction; SUVA, 1.8; pH 10	27		
		Hydrophobic fraction; SUVA, 3.0; pH 4	30		
		Hydrophobic fraction; SUVA, 3.0; pH 7	40		
		Hydrophobic fraction; SUVA, 3.0; pH 10	15		
		Transphilic fraction; SUVA, 2.6; pH 4	63		
Surface water; DOC, 3.9 mg L^{-1}; SUVA, 4.5	GM, PA TFC, 8 kDa, negative charge, flat sheet, cross-flow, bench scale		72		Cho et al. (1999)

Different surface water and groundwater	GM, PA TFC, 8 kDa, flat sheet, tangential cross-flow, bench scale	DOC, 2.0 mg L^{-1}; SUVA, 2.4 DOC, 3.9 mg L^{-1}; SUVA, 4.4 DOC, 9.8 mg L^{-1}; SUVA, 4.9 DOC, 6.8 mg L^{-1}; SUVA, 5.7 DOC, 9.8 mg L^{-1}; SUVA, 4.9 DOC, 47.80 mg L^{-1}; SUVA, 3.7	38 60 85 87 84 60	UV254, 64% UV254, 84% UV254, 93% UV254, 93%	Cho et al. (2000b)

CA, cellulose acetate; *DIC*, diisocyanate; *DOC*, dissolved organic carbon; *HAA*, haloacetic acid; *HAAFP*, haloacetic acid formation potential; *HMM*, high molar mass; *HPC*, hydroxypropyl cellulose; *IHSS*, international humic substance society; *LMM*, low molar mass; *MF*, microfiltration; *NOM*, natural organic matter; *PA*, polyamide; *PE*, polyester; *PES*, polyethersulfone; *PSu*, polysulfone; *PVDF*, polyvinylidene fluoride; *RC*, regenerated cellulose; *RCA*, regenerated cellulose acetate; *S-PES*, sulfonated polyethersulfone; *S-PSu*, sulfonated polysulfone; *SUVA*, specific ultraviolet absorbance; *TFC*, thin-film composite; *THM*, trihalomethane; *THMFP*, trihalomethane formation potential; *TMP*, transmembrane pressure; *TOC*, total organic carbon; *UF*, ultrafiltration; zeta, zeta potential. SUVA values are in liters per meter per milligram.

increase NOM removal significantly. Yoon et al. (2005) reported NOM removal with hydrophobic PES membranes eight times greater than expected, considering the average MM of NOM and membrane pore sizes. These results clearly show that much smaller molecules are rejected than the membrane cutoff value presumes.

6.3.1 Effect of NOM characteristics

Some of the important properties of NOM that enhance its removal by membranes include high MM, negative charge, and a hydrophobic and aromatic structure. All these characteristics are associated with the hydrophobic NOM fraction. This is consistent with the fact that the studies performed with humic acids show greater DOC removal than those performed with surface waters or NOM solutions (Table 6.1). Lamsal et al. (2012) noted an overall DOC removal of 66%, whereas the removal of hydrophobic acids was as high as 93% in river water UF.

Several researchers have applied a ratio of UV_{254} to DOC (specific ultraviolet absorbance (SUVA)) in the identification of NOM removal and reported that the majority of DOC removed from raw waters by UF was aromatic or hydrophobic compounds (humic and fulvic acids (de la Rubia et al., 2008; Amy and Cho, 1999; Cho et al., 1999, 2000a,b; Ates et al., 2009)). The high SUVA value corresponds to high hydrophobicity and to a high concentration of humic substances (Edzwald and Tobiason, 1999). Amy and Cho (1999) reported that DOC rejection increased from 38% to 87% when SUVA increased from 2.4 to 5.7 $L\ m^{-1}\ mg^{-1}$. In conclusion, NOM rejection increases with an increase in the proportion of humic substances in NOM, and thus with its SUVA value.

Hydrophobic NOM fractions have a significantly greater charge density than hydrophilic NOM. Acidic contents of 5.1, 3.6, and 1.0 $meq\ g^{-1}$ DOC were noted for the humic, fulvic, and hydrophilic acid fractions, respectively (Sharp et al., 2006). Furthermore, zeta potential values of about −25 and −13 mV have been observed at pH greater than four for the humic and fulvic acid fractions, respectively (Sharp et al., 2006). Hence, humic and fulvic acids are removed more favorably by charge repulsion than by hydrophilic acids.

This corresponds to the results of Zularisam et al. (2007a,b). They isolated surface water into different NOM fractions and investigated their behavior in UF equipped with a 68-kDa polysulfone membrane. The hydrophobic fraction showed the greatest DOC removal (45%) of all the fractions. This phenomenon was presumed to occur as a result of charge repulsion. DOC removal of hydrophilic NOM was low, only 14%, as was expected because of its relatively low MM, aliphatic molecular structure, and lesser hydrophobic interactions compared with hydrophobic NOM.

The charge of the hydrophobic fraction of NOM is associated largely with the carboxylic and phenolic groups (Edwards et al., 1996; van Loon and Duffy, 2005). The former have pKa values in the range of 2.5–5, whereas the phenolic hydrogens have pKa values around 9 or 10 (van Loon and Duffy, 2005). As a result, humic and fulvic acids are negatively charged as a result of the ionization of carboxylic groups in the pH range of natural waters. The usual concentration of carboxyl groups in humic substances ranges from 2 to 6 mmol g^{-1} (Leenheer et al., 1995).

6.3.2 Effect of solution composition

The importance of pH in membrane filtration performance has been demonstrated in various research studies, but mainly in relation to membrane fouling rather than NOM removal. Yet, the degree of DOC removal is influenced strongly by the feed solution pH and ionic content because they affect all removal mechanisms by changing the surface properties of both the membrane and NOM molecules.

The volumetric size of humic and fulvic acids changes according to the solution conditions. Low concentrations, relatively low ionic strength, and high pH renders a large, flexible, and linear shape, which changes to a small, rigid, and spherical shape at high humic concentration, high ionic strength, and low pH (Ghosh and Schnitzer, 1980).

NOM removal has been shown to increase with pH (de la Rubia et al., 2008; Shao et al., 2011; Hong and Elimelech, 1997), although greater removal at lower pH has also been reported (Lin et al., 2001; Dong et al., 2006). Removal is expected to be greater at higher pH because, in this case, humic molecules have a more linear configuration and a larger radius than at low pH, and because the ionization and charge repulsion between the membrane and the molecules are also greater. Thus, the effect of pH on NOM removal is much greater for charged membranes than for neutral membranes.

A very high removal of humic acid (99%) has been observed at pH 6.9 and pH 7.9 compared with 59% at pH 2.4 (de la Rubia et al., 2006). This was the result of the high charge density of the ceramic membrane used; its zeta potentials were −2.9 and −15.6 mV at a pH of 2.4 and 7.9, respectively.

Lin et al. (2001) observed greater DOC removal (~60%) at pH four than at pH 7 (~25%), but this was the result of the fouling of a hydrophobic poly-sulfone membrane at pH 4. In addition, the poly-sulfone membrane may have a positive net charge at this pH, so charge attraction—rather than repulsion—occurs between the membrane and the molecules. Charge attraction and hydrophobic interactions might lead to fouling of the membrane and, as a consequence, to greater retention.

Removal mechanisms by hydrophobic interactions are more obvious at low pH because the protonation of the phenolic and carboxylic hydroxyl groups of humic and fulvic acids leads to lower water solubility and greater hydrophobicity (Jucker and Clark, 1994). This phenomenon is more pronounced for humic acids than for fulvic acids because they have more phenolic groups and relatively fewer carboxylic groups, and hence are relatively more hydrophobic than fulvic acids (Jucker and Clark, 1994).

High ionic strength decreases the thickness of the double layer around the molecules and of the membrane pore walls (shielding of electrostatic charges) and, as a result, decreases charge repulsion. Furthermore, an increase in ionic strength would decrease intra- and intermolecular repulsion of the humic acid molecules, leading to more coiled structure. As a consequence, NOM removal decreases (Cho et al., 1999; Shao et al., 2011; Lin et al., 2001; de la Rubia et al., 2006). For example, de la Rubia et al. (2006) noted that increasing ionic strength from 0.001 to 0.152 mol KCl L^{-1} decreased the average DOC reduction from 98% to 74%.

Calcium ions do more than just affect ionic strength; they can bind to the acidic functional groups of NOM and the membrane surface, significantly neutralizing the charges of both (Lee et al., 2005; Shao et al., 2011; Hong and Elimelech, 1997). Therefore, NOM rejection decreases when calcium ions are in the solution, but at greater concentrations it can also enhance NOM rejection by causing NOM aggregation (Cho et al., 1999; Shao et al., 2011; Schafer et al., 2001).

DOC concentration of raw water seems to be less significant to NOM removal than the aromatic content of NOM. This can be seen clearly in the results of Cho et al. (2000b). They filtered two different surface waters containing 3.9 DOC L^{-1} and 47.8 mg DOC L^{-1} with SUVA values of 4.5 and 3.7 L mg^{-1} m^{-1}, respectively. Despite the differences between these two waters, they reported that NOM rejection was similar for both (60%), whereas the greater DOC rejection, 84%, was achieved with surface water characterized by 9.8 mg DOC L^{-1} and SUVA of 4.9 L mg^{-1} m^{-1} (third surface water).

6.3.3 Effect of hydrodynamic conditions

Hydrodynamic operation conditions such as transmembrane pressure, permeate flux, cross-flow velocity, aeration, filtration mode, and membrane module configuration can all affect NOM removal via changes in the concentration and transport of NOM molecules at the membrane interface.

Amy and Cho (1999) used a ratio of permeate flux (J) to a diffusional back-transport coefficient (k) to describe the hydrodynamic conditions at the membrane interface. The greater J/k value corresponds to greater concentration polarization, and the greater cross-flow velocity corresponds to the greater back-transport velocity of the retained solutes and lower concentration polarization. Particle back-transport velocity depends also on its size; it is lowest for particles with a diameter of approximately 0.2 (*x*m) (Lahoussine-Turcaud et al., 1990). Smaller than this, as in the case of dissolved NOM, back-transport is dominated by Brownian diffusion, and the smaller the particle diameter, the greater its particle back-transport.

Furthermore, solution chemistry and NOM membrane interactions have an impact on NOM diffusivity (Cheryan, 1998; Clark and Lucas, 1998; Lee et al., 2004b; Wang et al., 2001).

Lee et al. (2005) reported that NOM removal decreased with increasing J/k values, and transmission of the greater MM compounds was also found to increase. Cho et al. (2000a) did not notice significant differences in DOC rejection between the J/k values 2 and 1 with the 8-kDa polyamide membrane, whereas with the 10-kDa PES membrane, NOM rejection decreased and flux decline increased with increasing J/k values. Yet, the J/k value had no effect on the rejection of NOM containing more fulvic/humic acids (Cho et al., 2000a).

6.3.4 NOM fractions and DBPs

Some researchers have isolated different NOM fractions to understand their removal and, especially, DBP precursors more completely. The proportions of the hydrophobic, transphilic, and hydrophilic fractions in raw waters have been observed to vary in the ranges of 38%–79%, 11%–31%, and 9%–38%, respectively (Zularisam et al., 2007a; Kennedy et al., 2005; Yoon et al., 2005; Lee et al., 2004b).

Yee et al. (2009) investigated the DBP formation potentials of the different NOM fractions. They found that the hydrophobic and hydrophilic bases were the most reactive fractions to form trihalomethanes (THMs), but still they contributed the least to the total THM formation potential because of their low proportion (14% of DOC) in the raw water.

The greatest THM formation potential was associated with the major fractions of NOM: hydrophobic acids (40% of DOC) and hydrophilic neutrals (22% of DOC).

Lin et al. (1999) reported that a 100-kDa UF membrane was able to remove 95% of DOC and THM formation potential in the greatest MM fraction of humic acid. The concentration of THM formation potential in the permeate was reduced to less than the European (EU) regulation limit of 100 ($\varkappa$g L^{-1}) (The Council of the European Union, 1998). Yet, only 10% of THM formation potential from the smallest MM fraction was reduced, and its concentration in the permeate was still ~480 ($\varkappa$g L^{-1}).

In relation to their relatively high cutoff values, some UF membranes have been shown to be quite effective in the removal of DBP precursors from surface waters. Removals of DOC, trihalomethane formation potentials, and haloacetic acid (HAA) with formation potentials of 27%–85%, 47%–82%, and 28%–63%, respectively, have been observed in lab-scale experiments with 1–60-kDa membranes (Table 6.1). The corresponding removals of 66%, 54%, and 77% have been observed in full-scale plants operating with 20-nm pore size membranes (Lamsal et al., 2012).

In contrast, the removal of assimilable organic carbon (AOC) with UF membranes has not been successful. Hem and Efraimsen (2001) noted that AOC is related mainly to NOM with an MM less than 1 kDa, which corresponded to 16%–38% of the total organic carbon in the tested waters. AOC removal rates of less than 30% and 49% have been observed with 10-kDa and 1-kDa membranes, respectively. Hence, reduction in biofilm formation may be difficult without combining UF with a tight nanofiltration (NF)/reverse osmosis (RO) or biological treatment.

Tight UF membranes can be used for the removal of greater fractions of NOM from raw water, including high MM DBP precursors, whereas these should be combined with other processes to remove low and medium MM compounds (Fig. 6.3).

6.4 Nanofiltration

NF was developed in the United States at the end of 1970s for the treatment of salt-containing groundwater and was used to soften surface waters during the 1980s (Schafer et al., 2005). Since the 1990s, NF has been used to remove DBP precursors and micropollutants (Bodzek et al., 2011). NF membrane cutoff values are in the range of 100–500 Da and are able to remove solution components with diameters of about 1–3 nm, but at a greater transmembrane

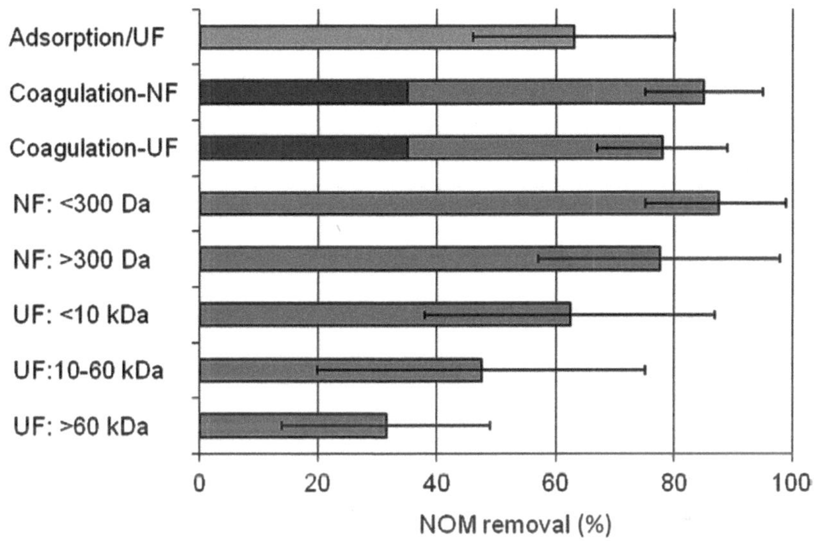

Figure 6.3 Natural organic matter (NOM) removal from raw water applying different membrane technology methods. *NF*, nanofiltration; *UF*, ultrafiltration.

pressure as UF. In general, NF operates at a transmembrane pressure range of 1–3 MPa (Bodzek et al., 2011), but operation at 0.8 MPa pressure is common in several full-scale plants (Lopes et al., 2013).

Size exclusion and charge repulsion are the major removal mechanisms for humic substances in NF (Ates et al., 2009; Cho et al., 1999; Hong and Elimelech, 1997; Nystrom et al., 1996; Visvanathan et al., 1998; Schafer et al., 1998; Ribau Teixeira and Rosa, 2006). The charge effect becomes less important when the molecules are larger than the membrane pores (van der Bruggen et al., 1999). Braeken et al. (2005) noticed that hydrophobicity is the most significant parameter determining the retention of molecules with an MM less than the membrane cutoff value.

The transport of NOM through NF membrane pores is influenced either by convection or diffusion, as in RO, depending on the hydrodynamic conditions and electrostatic interactions between the membrane surface and NOM molecules. Moreover, different mechanisms influencing the removal of small organic compounds have been suggested. The polarity and differences in diffusion rates in a nonporous structure may influence the rejection of small organic compounds in NF (van der Bruggen et al., 1999; van der Bruggen and Vandecasteele, 2003). The molecules with high dipole moments are reported to show a lower rejection than molecules with a lower dipole moment, although the sizes of the molecules have been approximately the same (van der Bruggen et al., 1999). This indicates that a lower fraction of polar molecules is removed compared with nonpolar molecules of equivalent size. The transport and rejection mechanisms by NF membranes are still partly unclear, however.

Almost total NOM rejection (>90%) has been achieved in lab- and bench-scale studies using NF membranes with cutoff values in the range of 200−400 Da (de la Rubia et al., 2008; Cho et al., 1999; Yoon et al., 2005; Costa and de Pinho, 2006a; Meylan et al., 2007) (Table 6.2). Large NOM compounds like polysaccharides and humic substances are clearly retained by NF membranes, whereas low-MM organic compounds permeate through the membrane (Schafer et al., 2000). Meylan et al. (2007) noted 98%−99% DOC removal with 300-Da NF membranes and an RO membrane. The high-MM polysaccharides and humic substances were almost completely rejected, charged low-MM organic compounds had 97%. rejection, and low-MM neutrals and hydrophobic organic compounds had rejections of 94% and 88%, respectively.

In general, the hydrophilic fraction of NOM is removed less easily than the hydrophobic fraction (Siddiqui et al., 2000; Ribau Teixeira and Rosa, 2006). Escobar et al. (2000) observed that AOC removal by NF membranes is strongly dependent on charge repulsion, and that biodegradable dissolved organic carbon (BDOC) removal was determined by both size exclusion and charge repulsion. In a full-scale drinking water treatment plant, they found NF a very effective means of reducing BDOC but not AOC.

Unlike UF permeates, several researchers have reported that NF permeates consist of NOM that appears to have an MM larger than the cutoff value of the membrane. Lee et al. (2005) noted 93%−96% DOC rejections with a negatively charged 250-Da membrane, but still the apparent MMs of the permeated NOMs were about twice the membrane cutoff. Furthermore, the proportions of BDOC were the same—10% and 20% of NOM—as in the two raw waters filtered.

Ates et al. (2009) noted an 85% DOC rejection with tight NF membranes for natural water. The high MM fraction (average MM, ~12 kDa) was rejected by 98%, but the low MM fraction (average MM, 1.8 kDa) was rejected by only 30%. Because of the broad range of molecular sizes found in NOM, a membrane with a low cutoff is needed for complete removal of small organic compounds.

Table 6.2 Studies concerning NOM removal from natural waters and surrogate waters containing humic substances with nanofiltration (NF) and RO membranes.

Solution	Membrane, material, cutoff/pore diameter, module type, TMP	Variable	DOC removal (%)	Quality of permeate, content and/or removal (%)	References
Surface water; TOC, 2.3 mg L^{-1}; SUVA, 1.7; THMFP, 70 μg L^{-1}	Different flat-sheet membranes, stirred cell, lab scale	NTR-7450, S-PSu, 1 kDa	38	UV_{254}, 67%	Lee and Lee (2007)
		MPF-34, unknown, 200 Da	75	UV_{254}, 85%	
		NF-45, PPiA, 400 Da	73	UV_{254}, 86%	
		NTR-7250, PPiA, 300–400 Da	60	UV_{254}, 81%	
		CA865, CA; NaCl, 60%	47	UV_{254}, 52%	
		NTR-729HF, PVA, 700 Da	82	UV_{254}, 87%	
6.3% NOM, Australia	Different membranes, stirred batch cell, lab scale, 500 kPa	TFC-SR, <180 Da/<0.64 nm	69		Schafer et al. (2000)
		TFC-S, <180 Da/<0.64 nm	95		
		TFC-ULP, <180 Da/<0.64 nm	95		
Aquatic NOM; DOC, 10 mg L^{-1}	Different flat-sheet membranes, cross-flow, bench scale	NF-PES-10, PES, 1 kDa	58		Kilduff et al. (2000)
		NTR-7450, S-PSu, 1 kDa	83		

Solution	Membrane, material, cutoff/pore diameter, module type, TMP	Variable	DOC removal (%)	Quality of permeate, content and/or removal (%)	References
Surface water; DOC, 1.1 mg L^{-1}; AOC, 103 ug L^{-1}	Different flat-sheet membranes, cross-flow, lab scale, 2000 kPa	HL, ~300 Da	98	DOC, 68 ug L^{-1}; AOC, 81 ug L^{-1}	Meylan et al. (2007)
		NF270, PA TFC, ~300 Da	98	DOC, 55 ug L^{-1}; AOC, 84 ug L^{-1}	
		XLE, NaCl, 98.6%	99	DOC, 42 ug L^{-1}; AOC, 19 ug L^{-1}	
Surface water; DOC, 3.2 mg L^{-1}; AOC, 229 ug L^{-1}		HL, ~300 Da	99	DOC, 1.2 mg L^{-1}; AOC, 153 ug L^{-1}	Meylan et al. (2007)
		NF270, PA TFC,	99	DOC, 129 ug L^{-1};	
		~300 Da XLE; NaCl, 98.6%	99	AOC, 92 ug L^{-1} DOC, 110 ug L^{-1}; AOC, 31 ug L^{-1}	
Surface water; DOC, 11.7 mg L^{-1}; SUVA, 2.0	NF90, aromatic PA, 200–400 Da		92–96		Byun et al. (2011)

Surface water; DOC, mg L^{-1}; SUVA, 2.7	HL, PA TFC, 150–300 Da, zeta, −9.9 mV, cross-flow, bench scale	Colloidal fraction; pH 4, 7, and 10 Hydrophobic fraction; pH 4 Hydrophobic fraction; pH 7 Hydrophobic fraction; pH 10 Transphilic fraction; pH 4, 7, and 10	75–80 70 80 85 ~80		
Surface water; DOC, 2.2 mg L−1; SUVA, 2.9; THMFP, 75 μg L^{-1}; HAAFP, 80 μg L^{-1}	NF70, NF45, and Desal 5; TFC; 100–400 Da; all spiral wound	Averages of all 4 membranes	88	DOC, 0.27 mg L−1; THMFP, 3 μg L−1, 96%; HAAFP, 5 μg L −1, 94%	Siddiqui et al. (2000)
Surface water; TOC, 3.2 mg L^{-1}	NF200–400B, flat sheet, cross-flow, lab scale, 1500 kPa		>90	UV254, >97%	Costa and de Pinho (2006a)
Surface water; DOC, 3.8 mg L^{-1}	ESNA, PA TFC, 250 Da, flat sheet, cross-flow, bench scale		~78	UV254, ~82%; HAAFP, 79%	Lee and Cho (2004)

AOC, assimilable organic carbon; *CA*, cellulose acetate; *DAF*, dissolved air flotation; *DOC*, dissolved organic carbon; *HAA*, haloacetic acid; *HAAFP*, haloacetic acid formation potential; *HMM*, high molar mass; *LMM*, low molar mass; *NOM*, natural organic matter; *PA*, polyamide; *PES*, polyethersulfone; *PPi*, polypiperazine; *PPiA*, polypiperazine amide; *PVA*, polyvinyl alcohol; *RO*, reverse osmosis; *S-PES*, sulfonated polyethersulfone; *S-PSu*, sulfonated polysulfone; *SUVA*, specific ultraviolet absorbance; *TFC*, thin-film composite; *THM*, trihalomethane; *THMFP*, trihalomethane formation potential; *TMP*, transmembrane pressure; *TOC*, total organic carbon; *UF*, ultrafiltration. SUVA values are in liters per meter per milligram.

6.4.1 Effect of solution composition

It has been reported that NOM removal by NF membranes is not influenced significantly by factors such as DOC levels, humic content as identified by SUVA, pH, and ionic strength (Cho et al., 1999; Siddiqui et al., 2000; Glucina et al., 2000; Ventresque et al., 2000). Yet, NOM removal increases with water recovery rate (Costa and de Pinho, 2006b) (i.e., with increasing concentration). As a result, a stronger concentration polarization layer may form on the membrane surface (Visvanathan et al., 1998; Jarusutthirak et al., 2007).

The effect of concentration polarization is more pronounced when the raw water has a high NOM concentration, especially when the rejection of a membrane is high. This layer may become so dense that it prevents permeability and thus enhances rejection, as shown by Jarusutthirak et al. (2007). They found that an increase in NOM concentration increased the NOM and salt rejection and permeate flux decline by enhancing accumulation on the membrane surface.

Ribau Teixeira and Rosa (2006) noted that NF performance was influenced by the pH and calcium hardness, rather than by the type of NOM. In their study, DOC rejection increased with recovery rate and pH, whereas the UV_{254} rejection was almost complete (90%–100%) at all pH values and recovery rates. They also found that the rejection was reduced at the membrane isoelectric point. This is a result of the lower membrane surface and pore charges, which result in a lower charge repulsion between the membrane and NOM, and larger pores resulting from the lower electrical double layer (Al-Amoudi, 2010). The membrane becomes negatively charged above its isoelectric point, and with the increase in pH, the charge repulsion between the membrane surface and NOM molecules increases. This is in accordance with the results of Hong and Elimelech (1997), who found that humic acid is better removed at pH 8 than at pH 4.

In contrast, Kuchler and Miekeley (1994) observed that the retention of humic acid decreased with increasing pH in the pH range of 4.3–9.0, but that of fulvic acid increased. In their study, humic and fulvic acid retention with the 1-kDa membrane varied in the range of 80%–90% and 60%–70%, respectively. The differences between humic and fulvic acids may be explained by their different pKa and solubility values and sizes.-Visvanathan et al. (1998) found that rejection increases with increasing pH up to a certain level, but above this level the solubility of the humic substances increases so that they can again pass through the membrane easily, thereby decreasing rejection. Optimum rejection was obtained at near-neutral conditions, between pH 7.5 and 8.5.

The salt concentration of raw water may have a greater influence in NF than in UF as a result of partial salt rejection and, hence, increasing salt concentration on the membrane surface. As a consequence, osmotic pressure increases and, because it opposes the transmembrane pressure, the permeate flux decreases.

Jarusutthirak et al. (2007) reported that increased ionic strength ranging from 0.004 to 0.1 M led to a greater flux decline, possibly as a result of increasing osmotic pressure. The same effect was observed when the solution pH was increased from 4 to 10. They also reported that a solution with high ionic strength (0.05 M) showed greater NOM rejection than that of low ionic strength (0.01 M), which is contradictory to the observation of Hong and Elimelech (1997), who noted that both permeate flux and humic acid rejection decreased drastically when NF was carried out at low pH and high ionic strength.

The presence of calcium and other divalent ions reduces the rejection capacity by shielding the interchain charges of the humic molecules (Visvanathan et al., 1998; Ribau Teixeira and Rosa, 2006). Another possible explanation proffered by Visvanathan et al. (1998) is the Donnan effect: calcium and magnesium cations tend to pass through the negatively charged membrane and, to maintain electroneutrality on both sides of the membrane, the negatively charged humic acid passes through the membrane, thus reducing rejection. A tight NF membrane (NF200B) with a high passage of calcium, more than 55% as $CaCl_2$, demonstrated stable rejections (>95%) of DOC, AOC, and THM formation potential at water recovery rates from 6% to 85%, although the DOC content in the feed water increased up to 20 mg L^{-1} (Gorenflo et al., 2003).

6.4.2 Effect of hydrodynamic conditions

Studies of how permeate flux affects NOM rejection achieved contradictory results. On the one hand, Lee and Lee (2007) observed that the greater the flux, the lower the DOC rejection when they filtered surface water with a range of NF membranes using small stirred-cell equipment. Thus, they concluded that it is difficult to achieve high permeate flux and high NOM rejection simultaneously. On the other hand, several researchers have found that an increase in the transmembrane pressure produces a greater flux but does not have a significant effect on rejection (Visvanathan et al., 1998; Chang et al., 2009; Uyak et al., 2008). The rejection of the low

MM fraction may increase with an increase in the transmembrane pressure, however (Ates et al., 2009).

Alborzfar et al. (1998) found that high cross-flow velocity and low trans-membrane pressure reduced the flux decline of tight NF membranes, but NOM removal remained at 93%. Lee et al. (2004a,b) observed that NOM removal improved with an increasing cross-flow rate, which was attributed to an increase in the back diffusive transport away from the membrane surface and, hence, to a reduced-concentration polarization layer. Furthermore, as the concentration at the membrane surface decreased, the driving force for the NOM diffusive transport through the membrane decreased. Although J/k values are not available, it seems that a low J/k value is favorable for NOM removal.

High permeate flux may result in severe fouling resulting from high solute transport toward the membrane and more compressed fouling layers (Cho et al., 1999; Hong and Elimelech, 1997). Thus, although greater operating pressures allow greater initial permeate flux, the following rapid decline in flux may counteract this advantage. Although NF of surface waters generally achieves a high removal of organics and salts, high flux decline is common as a result of colloidal, organic, and biological fouling and scaling. The most practical way to mitigate NF fouling is to treat raw water before NF. Various appropriate pretreatment processes have been tested, including coagulation, adsorption, ozonation, chlorination, dissolved air flotation, and MF/UF treatment (Lee and Lee, 2007).

6.4.3 DBP removal

NF and tight UF membranes with a cutoff ranging from 150 to 1000 Da have proved to be effective in DBP formation potential removal without any chemical treatment processes (de la Rubia et al., 2008; Ates et al., 2009; Siddiqui et al., 2000; Glucina et al., 2000).

THM formation potential removal is supposed to be proportional to NOM removal (Lee et al., 2005).Yet, THM formation potential per DOC can be even greater in the NF permeate than in raw water (de la Rubia et al., 2008). Siddiqui et al. (2000) observed average reductions of 90%, 97%, and 94% in DOC, THM, and HAA formation potentials, respectively, with four 100–300-Da spiral-wound NF membranes. The THM formation potential in permeates was well below the EU regulation limit of 100 ng L^{-1} (The Council of the European Union, 1998). The 500-Da membrane was less effective, providing THM and HAA formation potential reductions of 79% and 72%, respectively. In their study, prefiltration by a 60-kDa UF membrane removed DOC, THM, and HAA formation potentials by ~30, 50%, and 32%, respectively, whereas MF was virtually ineffective in precursor removal.

More than 90% removal in THM and HAA formation potentials were also observed by other researchers with the 100-300-Da membranes (Ates et al., 2009; Glucina et al., 2000). Removal of 85%, 82%, and 60% of DOC, THMs, and HAAs, respectively, was noticed with the 1-kDa UF membrane (Ates et al., 2009). In general, THM formation potential was slightly better removed than HAA formation potential, suggesting that lower MM and/or hydrophilic DOC may be more responsible for HAA formation.

6.5 Reverse osmosis

RO has been used mainly for the production of drinking water from seawater by desalination. It is being applied increasingly to drinking water production from high-hardness and NOM-rich surface water and groundwater. RO plants generally operate with polyamide-based hollow fiber and spiral-wound membranes (Schrotter and Bozkaya-Schrotter, 2010).

In general, NOM concentration in seawater is low (DOC, <3 mg L^{-1}) (Jeong et al., 2013), and it contains more hydrophilic compounds than surface waters. Hydrophilic, transphilic, and hydrophobic fractions of 60%, 26%, and 14% of DOC, respectively, have been reported in raw seawater (Jeong et al., 2013).

Currently, new desalination plants are being designed to use MF or UF systems upstream from the RO membranes for the removal of turbidity; nonsoluble, colloidal organics; and pathogens such as Giardia and Cryptosporidium from raw seawater. Yet, they are not able to remove significant amounts of dissolved organics and marine microorganisms, which typically are responsible for RO membrane biofouling (Voutchkov, 2010).

Jeong et al. (2013) studied submerged MF with coagulation and with coagulation and adsorption as a pretreatment to seawater RO. A DOC removal of 23% was reported for MF pretreatment with coagulant and 48% for MF pretreatment with coagulant and adsorbent. The hydrophilic fraction was removed by 19% and 38%, respectively. Both pretreatments were able to remove more than 900-Da MM compounds and reduce flux decline from 49% to 24%—26%.

The flotation UF pretreatment method reduced low-MM neutrals and biopolymers by 33%—40% and 18%—19%, respectively. RO membranes acted as an almost complete barrier for humic compounds, building blocks, and low-MM neutrals, with rejections of 97%–100% (Simon et al., 2013).

Apart from seawater desalination, RO is often used to remove THMs, HAAs, and other halogenated hydrocarbons from water (Bodzek et al., 2011). A DOC removal of 99% (Meylan et al., 2007), and THM and HAA formation potential removals of more than 95% (Kruithof et al., 1998) from surface waters have been reported.

The main challenges in NF and RO plants are the concentration of scale-forming minerals and organic materials that decrease recovery rates, and treatment and disposal of membrane concentrates. Concentrate treatment may be necessary before its final disposal to a wastewater treatment plant or surface waters, or via deep-well injection. Comstock et al. (2011) investigated treatment of NF/RO concentrate by precipitative softening with lime, ferric sulfate coagulation, and anion exchange using MIEX-Cl resin. Both coagulation and anion exchange were able to reduce DOC concentration to the raw water level, whereas coagulation was not recommended for treating the membrane concentrate because the method is sensitive to alkalinity and pH, and does not remove inorganic chemicals. Anion exchange removed organic and inorganic chemicals, except calcium, which was removed by lime softening only.

6.6 Membrane fouling

The major problem in drinking water production by membrane processes is the decrease of flux and yield over time resulting from concentration polarization and fouling. Periodic cleaning and renewal of the membranes are needed, which considerably increase the price of produced water (Guizard and Amblard, 2009). Therefore, a number of methods for the removal of matter accumulated on the membrane surface have been introduced to prevent fouling and to lengthen operation time between cleanings.

Concentration polarization depends on the selectivity of a membrane. When particles or solutes are retained by the membrane, they accumulate in a boundary layer adjacent to the membrane surface. This leads to a reduction in solvent flow through the membrane, which results in a reduction in the effective transmembrane pressure resulting from an osmotic pressure difference between the filtrate and retentate sides of the membrane surface. This phenomenon is reversible and it can be reduced by simply releasing the transmembrane pressure or by using methods that increase particle back-transport away from the membrane surface (Fane and Chang, 2009).

Membrane fouling is a buildup of material on the membrane surface or within the membrane structure. Different fouling mechanisms, such as adsorption, pore blocking, pore constriction, or gel/cake layer formation, may occur, depending on the process (Fig. 6.4).

The typical fouling mechanisms for MF and UF are pore constriction, pore blocking, and gel/cake formation. If particles are smaller than membrane pores, they can enter and stick to the pores. If particles are about the same size as membrane pores, deposition of the particles onto the membrane surface may cause pore blocking. In a membrane filtration process, fouling may be a combination of different mechanisms resulting from the distribution of membrane pore size and component sizes of the feed water. During the initial stages, membrane fouling is thought to be dominated by pore constriction and pore blocking (Fane and Chang, 2009).

Fouling is a very complex phenomenon, and it has to be investigated for every single application to avoid it. It depends on the composition of the feed solution (nature, concentration and interactions of components, pH, and ionic strength), membrane properties (hydrophobicity, charge, roughness, pore size, pore size distribution, and porosity), and operating conditions (temperature, transmembrane pressure, and cross-flow velocity) (Fane et al., 2008). Membrane-fouling mitigation may involve pretreatment of the raw water (e.g., coagulation, adsorption, oxidation, and prefiltration) and the development of fouling-resistant membranes (Zhou et al., 2010; Huang et al., 2011).

Fouling can be classified according to the type of fouling material: colloidal, organic, inorganic (scaling), and microbial/biological (biofouling). The fouling caused by organic macromolecules may overlap organic, colloidal, and microbial fouling. Only the fouling caused by NOM molecules is discussed here. Aquatic colloidal, organic molecules are classified into rigid biopolymers, humic and fulvic acids, and flexible biopolymers (Buffle et al., 1998; Tang et al., 2011). Rigid biopolymers consist mainly of large MM polysaccharides with a long persistence length; flexible biopolymers consist of proteins and some low-molecular weight organic molecules (molecular weight, <1000 Da).

NOM is believed to be the major membrane foulant in drinking water treatment plants using surface water. A propensity for fouling depends on the composition of NOM. Of the two NOM types—allochthonous (terrestrially derived, mainly humic substances) and autochthonous or algal (microbially derived) NOMs—the autochthonous types have greater fouling potential (Amy, 2008). They consist of more extracellular and intracel-lular macromolecules and cellular debris. Surface water dominated by autochthonous NOM can cause serious fouling when algae bloom, especially in spring (Plottu et al., 2003).

Several authors observed that biopolymers cause more fouling than humic substances (Amy, 2008; Kennedy et al., 2008b; Lee et al., 2006; Jermann et al., 2008; Her et al., 2008; Zazouli et al., 2010).Jermann et al. (2008) investigated the fouling of a 100-kDa PES membrane using alginate, which is a polysaccharide secreted by brown algae, and humic acid in a dead-end UF. They reported that most alginate fouling was reversible, whereas humic acid fouling was more irreversible. Zazouli et al. (2010) investigated the fouling effects of humic acid and alginate on 150–300-Da polyamide NF membranes. Increasing organic concentration or ionic concentration, or decreasing pH increased the rate and extent of flux reduction. Alginate exhibited greater flux decline than humic acid fouling under the same conditions.

Lee et al. (2006) investigated the fouling of 100-kDa membranes made of regenerated cellulose and PES, and 0.22-m̂ pore size MF membranes made of mixed cellulose ester and PVDF with different isolated fractions of surface waters dominated by both autochthonous and allochthonous NOM. They found that colloidal NOM and autochthonous NOM derived from algae, both containing polysaccharides and protein compounds, were responsible for significant MF membrane fouling. Hydrophobic fractions also fouled the MF membranes. The UF membranes were more fouling

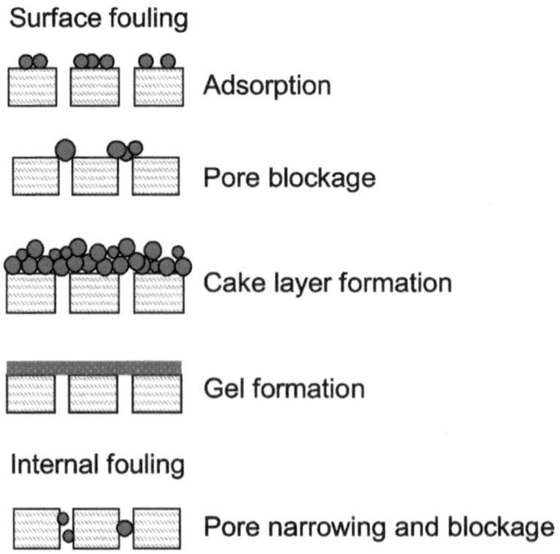

Figure 6.4 Major fouling mechanisms.

resistant, and the PES membrane was fouled significantly only by the NOM derived from algae. The hypothesized fouling mechanisms suggested by atomic force microscope (AFM) were (1) a largely hydraulically reversible surface (gel layer) fouling that was dominant for UF and (2) less hydraulically reversible pore blockage, which was dominant for MF.

Amy (2008) reported that the most severe foulants correspond to polysaccharide and protein macromolecules and organic colloids (peptidoglycans and amino sugars). It was hypothesized that polysaccharidelike and proteinlike foulants interact with a membrane whether by hydrogen bonding and dipole interactions, respectively, or, in a colloidal form, contributing to a cake/gel layer. Foulants larger than the pore size—for example, the high-MM polysaccharides, proteins, and organic colloids that cause only surface fouling—may form a gel layer that is largely hydraulically reversible, and hydraulic backwashing alone may be adequate to remove it (Amy, 2008).

6.6.1 Effects of pH, ionic strength, and divalent ion concentration

Many existing studies have shown that fouling by humic acid is promoted by low pH, high divalent ion concentration (Ca^{2+} and Mg^{2+}), high ionic strength, and high foulant concentration (Kim et al., 2011; Hong and Elimelech, 1997; Zazouli et al., 2010; Seidel and Elimelech, 2002). The adverse effect of low pH and high ionic strength can be attributed to the reduced electrostatic repulsion between humic acid molecules and the membrane surface, as well as between humic acid molecules. Calcium has a strong affinity for the carboxylic moieties of humic acid (Hong and Elimelech, 1997; Wang et al., 2001; Hering and Morel, 1988). Calcium ions can complex with humic acid and thus partially neutralize humic acid molecules. Shao et al. (2011) reported that the filtrate flux and the initial removal rate of humic acid decreased sharply as the calcium concentration increased from 0 to 0.5 mM; when the calcium concentration was increased further to 10 mM, both the flux and retention increased. The most likely explanation for these phenomena is electrostatic interactions and coagulation of humic acid, respectively.

Many polysaccharides can form an extensive gel in the presence of some multivalent metal ions. For example, calcium has a strong affinity to carboxylic groups (Hering and Morel, 1988) and can interact with some olysaccharides by crosslinking the biopolymers into a solidlike gel layer via metal-ligand complex formation (Wang and Waite, 2008; Wang et al., 2009).

Amy (2008) found that the addition of calcium (Ca^{2+}) to the hydrophobic NOM fraction did not increase fouling, because the resultant solute was still small enough to pass through the pores of 100-kDa UF and MF membranes. The negative impact of Ca^{2+} is clearer when the membrane pore size is smaller. Flux decline increased with the J/k value and, finally, a significant flux decline was observed. This result was attributed to greater concentration polarization and the aggregation of humic substances in the boundary layer. It was assumed that the addition of Ca^{2+} enhanced fouling by neutralizing the negative charge of the humic and fulvic acids, thus promoting aggregation.

Jermann et al. (2007) studied fouling of a 100-kDa PES membrane by humic acid and alginate with and without calcium. Calcium increased humic acid adsorption onto the membrane surface and in the pores, whereas alginate formed a cake layer, which was relatively reversible because of strong electrostatic repulsion. Only marginal alginate adsorption was noticed, and the addition of calcium let to complexation of alginate and gel formation.

Wang and Waite (2008) also observed that alginate gel formed a very compressible and porous cake layer (water content, >96%) on a 100-kDa UF membrane. In another study conducted with a membrane bioreactor, they noticed significant gel formation associated mainly with the polysaccharides through metal-ligand complexation provoked by multivalent metals (Ca^2+, Mg^2+) (Wang and Waite, 2009).

6.6.2 Biofouling

Microbial fouling is a result of the formation of biofilms on membrane surfaces. Wet membrane surfaces are ideal areas for microbial (bacterial, algal, or fungal) attachment and colonization, with a steady nutrient supply by convective flow (Chen et al., 2010). After attachment to the membrane, bacteria start to multiply and secrete extracellular polymeric substances such as heteropolysaccharides, proteins, and amino sugars, which form a viscous, slimy, hydrated gel (Kennedy et al., 2008a,b). This gel structure protects bacterial cells from hydraulic shearing and from chemical attacks by biocides such as chlorine (Kennedy et al., 2008a,b). In contrast to other foulants, microorganisms can repopulate soon after a cleaning procedure, if they are not removed completely from the system.

Microbial fouling is closely related to the characteristics of the feed water and is more severe in warm environments and at plants using surface waters that provide great potential for biological activity from numerous species of microorganisms. The growth of microorganisms is most pronounced in the temperature range of 30–45°C (Koltuniewich and Drioli, 2008).

Microcystins, the most commonly occurring cyanotoxins in surface water reservoirs used for water supply, are the first to be adsorbed on the membrane surface as a result of their hydrophobic character (Ribau Teixeira and Sousa, 2013). The parts of the treatment chain most vulnerable to bio-fouling are UF and MF pretreatments when situated downstream of the cartridge filters or sand filters, which are potential breeding grounds for microorganisms (Chen et al., 2010).

6.7 Conclusions

All surface waters contain NOM, which is a heterogeneous mixture of biopolymers and their different-size degradation products. NOM causes problems in drinking water treatment and distribution systems, and is a health risk when it reacts with disinfectants. The great variability of NOM composition makes it difficult to remove it from drinking water completely, particularly because its concentration and quality are influenced by seasonal variation. The characteristics of NOM that influence membrane filtration most greatly are its size distribution and acidic functional group content. The MM of NOM is in the colloidal size range from a few 100 Da to more than 100 kDa, covering the pore sizes of NF, UF, and MF membranes. Although the size distribution of NOM varies among different water sources, in general very high proportions of small molecules (<1 kDa) have been observed.

During disinfection, NOM can react with chlorine to form chlorinated DBPs, such as THMs and HAAs, which are recognized human carcinogens. The formation of DBPs in treated water depends on the raw water quality, especially the concentration of NOM. Hydrophobic acids and hydrophilic neutrals are the major precursors to the formation of DBPs and, therefore, are the fractions that should be removed primarily to minimize DBP formation. Biodegradable dissolved organic matter that causes bacterial growth and biofilm formation in drinking water distribution systems, measured as biodegradable DOC or AOC, also needs to be minimized. AOC is observed to be related mainly to NOM with an MM less than 1 kDa.

Molecules larger than the membrane pores are rejected by size exclusion; smaller compounds may be rejected by charge repulsion if they contain negatively charged groups and if a negatively charged membrane is used. Hence, NOM removal increases with decreasing membrane pore size and with increasing charge density on the membrane surface and in the pores. These conditions are favorable at higher pH and low ionic strength.

MF membranes have large pores and, hence, are unable to reject most NOM compounds unless they are coagulated into bigger flocs. The advantages of these membranes are that they remove suspended solids and colloidal materials, such as pathogens and algae, more economically than tighter membranes. Negatively charged UF membranes are able to remove larger hydrophobic NOM fractions, even more than would be expected on the basis of the membrane cutoff value and NOM size. Yet, NOM contains plenty of small hydrophilic compounds with a low acidic content that are difficult to remove by size or charge exclusion through UF membranes. Hence, UF is ineffective for removing AOC and for preventing the bacterial regrowth potential in drinking water distribution systems. Therefore, to achieve the required water quality, MF and UF need to be combined with other processes such as coagulation, adsorption, oxidation/biological activated carbon (BAC), or tighter membranes (NF or RO).

NF membranes are able to remove compounds from macromolecular size to multivalent ions, but at higher transmembrane pressure than UF. Almost complete NOM rejection was achieved by NF membranes with cutoff values in the range of 100–400 Da. DOC and THM formation potential removals increase with decreasing pore size in UF and NF as follows: 60-kDa UF, >30%; 1-kDa UF, >75%; 500-Da NF, >80%; and 100–300 DaNF, >90%. HAA formation potential removal has been found to be slightly lower. Yet, AOC removal is incomplete, so the disinfection process is still necessary. NF membranes are very susceptible to fouling in surface water treatment and, in general, require extensive pretreatment to control colloidal, organic, and biological fouling and scaling. Pretreatment steps may include MF/UF, conventional treatment, or slow sand or dual-media filtration.

Membrane technology has a great potential to shorten and simplify the long treatment chains of physical-chemical and biological unit processes needed to satisfy stringent water quality criteria. Yet, disinfection is still needed because membranes are not able to remove AOC compounds economically if high quality water is required.

References

A.S. Al-Amoudi, Factors affecting natural organic matter (NOM) and scaling fouling in NF membranes: a review, Desalination 259 (2010) 1–10.

M. Alborzfar, G. Jonsson, C. Gr0n, Removal of natural organic matter from two types of humic ground waters by nanofiltration, Water Res. 32 (1998) 2983–2994.

G. Amy, Fundamental understanding of organic matter fouling of membranes, Desalination 231 (2008) 44–51.

G. Amy, J. Cho, Interactions between natural organic matter (NOM) and membranes: rejection and fouling, in: H. Odegaard (Ed.), International IAWQ-IWSA Joint

Specialist Conference on Removal of Humic Substances from Water. June 24–26, 1999, 1999, p. 141.
E. Aoustin, A.I. Schafer, A.G. Fane, T.D. Waite, Ultrafiltration of natural organic matter, Separ. Purif. Technol. 22–23 (2001) 63–78.
N. Ates, L. Yilmaz, M. Kitis, U. Yetis, Removal of disinfection by-product precursors by UF and NF membranes in low-SUVA waters, J. Membr. Sci. 328 (2009) 104–112.
M. Ben-Sasson, Y. Zidon, R. Calvo, A. Adin, Enhanced removal of natural organic matter by hybrid process of electrocoagulation and dead-end microfiltration, Chem. Eng. J. 232 (2013) 338–345.
Y. Bessiere, B. Jefferson, E. Goslan, P. Bacchin, Effect of hydrophilic/hydrophobic fractions of natural organic matter on irreversible fouling of membranes, Desalination 249 (2009) 182–187.
M. Bodzek, K. Konieczny, A. Kwiecinska, Application of membrane processes in drinking water treatment—state of art, Desalination Water Treat. 35 (2011) 164–184.
L. Braeken, R. Ramaekers, Y. Zhang, G. Maes, B.V.d. Bruggen, C. Vandecasteele, Influence of hydrophobicity on retention in nanofiltration of aqueous solutions containing organic compounds, J. Membr. Sci. 252 (2005) 195–203.
B. van der Bruggen, J. Schaep, D. Wilms, C. Vandecasteele, Influence of molecular size, polarity and charge on the retention of organic molecules by nanofiltration, J. Membr. Sci. 156 (1999) 29–41.
B. van der Bruggen, C. Vandecasteele, Removal of pollutants from surface water and groundwater by nanofiltration: overview of possible applications in the drinking water industry, Environ. Pollut. 122 (2003) 435–445.
J. Buffle, K.J. Wilkinson, S. Stoll, M. Filella, J. Zhang, A generalized description of aquatic colloidal interactions: the three-colloidal component approach, Environ. Sci. Technol. 32 (1998) 2887–2899.
S. Byun, J.S. Taurozzi, A.L. Alpatova, F. Wang, V.V. Tarabara, Performance of polymeric membranes treating ozonated surface water: effect of ozone dosage, Separ. Purif. Technol. 81 (2011) 270–278.
S. Chae, H. Yamamura, B. Choi, Y. Watanabe, Fouling characteristics of pressurized and submerged PVDF (polyvinylidene fluoride) microfiltration membranes in a pilot- scale drinking water treatment system under low and high turbidity conditions, Desalination 244 (2009) 215–226.
E.E. Chang, Y. Chen, Y. Lin, P. Chiang, Reduction of natural organic matter by nanofiltration process, Chemosphere 76 (2009) 1265–1272.
V. Chen, J. Mansouri, T. Charlton, Biofouling in membrane systems, in: K. Peinemann, S. Pereira Nunes (Eds.), Membranes for Water Treatment, Wiley-VCH Verlag, Weinheim, 2010, pp. 25–49.
M. Cheryan, Ultrafiltration and Microfiltration Handbook, Technomic Publishing, Lancaster, USA, 1998.
Y. Chin, G. Aiken, E. O'Loughlin, Molecular weight, polydispersity, and spectroscopic properties of aquatic humic substances, Environ. Sci. Technol. 28 (1994) 1853–1858.
J. Cho, G. Amy, J. Pellegrino, Membrane filtration of natural organic matter: comparison of flux decline, NOM rejection, and foulants during filtration with three UF membranes, Desalination 127 (2000a) 283–298.
J. Cho, G. Amy, J. Pellegrino, Membrane filtration of natural organic matter: factors and mechanisms affecting rejection and flux decline with charged ultrafiltration (UF) membrane, J. Membr. Sci. 164 (2000b) 89–110.
J. Cho, G. Amy, J. Pellegrino, Membrane filtration of natural organic matter: initial comparison of rejection and flux decline characteristics with ultrafiltration and nanofiltration membranes, Water Res. 33 (1999) 2517–2526.

M.M. Clark, P. Lucas, Diffusion and partitioning of humic acid in a porous ultrafiltration membrane, J. Membr. Sci. 143 (1998) 13–25.
S.E.H. Comstock, T.H. Boyer, K.C. Graf, Treatment of nanofiltration and reverse osmosis concentrates: comparison of precipitative softening, coagulation, and anion exchange, Water Res. 45 (2011) 4855–4865.
A.R. Costa, M.N. de Pinho, Comparison of the performance of ultrafiltration and nanofiltration in surface water treatment, Desalination 199 (2006a) 73–75.
A.R. Costa, M.N. de Pinho, Performance and cost estimation of nanofiltration for surface water treatment in drinking water production, Desalination 196 (2006b) 55–65.
B.Z. Dong, Y. Chen, N.Y. Gao, J.C. Fan, Effect of pH on UF membrane fouling, Desalination 195 (2006) 201–208.
M. Edwards, M.M. Benjamin, J.N. Ryan, Role of organic acidity in sorption of natural organic matter (NOM) to oxide surfaces, Colloids Surf. A Physicochem. Eng. Aspects 107 (1996) 297–307.
J.K. Edzwald, J.E. Tobiason, Enhanced coagulation: US requirements and a broader view, Water Sci. Technol. 40 (1999) 63–70.
B. Ericsson, M. Hallberg, J. Wachenfeldt, Nanofiltration of highly colored raw water for drinking water production, Desalination 108 (1997) 129–141.
B. Ericsson, G. Tragardh, Treatment of surface water rich in humus — membrane filtration vs. conventional treatment, Desalination 108 (1997) 117–128.
I.C. Escobar, S. Hong, A.A. Randall, Removal of assimilable organic carbon and biodegradable dissolved organic carbon by reverse osmosis and nanofiltration membranes, J. Membr. Sci. 175 (2000) 1–17.
R. Fabris, C.W.K. Chow, M. Drikas, B. Eikebrokk, Comparison of NOM character in selected Australian and Norwegian drinking waters, Water Res. 42 (2008) 4188–4196.
L. Fan, J.L. Harris, F.A. Roddick, N.A. Booker, Influence of the characteristics of natural organic matter on the fouling of microfiltration membranes, Water Res. 35 (2001) 4455–4463.
A.G. Fane, S. Chang, Techniques to enhance performance of membrane process, in: A.K. Paddy, S.S.H. Rizvi, A.M. Sastre (Eds.), Handbook of Membrane Separations, Chemical, Pharmaceutical, Food, and Biotechnological Applications, CRC Press, Taylor & Francis Group, Boca Raton, USA, 2009, pp. 193–232.
A.G. Fane, R. Wang, Y. Jia, Membrane technology: past, present and future, in: L.K. Wang, J.P. Chen, Y. Hung, N.K. Shammas (Eds.), Handbook of Environmental Engineering. Membrane and Desalination Technologies vol. 13, Springer Science & Business Media, New York, 2008, pp. 1–45.
K. Ghosh, M. Schnitzer, Macromolecular structures of humic substances, Soil Sci. 129 (1980) 266–276.
K. Glucina, A. Alvarez, J.M. Laine, Assessment of an integrated membrane system for surface water treatment, Desalination 132 (2000) 73–82.
A. Gorenflo, D. Velazquez-Padron, F.H. Frimmel, Nanofiltration of a German groundwater of high hardness and NOM content: performance and costs, Desalination 151 (2003) 253–265.
C. Guizard, P. Amblard, Current status and prospects for ceramic membrane applications, in: A.K. Paddy, S.S.H. Rizvi, A.M. Sastre (Eds.), Handbook of Membrane Separations, Chemical, Pharmaceutical, Food, and Biotechnological Applications, CRC Press, Taylor & Francis Group, Boca Raton, USA, 2009, pp. 139–179.
B.I. Harman, H. Koseoglu, N.O. Yigit, M. Beyhan, M. Kitis, The use of iron oxide- coated ceramic membranes in removing natural organic matter and phenol from waters, Desalination 261 (2010) 27–33.
L.J. Hem, H. Efraimsen, Assimilable organic carbon in molecular weight fractions of natural organic matter, Water Res. 35 (2001) 1106–1110.

N. Her, G. Amy, J. Chung, J. Yoon, Y. Yoon, Characterizing dissolved organic matter and evaluating associated nanofiltration membrane fouling, Chemosphere 70 (2008) 495–502.

J.G. Hering, F.M.M. Morel, Humic acid complexation of calcium and copper, Environ. Sci. Technol. 22 (1988) 1234–1237.

N. Hilal, M. Al-Abri, H. Al-Hinai, Enhanced membrane pre-treatment processes using macromolecular adsorption and coagulation in desalination plants: a review, Separ. Sci. Technol. 41 (2006) 403–453.

S. Hong, M. Elimelech, Chemical and physical aspects of natural organic matter (NOM) fouling of nanofiltration membranes, J. Membr. Sci. 132 (1997) 159–181.

G. Huang, F. Meng, X. Zheng, Y. Wang, Z. Wang, H. Liu, M. Jekel, Biodegradation behavior of natural organic matter (NOM) in a biological aerated filter (BAF) as a pretreatment for ultrafiltration (UF) of river water, Appl. Microbiol. Biotechnol. 90 (2011) 1795–1803.

S.A. Huber, A. Balz, M. Abert, W. Pronk, Characterization of aquatic humic and non-humic matter with size-exclusion chromatography–organic carbon detection–organic nitrogen detection (LC-OCD-OND), Water Res. 45 (2011) 879–885.

J.C. Jacangelo, R. Rhodes Trussell, M. Watson, Role of membrane technology in drinking water treatment in the United States, Desalination 113 (1997) 119–127.

C. Jarusutthirak, S. Mattaraj, R. Jiraratananon, Factors affecting nanofiltration performances in natural organic matter rejection and flux decline, Separ. Purif. Technol. 58 (2007) 68–75.

S. Jeong, S. Kim, C. Min Kim, S. Vigneswaran, T. Vinh Nguyen, H. Shon, J. Kandasamy, I.S. Kim, A detailed organic matter characterization of pretreated seawater using low pressure microfiltration hybrid systems, J. Membr. Sci. 428 (2013) 290–300.

D. Jermann, W. Pronk, R. Kagi, M. Halbeisen, M. Boller, Influence of interactions between NOM and particles on UF fouling mechanisms, Water Res. 42 (2008) 3870–3878.

D. Jermann, W. Pronk, S. Meylan, M. Boller, Interplay of different NOM fouling mechanisms during ultrafiltration for drinking water production, Water Res. 41 (2007) 1713–1722.

C. Jucker, M.M. Clark, Adsorption of aquatic humic substances on hydrophobic ultrafiltration membranes, J. Membr. Sci. 97 (1994) 37–52.

J. Kaewsuk, G.T. Seo, Verification of NOM removal in MIEX-NF system for advanced water treatment, Separ. Purif. Technol. 80 (2011) 11–19.

P. Kajitvichyanukul, N.K. Shammas, Y.-. Hung, L.K. Wang, J. Ananpattarachai, Portable water biotechnology, membrane filtration and biofiltration, in: L.K. Wang, J.P. Chen, Y. Hung, N.K. Shammas (Eds.), Handbook of Environmental Engineering. Membrane and Desalination Technologies vol. 13, Springer Science & Business Media, New York, 2008, pp. 477–523.

M.D. Kennedy, J. Kamanyi, S.G. Salinas Rodriguez, N.H. Lee, J.C. Schippers, G. Amy, Water treatment by microfiltration and ultrafiltration, in: N.N. Li, A.G. Fane, W.S.W. Ho, T. Matsuura (Eds.), Advanced Membrane Technology and Applications, John Wiley & Sons, Hoboken, USA, 2008a, pp. 131–170.

M.D. Kennedy, H.K. Chun, V.A. Quintanilla Yangali, B.G.J. Heijman, J.C. Schippers, Natural organic matter (NOM) fouling of ultrafiltration membranes: fractionation of NOM in surface water and characterization by LC-OCD, Desalination 178 (2005) 73–83.

M.D. Kennedy, J. Kamanyi, B.G.J. Heijman, G. Amy, Colloidal organic matter fouling of UF membranes: role of NOM composition & size, Desalination 220 (2008b) 200–213.

J.E. Kilduff, S. Mattaraj, J.P. Pieracci, G. Belfort, Photochemical modification of poly(ether sulfone) and sulfonated poly(sulfone) nanofiltration membranes for control of fouling by natural organic matter, Desalination 132 (2000) 133–142.

D.H. Kim, H.K. Shon, G. Sharma, J. Cho, Charge effect of natural organic matter for ultrafiltration and nanofiltration membranes, J. Ind. Eng. Chem. 17 (2011) 109–113.

M.H. Kim, M.J. Yu, Characterization of NOM in the Han River and evaluation of treatability using UF-NF membrane, Environ. Res. 97 (2005) 116–123.

A.B. Koltuniewich, E. Drioli, Membranes in Clean Technologies, Theory and Practice, Wiley-VCH Verlag, Weinheim, 2008.

J.C. Kruithof, J.C. Schippers, P.C. Kamp, H.C. Folmer, J.A.M.H. Hofman, Integrated multi-objective membrane systems for surface water treatment: pretreatment of reverse osmosis by conventional treatment and ultrafiltration, Desalination 117 (1998) 37–48.

I.L. Kuchler, N. Miekeley, Ultrafiltration of humic compounds through low molecular mass cut-off level membranes, Sci. Total Environ. 154 (1994) 23–28.

V. Lahoussine-Turcaud, M.R. Wiesner, J. Bottero, Fouling in tangential-flow ultrafiltration: the effect of colloid size and coagulation pretreatment, J. Membr. Sci. 52 (1990) 173–190.

J.M. Laine, D. Vial, P. Moulart, Status after 10 years of operation — overview of UF technology today, Desalination 131 (2000) 17–25.

R. Lamsal, K.R. Montreuil, F.C. Kent, M.E. Walsh, G.A. Gagnon, Characterization and removal of natural organic matter by an integrated membrane system, Desalination 303 (2012) 12–16.

U. Lankes, H.D. Ludemann, F.H. Frimmel, Search for basic relationships between "molecular size" and "chemical structure" of aquatic natural organic matter—answers from ^{13}C and ^{15}N CPMAS NMR spectroscopy, Water Res. 42 (2008) 1051–1060.

N. Lee, G. Amy, J. Croue, Low-pressure membrane (MF/UF) fouling associated with allochthonous versus autochthonous natural organic matter, Water Res. 40 (2006) 2357–2368.

N. Lee, G. Amy, J. Croue, H. Buisson, Identification and understanding of fouling in low-pressure membrane (MF/UF) filtration by natural organic matter (NOM), Water Res. 38 (2004a) 4511–4523.

S. Lee, C. Lee, Effect of membrane properties and pretreatment on flux and NOM rejection in surface water nanofiltration, Separ. Purif. Technol. 56 (2007) 1–8.

S. Lee, G. Amy, J. Cho, Applicability of sherwood correlations for natural organic matter (NOM) transport in nanofiltration (NF) membranes, J. Membr. Sci. 240 (2004b) 49–65.

S. Lee, J. Cho, Comparison of ceramic and polymeric membranes for natural organic matter (NOM) removal, Desalination 160 (2004) 223–232.

S. Lee, B. Kwon, M. Sun, J. Cho, Characterizations of NOM included in NF and UF membrane permeates, Desalination 173 (2005) 131–142.

J.A. Leenheer, R.L. Wershaw, M.M. Reddy, Strong-acid, carboxyl-group structures in fulvic acid from the Suwannee River, Georgia. 1. Minor structures, Environ. Sci. Technol. 29 (1995) 393–398.

Y. Li, J. Wang, W. Zhang, X. Zhang, C. Chen, Effects of coagulation on submerged ultrafiltration membrane fouling caused by particles and natural organic matter (NOM), Chin. Sci. Bull. 56 (2011) 584–590.

C. Lin, Y. Huang, O.J. Hao, Ultrafiltration processes for removing humic substances: effect of molecular weight fractions and PAC treatment, Water Res. 33 (1999) 1252–1264.

C. Lin, S. Liu, O.J. Hao, Effect of functional groups of humic substances on uf performance, Water Res. 35 (2001) 2395–2402.

T. Liu, Z. Chen, W. Yu, J. Shen, J. Gregory, Effect of two-stage coagulant addition on coagulation-ultrafiltration process for treatment of humic-rich water, Water Res. 45 (2011) 4260–4268.

G.W. van Loon, S.J. Duffy, Environmental Chemistry: A Global Perspective, second ed., Oxford University Press, USA, 2005.

M.P. Lopes, C.T. Matos, V.J. Pereira, M.J. Benoliel, M.E. Valerio, L.B. Bucha, A. Rodrigues, A.I. Penetra, E. Ferreira, V.V. Cardoso, M.A.M. Reis, J.G. Crespo, Production of drinking water using a multi-barrier approach integrating nanofiltration: a pilot scale study, Separ. Purif. Technol. 119 (2013) 112–122.

S. Metsamuuronen, M. Sillanpaa, A. Bhatnagar, M. Manttari, Natural organic matter removal from drinking water by membrane technology, Separ. Purif. Rev. 43 (2014) 1–61.

S. Meylan, F. Hammes, J. Traber, E. Salhi, U. von Gunten, W. Pronk, Permeability of low molecular weight organics through nanofiltration membranes, Water Res. 41 (2007) 3968–3976.

J.C. Mierzwa, I. Hespanhol, M.C.C. da Silva, L.D.B. Rodrigues, C.F. Giorgi, Direct drinking water treatment by spiral-wound ultrafiltration membranes, Desalination 230 (2008) 41–50.

J.C. Mierzwa, M.C.C. da Silva, L.R.V. Veras, E.L. Subtil, R. Rodrigues, T. Li, K.R. Landenberger, Enhancing spiral-wound ultrafiltration performance for direct drinking water treatment through operational procedures improvement: a feasible option for the Sao Paulo Metropolitan Region, Desalination 307 (2012) 68–75.

M. Nystrom, K. Ruohomaki, L. Kaipia, Humic acid as a fouling agent in filtration, Desalination 106 (1996) 79–87.

A. Orecki, M. Tomaszewska, K. Karakulski, A.W. Morawski, Surface water treatment by the nanofiltration method, Desalination 162 (2004) 47–54.

A. Plottu, B. Houssais, C. Democrate, D. Gatel, J. Cavard, Autopsies of membranes fouled on Mery-sur-Oise pilot units: many lessons for the behaviour of the water treatment plant, Desalination 157 (2003), 367. 367.

P.J. Remize, C. Guigui, C. Cabassud, Evaluation of backwash efficiency, definition of remaining fouling and characterization of its contribution in irreversible fouling: case of drinking water production by air-assisted ultra-filtration, J. Membr. Sci. 355 (2010) 104–111.

M. Ribau Teixeira, M.J. Rosa, The impact of the water background inorganic matrix on the natural organic matter removal by nanofiltration, J. Membr. Sci. 279 (2006) 513–520.

M. Ribau Teixeira, V.S. Sousa, Fouling of nanofiltration membrane: effects of NOM molecular weight and microcystins, Desalination 315 (2013) 149–155.

A. de la Rubia, M. Rodriguez, D. Prats, pH, ionic strength and flow velocity effects on the NOM filtration with TiO2/ZrO2 membranes, Separ. Purif. Technol. 52 (2006) 325–331.

A. de la Rubia, M. Rodriguez, V.M. Leon, D. Prats, Removal of natural organic matter and THM formation potential by ultra- and nanofiltration of surface water, Water Res. 42 (2008) 714–722.

N.P. Sanchez, A.T. Skeriotis, C.M. Miller, Assessment of dissolved organic matter fluorescence PARAFAC components before and after coagulation-filtration in a full scale water treatment plant, Water Res. 47 (2013) 1679–1690.

A.I. Schafer, A.G. Fane, T.D. Waite, Nanofiltration: Principles and Applications, Elsevier Advanced Technology, Kidlington, UK, 2005.

A.I. Schafer, A.G. Fane, T.D. Waite, Cost factors and chemical pretreatment effects in the membrane filtration of waters containing natural organic matter, Water Res. 35 (2001) 1509–1517.

A.I. Schafer, A.G. Fane, T.D. Waite, Fouling effects on rejection in the membrane filtration of natural waters, Desalination 131 (2000) 215–224.

A.I. Schafer, A.G. Fane, T.D. Waite, Nanofiltration of natural organic matter: removal, fouling and the influence of multivalent ions, Desalination 118 (1998) 109–122.

J. Schrotter, B. Bozkaya-Schrotter, Current and emerging membrane processes for water treatment, in: K. Peinemann, S. Pereira Nunes (Eds.), Membranes for Water Treatment, Wiley-VCH Verlag, Weinheim, 2010, pp. 53–91.

A. Seidel, M. Elimelech, Coupling between chemical and physical interactions in natural organic matter (NOM) fouling of nanofiltration membranes: implications for fouling control, J. Membr. Sci. 203 (2002) 245–255.
J. Shao, J. Hou, H. Song, Comparison of humic acid rejection and flux decline during filtration with negatively charged and uncharged ultrafiltration membranes, Water Res. 45 (2011) 473–482.
E.L. Sharp, S.A. Parsons, B. Jefferson, The impact of seasonal variations in DOC arising from a moorland peat catchment on coagulation with iron and aluminium salts, Environ. Pollut. 140 (2006) 436–443.
M. Siddiqui, G. Amy, J. Ryan, W. Odem, Membranes for the control of natural organic matter from surface waters, Water Res. 34 (2000) 3355–3370.
F.X. Simon, Y. Penru, A.R. Guastalli, S. Esplugas, J. Llorens, S. Baig, NOM characterization by LC-OCD in a SWRO desalination line, Desalination Water Treat. 51 (2013) 1776–1780.
K. Sutherland, Drinking and pure water: filtration improvements progress global clean water provision, Filtr. Sep. 49 (2012) 12–16.
A.D. Syafei, C. Lin, C. Wu, Removal of natural organic matter by ultrafiltration with TiO2-coated membrane under UV irradiation, J. Colloid Interface Sci. 323 (2008) 112–119.
C.Y. Tang, T.H. Chong, A.G. Fane, Colloidal interactions and fouling of NF and RO membranes: a review, Adv. Colloid Interface Sci. 164 (2011) 126–143.
The Council of the European Union, Council directive 98/83/EC. Off, J. Eur. Communities L 330 (1998) 32–54.
T. Thorsen, Concentration polarization by natural organic matter (NOM) in NF and UF, J. Membr. Sci. 233 (2004) 79–91.
T. Thorsen, Flux and recovery influences on treatment efficiency in ultrafiltration of coloured drinking water, Desalination 117 (1998) 131–138.
V. Uyak, I. Koyuncu, I. Oktem, M. Cakmakci, I. Toroz, Removal of trihalomethanes from drinking water by nanofiltration membranes, J. Hazard Mater. 152 (2008) 789–794.
C. Ventresque, V. Gisclon, G. Bablon, G. Chagneau, An outstanding feat of modern technology: the Mery-sur-Oise nanofiltration treatment plant (340,000 m^3/d), Desalination 131 (2000) 1–16.
C. Visvanathan, B.D. Marsono, B. Basu, Removal ofTHMP by nanofiltration: effects of interference parameters, Water Res. 32 (1998) 3527–3538.
N. Voutchkov, Considerations for selection of seawater filtration pretreatment system, Desalination 261 (2010) 354–364.
X. Wang, T.D. Waite, Role of gelling soluble and colloidal microbial products in membrane fouling, Environ. Sci. Technol. 43 (2009) 9341–9347.
X. Wang, T.D. Waite, Gel layer formation and hollow fiber membrane filterability of polysaccharide dispersions, J. Membr. Sci. 322 (2008) 204–213.
Y.F. Wang, L. Shu, V. Jegatheesan, B.Y. Gao, Coagulation and nano-filtration: a hybrid system for the removal of lower molecular weight organic compounds (LMWOC), Desalination Water Treat. 11 (2009) 23–31.
Y. Wang, C. Combe, M.M. Clark, The effects of pH and calcium on the diffusion coefficient of humic acid, J. Membr. Sci. 183 (2001) 49–60.
L.F. Yee, P.M. Abdullah, A. Abdullah, B. Ishak, K.N.Z. Abidin, Hydrophobicity characteristics of natural organic matter and the formation of THM, Malays. J. Anal. Sci. 13 (2009) 94–99.
Y. Yoon, G. Amy, J. Cho, N. Her, Effects of retained natural organic matter (NOM) on NOM rejection and membrane flux decline with nanofiltration and ultrafiltration, Desalination 173 (2005) 209–221.

M.A. Zazouli, S. Nasseri, M. Ulbricht, Fouling effects of humic and alginic acids in nanofiltration and influence of solution composition, Desalination 250 (2010) 688–692.

J. Zhou, W. Li, J. Gu, H. Yu, Z. Tang, X. Wei, Development of a novel RAFT-UV grafting technique to modify polypropylene membrane used for NOM removal, Separ. Purif. Technol. 71 (2010) 233–240.

A.W. Zularisam, A.F. Ismail, M.R. Salim, M. Sakinah, O. Hiroaki, Fabrication, fouling and foulant analyses of asymmetric polysulfone (PSF) ultrafiltration membrane fouled with natural organic matter (NOM) source waters, J. Membr. Sci. 299 (2007a) 97–113.

A.W. Zularisam, A.F. Ismail, M.R. Salim, M. Sakinah, T. Matsuura, Application of coagulation-ultrafiltration hybrid process for drinking water treatment: optimization of operating conditions using experimental design, Separ. Purif. Technol. 65 (2009) 193–210.

A.W. Zularisam, A.F. Ismail, M.R. Salim, M. Sakinah, H. Ozaki, The effects of natural organic matter (NOM) fractions on fouling characteristics and flux recovery of ultrafiltration membranes, Desalination 212 (2007b) 191–208.

CHAPTER 7

NOM removal by biofiltration

Lindsay Anderson[1], Ryan Swinamer[1], Martin Earle[1], Amina Stoddart[1], Graham A. Gagnon[1], Yuri Park[2] and Mika Sillanpää[3,4,5,6]

[1]Center for Water Resources Studies, Dalhousie University, Halifax, Nova Scotia, Canada; [2]Institute of Environmental Technology, Department of Environmental Engineering, Seoul National University of Science and Technology, Seoul, South Korea; [3]Department of Chemical Engineering, School of Mining, Metallurgy and Chemical Engineering, University of Johannesburg, Doornfontein, South Africa; [4]Chemistry Department, College of Science, King Saud University, Riyadh, Saudi Arabia; [5]Department of Applied Physics, Faculty of Science and Technology, Universiti Kebangsaan Malaysia, Bangi, Selangor, Malaysia; [6]International Research Centre of Nanotechnology for Himalayan Sustainability (IRCNHS), Shoolini University, Solan, Himachal Pradesh, India

Abstract

Biofiltration is gaining attention in the drinking water treatment industry due to its ability to remove biodegradable organic matter and decrease disinfection by-product production by removing highly reactive Natural Organic Matter (NOM) species. The main removal mechanism of NOM in biofilters is biologically mediated through biodegradation and sorption to biofilms, although physical separation also occurs. When media with adsorption capacity (i.e., Granular Activated Carbon) are used over inert media types (e.g., sand, anthracite), NOM can also be removed through adsorption. GAC also demonstrates more stability under challenging operating conditions (e.g., under cold water temperature) compared to inert media types. Strategies such as preoxidation, pH control, filter media optimization, nutrient enhancement, and backwash optimization are used often to create an engineered environment within the filter media that will promote microbial growth and some studies have demonstrated their effectiveness in improving NOM removal. These parameters should be optimized on a site-specific basis to achieve maximum removal of NOM. In addition to conventional NOM metrics, tools such as photoelectrochemical oxygen demand, fluorescence spectroscopy (e.g., fluorescence excitation emission matrices) and liquid chromatography organic carbon detection (LC-OCD) have shown promise for monitoring the transformation and removal of NOM during biofiltration. Enzyme activity is also gaining attention as indicator for biodegradation and removal of NOM during biofiltration. This chapter provides an overview of biofiltration processes for NOM removal in drinking water treatment, as well as detailed information on types of biofiltration and their role in NOM removal, mechanisms for NOM removal in biofilters, operational aspects for targeting NOM removal, and various metrics for monitoring NOM removal biofilters.

Keywords: Anthracite; Biofilter; Biofiltration; Drinking water treatment; Granular activated carbon (GAC); Natural organic matter (NOM).

Natural Organic Matter in Water
ISBN 978-0-12-824274-2
https://doi.org/10.1016/B978-0-12-824274-2.00009-0

Abbreviations

AOC	Assimilable organic carbon
AOP	Advanced oxidation process
ATP	Adenosine triphosphate
BAC	Biologically active carbon
BDOC	Biodegradable dissolved organic carbon
BOM	Biodegradable organic matter
COD	Chemical oxygen demand
DBP	Disinfection by-product
DBPFP	Disinfection by-product formation potential
DOC	Dissolved organic carbon
EBCT	Empty bed contact time
EPS	Extracellular polymeric substance
FEEM	Fluorescence excitation-emission matrix
GAC	Granular activated carbon
IBM	Integrated Biofilm Model
LC-OCD	Liquid chromatography-organic carbon detection
PACl	Polyaluminum chloride
PeCOD	Photoelectrochemical oxygen demand
SUVA	Specific UV-absorbance
TOC	Total organic carbon

7.1 Introduction

In the past decade, biological filtration (biofiltration) has gained acceptance for drinking water treatment. In biofilters, filter media are covered in biofilms which break down (i.e., biodegrade) organic carbon through microbial degradation (e.g., Korotta-Gamage and Sathasivan, 2017; Terry and Summers, 2018). Biofilters pose many advantages when compared to conventional filtration practices with respect to NOM removal. Conventional filtration mainly relies on physical NOM removal mechanisms and sometimes removal via adsorption, depending on what type of media is used. Biofilters instead may provide multiple removal mechanisms, including biodegradation in addition to physical separation (e.g., Korotta-Gamage and Sathasivan, 2017). Their unique ability to remove NOM via biodegradation suggests the potential for biofiltration to be an increased barrier against NOM breakthrough. Additionally, biofiltration has been shown to limit DBP production by removing highly reactive NOM species (McKie et al., 2015).

Biofilters are like conventional filters in that they both use granular media such as sand, anthracite, or GAC. However, the environments that are created in biofilters present conditions that are more likely to facilitate microbial growth on the filter media. The use of prechlorination, or other disinfectant prior to filtration, usually dictates whether there will be appreciable biological activity in a filter (Stoddart and Gagnon, 2015). When biofiltration is targeted, enhancements (i.e., engineered biofiltration) can be made to the influent water to further promote microbial growth thus resulting in increased NOM removal. For example, as introduced by Lauderdale et al. (2012), engineered biofiltration with nutrient enhancement can improve biomass growth and consumption of organic material, while hydrogen peroxide addition can control excess biofilm development and minimize head loss.

This chapter provides an overview of biofiltration processes for NOM removal in drinking water treatment. The following sections provide detailed information on (1) the types of biofiltration and their role in NOM removal, (2) the mechanisms for NOM removal in biofilters, (3) the operational aspects of biofiltration with respect to targeting NOM removal as well as (4) an overview of metrics for monitoring NOM removal in biofilters used in bench, pilot, and full-scale studies.

7.2 Types of biofiltration and their role in NOM removal

Biofilters are sometimes characterized into two categories: engineered and passive. Engineered biofilters are optimized to accumulate active biomass on filter media. Strategies such as pH adjustment, filter media optimization, nutrient enhancement, and backwash optimization are used to create an engineered environment within the filter media that will promote microbial growth (Brown et al., 2020). Optimizing the environment for the microbial community (e.g., engineered biofiltration) has been shown to promote biological activity without generating excessive quantities of extracellular polymeric substances (EPS), resulting in potentially improved finished water quality without compromising filter hydraulics. In contrast, passive biofilters result from unintentional biomass growth when conventional filtration objectives are targeted. Kirstis et al. (2019) further categorizes the different types of biofilters as "classical" (conventional filtration operated in the absence of chlorine in the filter influent), in addition to customized or enhanced biofilters including "supplemented" (endogenous components are added to reach a higher concentration in the filter influent than is present naturally), "amended" (exogenous, components that are not naturally present are added), and "bioaugmented" (either endogenous or exogenous biological components are added).

Engineered or enhanced biofilters have been shown to remove on average 75% more DOC as compared to filters that act as passive biofilters (Lauderdale et al., 2012). Pilot scale studies have produced results showing higher DOC, UV_{254}, and SUVA removals in biofilters with nutrient enhancement. Furthermore these studies showed that as NOM removal was declining, nutrient addition was found to reverse this and temporarily increase NOM removal (Fu et al., 2017). Nyfennegger et al. (2013) conducted a pilot-scale study using nutrient enhancement for engineered biofiltration, tracking treatment performance via DOC among other parameters. The filters supplemented with PO_4 improved DOC removal compared to control filters. However, not all DOC is biodegradable and in waters with poorly biodegradable NOM, strategies used to enhance biofilter performance did not improve the effluent water quality in terms of DOC. McKie et al. (2015) showed that pH optimization, nutrient addition or in-line coagulant (aluminum sulfate) addition showed no impact on DOC concentrations in the biofilter effluent—only chemical interventions with polyaluminum chloride (PACl) were effective at improving NOM removal. Likewise, biofilter enhancements were not found to improve the removal of humic substances (McKie et al., 2015). More information on the various engineered biofiltration enhancements and their relation to NOM removal are provided in Section 7.4.

7.3 NOM removal mechanisms during biofiltration

The main NOM removal mechanism in biofilters is biologically mediated through biodegradation and sorption to biofilms, although physical separation (e.g., particle removal) can also occur. Like conventional filters, biofilters can remove particulate organic matter when particle charge has been stabilized through processes such as coagulation (Brown et al., 2020). NOM removal in biofilters can be a combination of sorption and biodegradation (Kaarela et al., 2021). When media with adsorption capacity (i.e., GAC) are used over inert media types (e.g., sand, anthracite), contaminants can also be removed through adsorption. However, the adsorption capacity of GAC media decreases with time and breakthrough can occur (Brown et al., 2020). For example, Servais et al. (1994) reported that due to exhaustion of GAC adsorption sites, the removal of the non-biodegradable fraction of NOM was reduced from 50% to 10% after 100 days. Regardless, GAC enables biodegradable DOC removal long after media becomes exhausted and continues to remove particles.

GAC media can also undergo regeneration, which is a process in which exhausted GAC media adsorption sites are made available for additional adsorption through biodegradation of adsorbed material (Brown et al., 2020). Biologically active carbon (BAC) is another term used to refer to GAC biofilters which specifically describes GAC where the adsorption sites are exhausted and a natural biofilm is established (e.g., Korotta-Gamage and Sathasivan, 2017).

Once steady state is achieved and biofilters are acclimated, biodegradable fractions of NOM compounds are removed through biodegradation and biotransformation (e.g., Korotta-Gamage and Sathasivan, 2017; Brown et al., 2020). Microorganisms in biofilters typically biodegrade organic compounds through primary substrate utilization, while compounds present at very low concentrations (e.g., algal metabolites like geosmin) are typically degraded through secondary substrate utilization because their concentrations are too low to support primary cell function (Brown et al., 2020). The process of biodegradation occurs for both inert (e.g., sand, anthracite) and adsorptive media such as GAC. According to Brown et al. (2020), for the first several months of operation of GAC biofilters the primary removal mechanism is adsorption, after which biodegradation dominates (i.e., BAC). For inert biofilters, biodegradation is always the primary removal mechanism (Brown et al., 2020). Biodegradation of NOM is a function of biomass growth rate and structure, water chemistry (e.g., temperature, pH, dissolved oxygen, nutrient and carbon availability, NOM characteristics) and media characteristics (pore size and volume, surface area, etc.) (Kaarela et al., 2021; Korotta-Gamage and Sathasivan, 2017).

It is common to employ coagulation processes prior to biofiltration. Coagulation removes hydrophobic NOM resulting in primarily hydrophilic NOM in biofilter influent. BAC is known to remove low molecular weight hydrophilic NOM, consisting of carbohydrates, carboxylic acids, amino acids, and sugars (Owen et al., 1995; Sillanpaa et al., 2015; Korotta-Gamage and Sathasivan, 2017). The hydrophilic fraction acts as a substrate and nutrient source for microorganisms in biofilms and is typically measured as BDOC or AOC (Owen et al., 1995; Sillanpaa et al., 2015). Chen et al. (2016) used different NOM characterization approaches to understand NOM removal during biofiltration and found that biofilters removed high molecular weight biopolymers and protein like materials but achieved only low removal of humic substances. Therefore, it is important to monitor NOM quality in biofilters to understand biodegradable and non-biodegradable components.

7.4 Factors that influence NOM removal during biofiltration

7.4.1 Media type

The filter media used for biofiltration can affect the growth and health of the microbial community, which can impact NOM removal. Wang et al. (1995) found that GAC media could hold more biomass compared to anthracite media. Similarly, de Vera et al. (2018) noted that GAC contained greater biomass development in comparison to anthracite due to disinfectant quenching by GAC. Sharma et al. (2018) also showed that GAC had nearly two times more biomass compared to anthracite through ATP measurements (244 vs. 110 ng g^{-1}, respectively). The denser microbial population offered by GAC in comparison to anthracite and sand may be attributed to a combination of factors such as surface area and charge, adsorption capacity, porosity and roughness (Basu et al., 2016), or pretreatment conditions (e.g., chlorination or pre-ozonation). Conversely, it has also been found (Emelko et al., 2006) that anthracite contained higher biomass in comparison to GAC, or that both media types had similar biomass concentrations (Pharand et al., 2014). Thus, the potential for greater biomass density provided by GAC could offer improved removal of organic matter through biodegradation (Basu et al., 2016). Brown et al. (2016) found that out of 38 North American utilities, approximately 37% used GAC/Sand, 37% used anthracite/sand, and that GAC alone and sand alone were 23% and 3%, respectively.

Initial NOM removal through GAC biofilters typically occurs through adsorption, but as adsorption sites on GAC media become exhausted, removal occurs through biological mechanisms (i.e., biodegradation) (Emelko et al., 2006). The biodegradation of adsorbed NOM can also renew the adsorption sites, prolonging the capacity of the GAC media (Brown et al., 2020). The adsorption capacity of GAC also can quench oxidants, providing a barrier for healthy biomass growth and biodegradation of NOM deeper in the filter media (de Vera et al., 2019). The additional roughness offered by GAC media can also aid in protecting the biomass from fluid shear stresses from the flow in the filter (Emelko et al., 2006). Inert media types, such as anthracite and sand are also used for biofiltration. These media are often less expensive and are noted for being less susceptible to abrasion which increases longevity (Brown et al., 2020). However, the lack of adsorption capacity in inert media biofilters makes them more susceptible to oxidants, which can lead to increased breakthrough and could reduce or even eliminate the microbial community in the media (de Vera et al., 2019), which may reduce biodegradable NOM removal.

Many studies have investigated the impact of media type, particularly GAC, on NOM removal through biofiltration. Wang et al. (1995) showed that after 6 months of operation, 29% DOC removal for GAC biofilters, compared to 16% and 20% removal for sand and anthracite, respectively. Liu et al. (2001) found that GAC biofilters were less affected by temperature in comparison to anthracite filters, and that GAC media could also provide better removal of less biodegradable compounds. Emelko et al. (2006) evaluated the effect of media on organic matter removal and found that at colder temperatures (e.g., 1–3 °C), GAC provided better removal of TOC and biodegradable organic matter (e.g., carboxylic acids such as oxalate) over anthracite media, however at warmer temperatures media type did not impact the removal of biodegradable organic matter. Liu et al. (2001) also demonstrated high levels of BOM removal at 5 °C compared to anthracite/sand filters. Selbes et al. (2017) evaluated the impact of media type on the removal of NOM across biofilters and found that the removal of DOC was slightly higher in the GAC media compared to anthracite media. Conversely, Granger et al. (2014) evaluated media type and DOC removal in a bench-scale direct biofiltration set up and found that media type (e.g., anthracite/sand vs. GAC/sand) had no significant impact on DOC removal, although they were operated at 19 °C. McKie et al. (2015) investigated both passive and engineered biofiltration for the removal of DBP precursors (e.g., NOM) and found that GAC media did not provide widespread benefits compared to anthracite, and the impacts of biofiltration on DOC removal was small because not all DOC was biodegradable. Table 7.1 provides more detail on NOM removal rates for various media types and configurations.

Accordingly, biofiltration can be effective with various media types (e.g., sand, anthracite, GAC), but GAC provides improved removal of biodegradable NOM because it supports higher surface area and biomass density, can adsorb NOM and quench oxidants, and improves biomass stability under challenging operating conditions (e.g., under cold water temperature) compared to inert media types (Basu et al., 2016; Terry and Summers, 2018).

7.4.2 Nutrient enhancement

Nutrients are crucial for microbial respiration and growth. These include biodegradable organic carbon (C), nitrogen (N), and phosphorous (P), which can be found in natural waters. Low availability of nutrients—typically

Table 7.1 Summary of bulk organic matter removal (%) in bench, pilot and full-scale biofiltration studies.

Study	Scale	Pretreatment	Filter configuration	EBCT (min)	Temp range (°C)	TOC/DOC*[1]	BDOC	AOC
						% Removal		
LeChevalier et al. (1992)	Pilot	Ozonation	ANT/Sand/Garnet	5	3.5–25.5	26	NA	75
		Coagulation/Flocculation	GAC/Sand	5–10	3.5–25.5	51	NA	86
		Sedimentation	Deep Bed GAC	14	3.5–25.5	62	NA	85
Servais et al. (1994)	Pilot	Coagulation/Flocculation	GAC (1 m, virgin)	10	9–22	5–49*	50–56	NA
		Sand Filtration	GAC (2 m, virgin)	10	9–22	7–47*	45–51	NA
		Ozonation	GAC (3 m, virgin)	10	9–22	5–43*	40–49	NA
			GAC (1m, precolonized)	10	9–22	0–3*	43–56	NA
Persson et al. (2006)	Pilot	None (Filters fed with raw water)	GAC	31	1.5–20	12*	34	22
			Expanded Clay (Fine)	31	1.5–20	5*	30	35
			Expanded Clay (Coarse)	31	1.5–20	3*	28	41
Moll et al. (1999)	Bench	None (Filters fed with raw water)	Sand	7	5	15*	38	43
			Sand	7	20	24*	60	55
			Sand	7	35	24*	60	57
Melin et al. (2002)	Full	Ozonation	GAC	26–30	2–14	14–33	55	NA

Emelko et al. (2006)	Full	Coagulation/Flocculation Sedimentation Ozonation	GAC/Sand	17–36	1–25	20–25	92 (Oxalate)	NA
			ANT/Sand	17–36	1–25	14–20	93 (Oxalate)	NA
Wang et al. (1995)	Pilot	Ozonation Coagulation/Flocculation Sedimentation	GAC	9.2	13–27	29*	NA	51
			Anthracite	9.2	13–27	16*	NA	39
			Sand	9.2	13–27	20*	NA	43
Nemani (2016)	Pilot	Ozonation	GAC (exhausted)	4	1.7–16.8	~4*	NA	NA
			GAC (exhausted)	8	1.7–16.8	~8*	NA	NA
			BAC (Full-Scale)	4–10	1.7–16.8	~9*	NA	NA
Liu et al. (2001)	Bench	Coagulation Flocculation Sedimentation	GAC/Sand	5.6	5	NA	~80 (Acetate)	NA
			GAC/Sand	5.6	20	NA	~90 (Acetate)	NA
			ANT/Sand	5.6	5	NA	~40 (Acetate)	NA
			ANT/Sand	5.6	20	NA	~90 (Acetate)	NA
Basu and Huck (2004)	Bench	Membrane filtration	ANT/Sand	22	NA	10–36	44–80 (Acetate) 75–83 (Formate)	NA
Theil et al. (2006)	Pilot	Magnetic ion exchange or enhanced coagulation	GAC	8	NA	11–14*	5–17	NA
			ANT	8	NA	1–4*	0–5	NA
			GAC	16	NA	15–20*	23	NA
			ANT	16	NA	2–7*	5–14	NA

Continued

Table 7.1 Summary of bulk organic matter removal (%) in bench, pilot and full-scale biofiltration studies.—cont'd

Study	Scale	Pretreatment	Filter configuration	EBCT (min)	Temp range (°C)	TOC/DOC*[1]	BDOC	AOC
						% Removal		
Wert et al. (2008)	Pilot	Ozonation Coagulation/ Flocculation	Shallow ANT/ Sand	3.2–9.7	13–19	NA	~80 (Oxalate)	~60
			Shallow ANT/ Sand	3.2–9.7	13–19	NA	~75 (Oxalate)	~50
			Deep ANT/Sand	8.3–25	13–19	NA	NA	~60
Ko et al. (2007)	Pilot	Ozonation	GAC/Sand	5	18–28.3	31*	NA	NA
			GAC/Sand	10	18–28.3	26*	NA	NA
			GAC/Sand	20	18–28.3	19*	NA	NA
Price and Bailey (1993) (Adapted from Dussert and Tramposch (1997)	Pilot	Sedimentation Ozonation	GAC	10	9	NA	NA	42
			GAC	10	16	NA	NA	57
			ANT/Sand	10	9	NA	NA	14
			ANT/Sand	10	16	NA	NA	57
Kramer et al. (1993)	Pilot	Ozonation Coagulation/ Flocculation Sedimentation	GAC	4.2	15–25	8–22 4–17	NA	70–75
			ANT	4.2	15–25	10–20	NA	70–75
Yang et al. (2011)	Pilot	Ozonation Sedimentation Coagulation Softening Filtration Ozonation	GAC	6.6	NA	NA	NA	82
			ANT/Sand	6.6	NA	NA	NA	70

Stoddart and Gagnon (2017)	Pilot	Coagulation Flocculation	GAC/ANT/Sand	12.5	9–12	~40	0.2	NA
Granger et al. (2014)	Bench	None (Filters fed with raw water)	ANT	15	6–19	11–21*	NA	NA
			GAC	15	6–19	10–19*	NA	NA
Hozalski et al. (1995)	Bench	Coagulation	Sand	4	22.5	~17–33	NA	NA
		Ozonation	Sand	10	22.5	~12–31	NA	NA
			Sand	20	22.5	~18–33	NA	NA
Rittmann et al. (2002)	Pilot	Ozonation	GAC	3–9	22	~40*	NA	NA
			ANT	3–9	22	~40*	NA	NA
			Sand	6	22	~40*	NA	NA
Zhang et al. (2017)	Bench	None (Filters fed with raw water)	GAC	10	NA	<5*	NA	0 –~47
			GAC	18	NA	<5–15*	NA	~44 –~53
			ANT/Sand	10	NA	<5–10*	NA	0 –~40
			ANT/Sand	18	NA	5–15*	NA	~27 –~66
		Ozonation	GAC	10	NA	<5*	NA	~40 –~60
			GAC	18	NA	<5–15*	NA	~55 –~85
			ANT/Sand	10	NA	<5–10*	NA	~40 –~50
			ANT/Sand	18	NA	<5–17*	NA	~60 –~85

[1]*Indicates DOC removal.

phosphorous—can result in reduced active biomass and increased production of EPS, which can lead to increased head loss (Brown et al., 2020). Coagulants can bind phosphorous which reduces its concentration in biofilter influent water, sometimes leading to limiting conditions for microbial growth and carbon removal (Brown et al., 2020; Lauderdale et al., 2012).

Many studies have shown the benefits of nutrient augmentation when phosphorus is the limiting nutrient. Maintaining a molar ratio of 100:10:1 for C:N:P is often done to limit the production of EPS while improving microbial activity (LeChevalier et al., 1991; Lauderdale et al., 2012); however, it has also been shown that this ratio is not always ideal (Wong, 2015). Lauderdale et al. (2012) found that by meeting this ratio, terminal head loss in biofilters decreased by 15%, allowing for longer run times, and provided substantial improvement in DOC removal. In addition, EPS production was decreased by more than 30% and ATP concentrations, indicating microbial activity, increased by more than 30% (Lauderdale et al., 2012). Dhawan et al. (2017) evaluated the influence of nutrient supplementation on DOC removal in biofilters, with C:N:P ratios of 546:24:1, 100:10:1, and 25:5:1 to simulate nutrient limitation and supplementation conditions and found that nutrient availability influenced DOC removal. The supplementation improved DOC removal to ~32% (25:5:1) and ~28% (100:10:1). Granger et al. (2014) found that an average of 23% removal of DOC occurred with phosphorus enhancement, in comparison to 12% for the nonenhanced filter suggesting that enhancement with phosphorus increases DOC removal. Conversely, Selbes et al. (2017) observed increased pilot-scale biofilter run time and reduced head loss with phosphate amendment but did not observe improved DOC or DBP precursor material removal. Several other studies (Azzeh et al., 2015; Fu et al., 2017; Nemani et al., 2018) have observed minor to no improvement in the removal of DBP precursor material. Stoddart and Gagnon (2017) evaluated water quality and filter performance on nutrient enhanced biofilters and determined that improvements in DOC, TOC, SUVA, and DBPfp through nutrient enhancement were not substantial enough to be considered of practical importance. Overall, the implementation of nutrient enhancement for improving NOM removal during biofiltration has resulted in mixed outcomes but demonstrates potential to maximize organic removal. Kiethley and Kristis (2019) noted that phosphorus supplementation could improve biofilter performance only if the biofilter is severely phosphorus limited, which highlights the importance of how phosphorus limitation is defined. An additional consideration is the aspects of upstream treatment that can impact the C:N:P ratios and their effectiveness. For example, phosphorous can be removed during upstream treatment and pre-ozonation can increase BDOC, both would alter the C and P components of the C:N:P ratio (Kiethley and Kristis, 2019). Therefore, additional research into nutrient enhancement and biofiltration optimization is necessary.

7.4.3 Temperature

Temperature impacts NOM removal in biofilters. For example, Wulfeck and Summers (1994) found that at temperatures below 14 °C, TOC removal in full-scale biofilters decreased by 34%. Moll et al. (1999) evaluated the impact of temperature on biofilter performance and microbial community structure and found that a biofilter operated at 5 °C had significantly lower NOM removal compared to filters operated at 20 and 35 °C. The filters operated at 20 and 35 °C had identical DOC and BDOC removal performance. The impact of low temperatures may be mitigated by increasing EBCT (Moll et al., 1999). Nishiima et al. (1998) found that there was no impact of DOC removal at operating temperatures between 5 and 30 °C, when the EBCT was above 24 min. According to Terry and Summers (2018), biofilters operated at temperatures greater than 20 °C removed twice as much TOC than those operated below 10 °C. Pharand et al. (2015) observed seasonal effects in biofiltration for the removal of NOM at the full scale. When temperatures increased from below 10 °C to above 10 °C, DOC removal increased by 6% and the removal of low molecular weight acids/humics increased by 12%.

Overall, higher water temperatures are typically associated with higher NOM removal rates during biofiltration. In general, many microorganisms grow faster in warmer conditions. Warmer conditions also improve the microbial kinetics by increasing the metabolism rate and nutrient utilization of the biomass (Korotta-Gamage and Sathasivan, 2017; Moll et al., 1999; Urfer et al., 1997). As described earlier, GAC appears to exhibit better removal of TOC over anthracite under cold temperature conditions (i.e., <5 °C) (Emelko et al., 2006); however, according to Basu et al. (2016), there is a lack of research on understanding the effects of cold-water temperature effects on biofilter performance.

7.4.4 pH control

pH is an important parameter when considering the growth, health, and biodegradation ability of the microbial community in biofilters. A lower pH (5.5–6.5) can promote the rapid growth of filamentous bacteria in biofilters (Scholz and Martin, 1997). An excessive amount of filamentous organisms can lead to clogging in the filter media, increasing head loss (Korotta-Gamage and Sathasivan, 2017). Increasing the pH to between 6.5 and 8.5 has been shown to increase TOC removal. This pH range is associated with the optimal growth of bacteria, specifically ciliated protozoa and rotifers which are essential for biodegradation processes (Scholz and Martin, 1997; Korotta-Gamage and Sathasivan, 2017; Brown et al., 2020). However, only few studies have investigated the use of pH control for improving NOM removal during biofiltration. Granger et al. (2014) investigated pH control in bench-scale biofilters and found that pH was not a significant factor in DOC removal. More recently, Hamidi et al. (2020) evaluated the influence of pH and alkalinity on TOC removal in biofilters and found that TOC removal was near 70% at lower pH ranges (6−7.5) and decreased with increasing pH reaching a minimum at pH 10 (TOC removal ~30%). Overall, a pH of 7.5 was recommended by Hamidi et al. (2020), resulting in 67% removal of TOC while providing the lowest head loss development. However, pH control for NOM removal in biofilters may not be practical when biofiltration follows coagulation and clarification processes, because mixing requirements for pH adjustment chemicals may break up flocs causing filter break through.

7.4.5 Pretreatment

Common pretreatment strategies for plants employing biofiltration include coagulation and oxidation. Most notably, using ozone as an oxidant can increase the biodegradable fraction of NOM (Terry and Summers, 2018). Ozone breaks down NOM into smaller and more biodegradable products (Amy and Carlson, 1997; Hozalski et al., 1999), which can be consumed by microorganisms. This results in less complex molecules which are more biodegradable and more easily removed by biofilters (Brown et al., 2020).

Several studies have evaluated ozonation as a pretreatment strategy to improve NOM removal during biofiltration. Seredynska-Sobecka et al. (2006) conducted an ozonation-biofiltration study with model solutions of humic acid and found that GAC biofilters with ozone pretreatment had higher removal of humic acids compared to biofiltration alone. Biofiltration processes are known to remove the rapidly biodegradable NOM fraction, leaving the slowly biodegradable NOM in the filter effluent. It has been determined experimentally that dosing a small amount of easily biodegradable carbon to ozonated water can stimulate the removal of slower fractions (Yavich et al., 2004). Lin and Hsien (2011) evaluated the transformation of humic acid during ozonation and biofiltration and noted that ozonation (1, 4, and 7 mg L^{-1}) cleaved high molecular weight humic acid into low molecular weight particles and increased the biodegradable fraction of organic carbon for biofiltration. In a pilot-scale study, Nguyen et al. (2020) integrated trickling filters followed by low-dose ozonation and biofiltration to reduce DBP precursor material. The ozone dose used ($\sim$3 mg L^{-1}) could not mineralize DOC but converted to biodegradable DOC and reduced DBP formation overall (Nguyen et al., 2020). Black and Berube (2014) evaluated the rate and extent of NOM removal during oxidation and biofiltration and found that oxidation prior to biofiltration increased NOM removal but did not affect the biodegradation rate; ozone and AOPs reacted with the biodegradable and non-biodegradable fractions equally (Black and Berube, 2014). One nuance regarding pre-oxidation is that it is not strictly used for enhancing biofiltration. Some water treatment facilities employ pre-oxidation strategies for the removal of other contaminants (e.g., taste and odor compounds, trace metals) and then implement additional treatment such as biofiltration to remove the oxidation by-products. In this case, facilities may be unoptimized for NOM removal.

Coagulation and clarification or coagulation alone (i.e., direct filtration) are often used in conjunction with ozonation as a pretreatment to biofiltration. Chaiket et al. (2002) investigated the effect of coagulation with ozonation and biofiltration for a water with low SUVA (e.g., hydrophilic NOM) at pilot-scale. Results showed that ozonation, coagulation and biofiltration provided significant reduction in UV absorbance, and pre-ozonation relative to coagulation was more effective than intermediate ozonation (Chaiket et al., 2002). Recently, Mirnasab et al. (2021) evaluated pre-ozonation, enhanced coagulation, and biofiltration with GAC as a treatment upgrade for improving NOM removal. With a pre-ozonation dose of 0.5 mg Ozone/mg TOC, a PACl dose of 15 mg L^{-1} and coagulation pH of 7.5 as well as a 5-minute EBCT, TOC removal was 78% and finished water TOC was less than 1 mg L^{-1}. Conversely, Collins et al. (2017) showed that although ozonation prior to biofiltration improved removal of NOM, precoagulation reduced biofiltration efficiency and hindered the positive effect that ozonation had on filter performance.

Other studies have considered coagulation without pre-oxidation (e.g., ozonation) as a pretreatment to biofiltration. Zhang et al. (1998) studied the synergistic combination of coagulation and biofiltration, where biofiltration was considered in two process sequences: before and after coagulation. Coagulation and biofiltration had synergistic effects, and biofiltration preceding coagulation provided substantially lower coagulant demand due to a reduction in zeta potential through biofiltration (Zhang et al., 1998). Abkar et al. (2019) investigated improvements in biofiltration through coagulation and sedimentation at pilot-scale and found that although sedimentation improved head loss and increased filter run times by approximately 30%, NOM removal was unchanged through coagulation and sedimentation compared to the direct-biofiltration operation.

Overall, pre-ozonation coupled with biofiltration can be an effective strategy to improve organics removal. Pre-ozone doses of 1–2 mg O_3/mg TOC may be effective for enhancing biodegradation of NOM during biofiltration (Basu et al., 2016). However, the optimal ozone dose is site specific and depends on influent NOM character and concentration, and it is recommended that pilot-scale studies be conducted to determine dose (Brown et al., 2020).

7.4.6 EBCT and hydraulic loading rate

It has been shown that the hydraulic loading rate (the rate of flow per unit area of filter media), or surface loading rate, does not affect NOM removal (Servais et al., 1994). Instead, EBCT is one of the most common parameters modified for improving NOM removal in biofilters. It is a measure of the time that water is in contact with the filter media is calculated as the volume of filter media divided by the flowrate (Terry and Summers, 2018). EBCT is changed by either changing the depth of the filter media or the flow rate of the filter.

Many studies have investigated the impact of EBCT on NOM removal during biofiltration, and there is no consensus on adequate EBCTs to optimize NOM removal. Some studies have reported increased NOM removal with EBCT but the relationship is not linear (Urfer et al., 1997). One study by Urfer et al. (1997) found an increase of 20% in the removal of DOC with GAC when the EBCT was changed from 5 to 20 min. LeChevalier et al. (1992) also reported improved TOC removal when the EBCT was increased from 5 to 20 min. A recent study by Moona et al. (2021) investigated four different EBCT to investigate its impact on NOM removal in biofilters. An EBCT of 80 min was used to represent normal operating conditions based on existing operations at the treatment plant where the study was conducted. EBCTs of 15, 30, and 50 min were investigated to reduce treatment time and increase production. An EBCT of 15 min resulted in 1.5% DOC removal and improved to 2.5%–4% for the three higher EBCTs. This is a result of increased EBCTs having more contact time with the biomass allowing for further degradation (Moona et al., 2021).

Other research has demonstrated that EBCT had minimal impact on NOM removal, although EBCTs reported were shorter (e.g., <20 min) (Wang et al., 1995). Hozalski et al. (1995) found no differences in TOC removal when EBCT was increased from 4 to 20 min; TOC removal was independent of EBCT when operated within this range. Accordingly, longer EBCTs may be desired (e.g., >20 min) for maximizing NOM removal although this may be impractical from an operational perspective—the flow rate may be too low for sufficient water production or media would need to be very deep. A survey conducted on 38 North American utilities that use biofiltration demonstrates that many utilities operate at low EBCTs—33% used an EBCT between 2 and 5 min, 20% used an EBCT between 5 and 7 min, 27% used an EBCT between 7 and 10 min, and the remaining 20% used EBCT longer than 10 min (Brown et al., 2020). According to Muller et al. (2019), more research needs to be performed to determine a minimum EBCT needed for adequate removals.

7.4.7 Backwashing

Backwashing needs to be carefully considered for biofiltration to control the loss of biomass, but backwashing must still be sufficient to remove the buildup and clogging of particles that cause head loss and reduce run times. Water alone may not be sufficient for backwashing biofilters; it is common to use air to achieve conditions optimal for particle detachment and removal. TOC removal has been found to return to previous levels when using air scouring during backwashing, regardless of whether the biofilter used GAC or anthracite media (Emelko et al., 2006). Laio et al. (2015) observed that if a significant amount of biomass was detached from the BAC media during backwashing, the biomass was able to regenerate, and DOC removal was usually highest directly after backwashing.

Collapse pulsing is another backwash strategy employed by treatment facilities with biofilters. Collapse pulsing describes when a combination of air and sub-fluidization water flows through the media simultaneously (Emelko et al., 2006). Ahmad et al. (1998) found that biofilters backwashed with collapse pulsing produced water with lower AOC concentrations compared to traditionally backwashed biofilters. Emelko et al. (2006) noted that collapse pulsing increases backwashing velocities and abrasion of filter media grains and noted that collapse pulsing for biofilters requires optimization. Ikhlef and Basu (2017) found that in comparison to a water only backwash procedure, collapse pulsing increased DOC removal, although bed expansions beyond 30% decreased DOC removal.

Many studies observed that backwashing biofilters with water having no residual chlorine or other oxidant results in better water quality post-backwash (Terry and Summers, 2018). The residual oxidant reacts with biomass, potentially decreasing the biodegradation capacity. Many studies have reported the impact of chlorine in backwash water (Wang et al., 1995; Miltner et al., 1995; McKie et al., 2015). Miltner et al. (1995) found loss of attached biomass with chlorinated backwashing (10 min, 2.5–3 mg L^{-1} chlorine). In some cases, having residual oxidant concentration in the backwash water may be favorable to control or reduce excess biofilm growth in the media, but typically is only suggested for biofilters with uncontrolled biomass growth (Brown et al., 2020). Overall, the removal of BOM appears to be improved when a disinfectant residual is not present in the backwash water, and this is reflected in a survey by Brown et al. (2016) of 38 North American water utilities with biofiltration that showed that nearly 80% of utilities did not use chlorinated backwash water.

The frequency, intensity, duration, and type of backwashing (collapse vs. air) as well as the use of chlorination vary by facility based on the water quality and temperature (Korotta-Gamage and Sathasivan, 2017), and should be optimized on a site-specific basis (Brown et al., 2020).

7.5 Metrics and monitoring tools for NOM removal in biofilters

7.5.1 Bulk organic carbon

TOC and DOC are conventional NOM metrics that are commonly used for monitoring of biofilter performance in terms of NOM removal. However, these metrics do not describe the character or biodegradability of NOM. Further, TOC removal on its own may not be sufficient for assessing biological removal processes given that particulate organic matter can be removed through physical separation.

The two major metrics for capturing the biodegradable organic matter (BOM) fraction of NOM include assimilable organic carbon (AOC) and biodegradable dissolved organic carbon (BDOC) (Huck, 1990; Volk and Lechevallier, 2002). AOC and BDOC are sometimes measured in biofilter effluent because they have implications for biological regrowth in distribution systems (Servais et al., 1989; Van Der Kooij et al., 1982), and can be used to ensure biofiltration conditions are optimized for minimizing regrowth. Carboxylic acids (i.e., acetate, formate, oxalate) are also measured in biofiltration studies and can correlate with AOC (Huck et al., 1991). At higher concentrations (>400 μg-C/L), the measurement of carboxylic acids over AOC may be advantageous due to lower cost and more rapid turnover time for analysis (Brown et al., 2020). Carboxylic acids have also shown high correlation with BDOC and are also used as a measure of potential BDOC formation (Gagnon et al., 1997). It has been recommended by Escobar and Randall (2001) that both AOC and BDOC are measured as complementary parameters given the high detection limit for BDOC ($0.1-0.2$ mg L^{-1}); a sample can have a BDOC concentration lower than detection with significant levels of AOC. Both AOC and BDOC should also be measured in conjunction with conventional NOM metrics such as DOC, as the relationship between these parameters can vary given changes in the type of organic matter present (Huck, 1990).

Removal of bulk organic matter can be achieved using biofiltration. However, removals are typically a function of temperature as well as operational factors (EBCT, hydraulic loading rate), media type, and can be improved through pretreatment such as ozonation. Table 7.1 presents a summary of bulk organic matter removal under various biofiltration treatment configurations for bench, pilot, and full-scale studies. As evidenced in Table 7.1, media type, pre-oxidation, EBCT and temperature impact the removal efficacy of NOM in terms of TOC/DOC, AOC, and BDOC. These parameters should be optimized on a site-specific basis to achieve maximum removal of NOM.

Mathematical models have also been developed to understand the removal of organic matter as a function of various input parameters for the design and operation of biofilters. Huck et al. (1994) developed a regression model for predicting BOM removal during biological treatment using data from pilot and full-scale studies. It was determined that a first-order relationship could successfully be used to model removal of biodegradable material (i.e., AOC and BDOC) as well as DBPs during biofiltration (i.e., removal rate is proportional to biofilter influent concentration) (Huck et al., 1994). Zhang and Huck (1996) used pilot scale data to develop a steady state model for the removal of AOC from drinking water via biofiltration and showed that overall removal is proportional to the influent concentration. It is important to note that although the models described by Huck et al. (1994) and Zhang and Huck (1996) are simple to use, they do not consider the non-steady state nature of biofiltration processes. Hozalski and Bouwer (2001) developed a numerical model (BIOFILT) to simulate non-steady state behavior of BOM in biofilters as a function of time. This model can also simulate the impact of biofilm loss due to backwashing on BOM removal performance. This model requires prior knowledge on backwashing effectiveness to achieve successful simulations. Other models include the CHABROL model, which can approximate BDOC removal during biofiltration (Billen et al., 1992), and the Integrated Biofilm Model (IBM) which is a spreadsheet-based modeling program designed for ease of use that can be used to analyze consumption of BOM during biofiltration (Rittman and Stillwell, 2002). Laurent et al. (1999) applied the CHABROL model that relates BOM consumption to biomass density for predicting BOM removal in biofilters using a dataset from pilot and full-scale studies conducted at temperatures ranging between 1 and 25°C. It was determined that BDOC removal was dependent on EBCT, and that at low temperatures 22% removal of BDOC could be achieved, while at a higher temperature 50% removal was achieved at the same EBCT (12 min) (Laurent et al., 1999).

7.5.2 Other

7.5.2.1 UV-absorbance

Specific UV-absorbance (SUVA) is often considered in biofilter monitoring strategies for its insight on DBP formation although it does not directly indicate biodegradadability of NOM or microbiological activity, which is the dominant mechanism for NOM removal in biologically active filters.

There has been mixed information on the relationship between SUVA and biofiltration performance. Selbes et al. (2017) found no distinct changes in SUVA values (reported between 0.5 and 0.9 L mg^{-1}-m) across biofilters with ozonation as pretreatment, given that UV absorbance and DOC levels decreased at similar levels. Similarly, Zhang et al. (2017) observed constant SUVA values (approx. 1.6 L mg^{-1}-m) across biologically active filters and attributed this to consistent removals of DOC and UV254. McKie et al. (2019) also found that bench and pilot scale biofilters showed no significant difference in influent and effluent SUVA values (1.7 L mg^{-1}-m). Stoddart and Gagnon (2017) also found no improvement in SUVA through pilot-scale biofiltration experiments with various nutrient, oxidant, and media enhancement strategies. Thompson (2019) observed a slight decrease in SUVA (from 1.2 to 1.5 to 1.1 L mg^{-1}-m) across GAC biofilters during stable operation and attributed this to low weight biodegradable organic matter in the filter influent. In contrast, Han et al. (2013) found that SUVA increased slightly across a biologically active filter (from 1.96 to ~2.1 L mg^{-1}-m). Hozalski et al. (1999) reported that the removal of

NOM through biodegradation was inversely proportional to the ratio of UV254 to TOC. Thus, in biofilter monitoring, SUVA should be used as an indicator for DBP formation in conjunction with parameters that provide information on the biodegradability of NOM (e.g., AOC and BDOC), and it is not expected that biofiltration would provide substantial changes in SUVA values.

7.5.2.2 Advanced NOM metrics

Advanced NOM metrics including chemical oxygen demand (COD), as well as the ratio of COD to TOC (referred to as mean oxidation state) have been used to monitor biofiltration performance. (Stoddart and Gagnon, 2014) used photoelectrochemical oxygen demand (peCOD), an alternative measurement of COD for monitoring NOM removal in biofilters, finding that peCOD reduction was 1.5 mg L^{-1} compared to 0.3 mg L^{-1} for DOC, improving the resolution for monitoring the biodegradable fraction of DOC—typically only 10%–30% of total DOC. Accordingly, peCOD has potential for use as a biofiltration performance indicator (Stoddart and Gagnon, 2014). Li et al. (2018) introduced a concept called "mean oxidation state of organic carbon," a metric which combines TOC/DOC and peCOD, for monitoring biofilter performance. In the case of Li et al. (2018), the mean oxidation state increased across biofilters in the study and when coupled with metrics such as ATP could provide insight on the transformation of NOM during biofiltration.

Fluorescence spectroscopy, or fluorescence excitation emission matrices (FEEM), and liquid chromatography coupled with organic carbon detection (LC-OCD) are additional tools that have also been applied to monitor NOM removal and transformation in biofilters. As an example, Peleato et al. (2017) used fluorescence spectroscopy to monitor NOM character and biofiltration performance on BAC filters that received different AOP pretreatments. Four independent humic-like components and one protein like component were identified, and reductions of most humic and the protein like component were observed in filters pretreated with both peroxide and ozone (Peleato et al., 2017). Vines and Terry (2020) used fluorescence spectroscopy to evaluate biodegradability of DOM via biofiltration and noted that protein like components were most effectively removed via biofiltration and that increasing temperature (6, 22, and 28°C) and EBCT (5, 15, and 30 min) improved removal efficiencies of these compounds. McKie et al. (2015) used LC-OCD to monitor DBP precursor material removal through engineered biofiltration and found that biofilters were able to remove biopolymers (28%–30%) and humic substances (4%–9%).

FEEM and LC-OCD are often used in conjunction to provide additional information on NOM in biofilters. Peleato et al. (2016) comparatively used fluorescence spectroscopy as well as LC-OCD for monitoring NOM reduction through pilot-scale biofilters. Through this analysis, five FEEM components were identified (four humic, one protein like); the humic component identified through LC-OCD analysis correlated well with humic components measured through FEEM (C1–C3), however there was a low correlation between the protein like FEEM component (C5) and the LC-OCD biopolymer component. One humic FEEM component (C4, terrestrial humic like) did not correlate well with organic measures including LC-OCD, DOC, and UV absorbance (Peleato et al., 2016). Chen et al. (2016)evaluated different NOM characterization approaches to evaluate NOM removal during biofiltration and noted that LC-OCD and FEEM are complementary for NOM characterization in biofiltration. Biofilter removal kinetics for FEEM protein like components were different than for LC-OCD biopolymers. Biofilters removed substantial amounts of biopolymers, and low quantities of humic substances, low molecular weight neutrals, and acids as well as building blocks (an LC-OCD component repressing degradation products of humic substances) identified by LC-OCD, and FEEM protein-like materials (Chen et al., 2016).

7.5.2.3 Bioactivity

Adenosine triphosphate (ATP) is the primary energy carrier in all living cells and is used to quantify viable or bulk active biomass when measured in water or biofilter media. ATP is used for biomass monitoring due to its availability in the form of rapid, sensitive, and easy to use commercial kits (Pharand et al., 2014). Although ATP is not a direct measurement or indicator of NOM concentration or quality, it provides valuable information on biomass in biofilters.

Previous studies have used ATP based measurements for biomass monitoring in biofilters although correlations between ATP and NOM removal are not typically observed. Pharand et al. (2014) assessed biomass in biofilters by ATP and summarized ATP concentration from acclimated biofilters (top 15 cm) and NOM removal from previous bench, pilot and full-scale studies. Pharand et al. (2014) found an increasing trend in ATP concentration with increasing influent DOC concentration (with no pre-ozonation), and no relationship between influent DOC and ATP concentration for studies with ozonation, suggesting that the biodegradable fraction is utilized for microbial growth. Pharand et al. (2014) also noted that ATP data was highly variable (10^2–10^3 ng ATP/cm^3 media) with a typical concentration in the top 15 cm of filter media of approximately 600 ng ATP/cm^3. ATP concentration also increased with increased EBCT and hydraulic loading rate. However, Pharand et al. (2014) did not observe a relationship between ATP concentration in the filter media and DOC removal during biofiltration when they reviewed 16 published sources of ATP data. Others (Elhadidy et al., 2017; Sharma et al., 2018; McKie et al., 2019) have noted that the relationship between NOM removal and ATP concentration is unclear, and few studies have shown a correlation. As an example, Wang et al. (1995) found lower biomass in an anthracite biofilter, and reduced TOC removal performance compared to a GAC biofilter. Lauderdale et al. (2012) evaluated enhanced biofilter performance through nutrient and peroxide addition (i.e., engineered biofiltration) in a pilot study and found that increased ATP correlated with higher DOC removal in nutrient enhanced biofilters. Conversely, Dhawan et al. (2017) evaluated the influence of nutrient addition on DOC removal in drinking water biofilters and found no correlation between DOC removal and ATP concentration.

Recently, Keon et al. (2021) proposed an additional parameter to ATP referred to as enzyme activity to be used as an indicator for biodegradation and subsequent removal of NOM. More specifically, the enzyme activity for esterase and chitinase were strongly and consistently correlated to DOC and DBP reductions across biofilters at both pilot and full-scale. Similarly, McKie et al. (2019) presented an "effective activity," which combines enzyme activity with EBCT, was strongly correlated to DOC reduction. In particular, the effective activity of enzyme esterase was strongly correlated to reductions in DOC concentration across biofilters. Accordingly, these parameters may be viable additions to biofilter monitoring programs for gaining insight on NOM removal through biodegradation although additional work is required to support these correlations given the limited about of research on this new parameter.

7.6 Conclusions

This chapter provides an overview of biofiltration processes for NOM removal in drinking water treatment, as well as detailed information on types of biofiltration and their role in NOM removal, mechanisms for

NOM removal in biofilters, operational aspects for targeting NOM removal, and various metrics for monitoring NOM removal biofilters.

NOM removal in biofilters is usually a combination of sorption and biodegradation. When media with adsorption capacity such as GAC (or BAC) are used over inert media types, contaminants can also be removed through adsorption. Although biofilters with inert media have shown to be effective at NOM removal, BAC can improve removal of biodegradable NOM due to its ability to support higher surface area and biomass density, added adsorption capacity, as well as its stability under challenging operating conditions such as cold-water temperature (e.g., $<5°C$).

Engineered biofilters are controlled for the purpose of biomass accumulation in the filter and have been shown to enhance hydraulic performance while producing improved effluent water quality. In some cases, engineered biofilters have been shown to remove more DOC as compared to passive biofilters, although since not all DOC is biodegradable, strategies used to enhance biofilter performance do not always improve the effluent water quality in terms of DOC.

Nutrient enhancement (i.e., maintaining a molar ratio of 100:10:1 for C:N:P) has demonstrated some success in improving DOC removal and improving hydraulic performance, although some have noted that improvements in NOM removal through nutrient enhancement were not substantial enough to be considered of practical importance. In contrast to nutrient enhancement, pH control for improving NOM removal in biofilters is less studied, although it has been shown that increasing the pH to a range between 6.5 and 8.5 can increase TOC removal. Further, pH control for NOM removal in biofilters may not be practical when biofiltration follows coagulation and clarification processes, as mixing requirements for pH adjustment chemicals may break up flocs causing filter break through. Instead, enhancement strategies with oxidants such as ozone are common for biofiltration facilities—treatment with ozone preceding biofiltration can increase BOM and has demonstrated good correlation with NOM removal in biofilters. The optimal ozone dose is site specific and depends on influent NOM character and concentration.

Longer EBCT (>20 min) improves NOM removal in biofilters. However, this may be impractical from an operational perspective because of production requirements and filter depth limits. Backwashing strategies should also be carefully considered to control the loss of biomass. Collapse pulsing (a combination of air and sub-fluidization water) backwashing may provide improved DOC removal in comparison to backwashing with water only. Additionally, the removal of BOM may be improved when there is no disinfectant residual in backwash water.

TOC and DOC commonly used for monitoring of biofilter performance in terms of NOM removal. However, these metrics do not provide any insight on the character or biodegradability of NOM. BOM should be measured in conjunction with conventional NOM metrics as the relationship between these parameters can vary given changes in the type of organic matter present, although the literature suggests that AOC and BDOC are measured less frequently. Carboxylic acids can be used as alternative measure for BOM as they have been shown to correlate with AOC and BDOC. PeCOD is a newer technology that has shown promise as it provides expanded scale of resolution for monitoring the biodegradable fraction of DOC. Recently, tools such as FEEM and LC-OCD have been used to provide information on the transformation and removal of NOM during biofiltration.

Overall, biofiltration is a promising technology for improving NOM removal during drinking water treatment, particularly when biodegradable NOM is present. Although this chapter provides an overview, the optimization of pretreatment (e.g., pre-oxidation, coagulation) and enhancement strategies (e.g., pH control, nutrient addition), operational aspects (i.e., EBCT, backwashing strategies), and media selection should be conducted on a site-specific basis given that NOM quality and quantity will vary by source.

References

L. Abkar, A.K. Stoddart, G.A. Gagnon, Sedimentation: hydraulic improvement of drinking water biofiltration, AWWA Water Sci. 1 (5) (2019) e1160.

R. Ahmad, A. Amirtharajah, Detachment of particles during biofilter backwashing, J. Am. Water Works Assoc. 90 (12) (1998) 74–85.

J. Azzeh, L. Taylor-Edmonds, R.C. Andrews, Engineered biofiltration for ultrafiltration fouling mitigation and disinfection by-product precursor control, Water Sci. Technol. Water Supply 15 (1) (2015) 124–133.

O.D. Basu, S. Dhawan, K. Black, Applications of biofiltration in drinking water treatment—a review, J. Appl. Chem. Biotechnol. 91 (3) (2016) 585–595.

O.D. Basu, P.M. Huck, Integrated biofilter-immersed membrane system for the treatment of humic waters, Water Res. 38 (3) (2004) 655–662, https://doi.org/10.1016/j.watres.2003.10.043.

G. Billen, P. Servais, P. Bouillot, C. Ventresque, Functioning of biological filters used in drinking water treatment-the Chabrol model, Aqua 41 (4) (1992) 231–241.

K.E. Black, P.R. Bérubé, Rate and extent NOM removal during oxidation and bio-filtration, Water Res. 52 (2014) 40–50.

J. Brown, Biofiltration Guidance Manual for Drinking Water Facilities, 2020. Retrieved from, https://www.waterrf.org/research/projects/biofiltration-guidance-manual-drinking-water-facilities.

J. Brown, G. Upadhyaya, J. Carter, T. Brown, C. Lauderdale, North American Biofiltration Knowledge Base: Project Number: 4459. Water Research Foundation, 2016. http://doi:10.1385/JCD:7:1:7.

K.H. Carlson, G.L. Amy, The formation of filter removable BOM during ozonation, Ozone Sci. Eng. 19 (2) (1997) 145.

T. Chaiket, P.C. Singer, A. Miles, M. Moran, C. Pallotta, Effectiveness of coagulation, ozonation, and biofiltration in controlling DBPs, J. Am. Water Works Assoc. 94 (12) (2002) 81−95.

F. Chen, S. Peldszus, A.M. Elhadidy, R.L. Legge, M.I. Van Dyke, P.M. Huck, Kinetics of natural organic matter (NOM) removal during drinking water biofiltration using different NOM characterization approaches, Water Res. 104 (2016) 361−370.

M.R. Collins, C.W. Vaughan, Characterization of NOM removal by biofiltration: impact of coagulation, ozonation, and sand media coating, in: Disinfection By-Products in Water Treatment, CRC Press, 2017, pp. 449−476.

G.A. de Vera, D. Gerrity, M. Stoker, W. Frehner, E.C. Wert, Impact of upstream chlorination on filter performance and microbial community structure of GAC and anthracite biofilters, Environ. Sci. Water Res. Technol. 4 (8) (2018) 1133−1144.

S. Dhawan, O.D. Basu, B. Banihashemi, Influence of nutrient supplementation on DOC removal in drinking water biofilters, Water Sci. Technol. Water Supply 17 (2) (2017) 422−432.

B.W. Dussert, W.G. Tramposch, Impact of support media on the biological treatment of ozonated drinking water, Ozone: Sci. Eng. 19 (2) (1997) 97−108, https://doi.org/10.1080/01919519708547308.

A.M. Elhadidy, M.I. Van Dyke, F. Chen, S. Peldszus, P.M. Huck, Development and application of an improved protocol to characterize biofilms in biologically active drinking water filters, Environ. Sci. Water Res. Technol. 3 (2) (2017) 249−261.

M.B. Emelko, P.M. Huck, B.M. Coffey, E.F. Smith, Effects of media, backwash, and temperature on full-scale biological filtration, J. Am. Water Works Assoc. 98 (12) (2006) 61−73.

I.C. Escobar, A.A. Randall, Assimilable organic carbon (AOC) and biodegradable dissolved organic carbon (BDOC): complementary measurements, Water Res. 35 (18) (2001) 4444−4454, https://doi.org/10.1016/s0043-1354(01)00173-7.

J. Fu, W.N. Lee, C. Coleman, K. Nowack, J. Carter, C.H. Huang, Removal of disinfection byproduct (DBP) precursors in water by two-stage biofiltration treatment, Water Res. 123 (2017) 224−235.

G.A. Gagnon, S.D. Booth, S. Peldszus, D. Mutti, F. Smith, P.M. Huck, Carboxylic acids: formation and removal in full-scale plants, J. Am. Water Works Assoc. 89 (8) (1997) 88−97.

H.C. Granger, A.K. Stoddart, G.A. Gagnon, Direct biofiltration for manganese removal from surface water, J. Environ. Eng. 140 (4) (2014) 04014006.

H.P. Hamidi, S.L.D. Kenari, O.D. Basu, Simultaneous TOC and ammonia removal in drinking-water biofilters: influence of pH and alkalinity, J. Environ. Eng. 146 (8) (2020) 04020080.

L. Han, W. Liu, M. Chen, M. Zhang, S. Liu, R. Sun, X. Fei, Comparison of NOM removal and microbial properties in up-flow/down-flow BAC filter, Water Res. 47 (14) (2013) 4861−4868.

R.M. Hozalski, S. Goel, E.J. Bouwer, TOC removal in biological filters, J. Am. Water Works Assoc. 87 (1995) 40−54.

R.M. Hozalski, E.J. Bouwer, Non-steady state simulation of BOM removal in drinking water biofilters: model development, Water Res. 35 (1) (2001) 198−210.

R.M. Hozalski, E.J. Bouwer, S. Goel, Removal of natural organic matter (NOM) from drinking water supplies by ozone-biofiltration, Water Sci. Technol. 40 (9) (1999) 157–163.
P.M. Huck, Measurement of biodegradable organic matter and bacterial growth potential in drinking water, J. Am. Water Works Assoc. 82 (7) (1990) 78–86, https://doi.org/10.1002/j.1551-8833.1990.tb06995.x.
P.M. Huck, P.M. Fedorak, W.B. Anderson, Formation and removal of assimilable organic carbon during biological treatment, J. AWWA 83 (12) (1991) 69–80, https://doi.org/10.1002/j.1551-8833.1991.tb07268.x.
P.M. Huck, S. Zhang, M.L. Price, BOM removal during biological treatment: a first-order model, J. Am. Water Works Assoc. 86 (6) (1994) 61–71.
S. Ikhlef, O.D. Basu, Influence of backwash regime on biofilter performance in drinking water treatment, J. Appl. Chem. Biotechnol. 92 (7) (2017) 1777–1784.
O. Kaarela, M. Koppanen, T. Kesti, R. Kettunen, M. Palmroth, J. Rintala, Natural organic matter removal in a full-scale drinking water treatment plant using ClO2 oxidation: performance of two virgin granular activated carbons, J. Water Proc. Eng. 41 (2021) 102001.
M.R. Keon, M.J. McKie, L. Taylor-Edmonds, R.C. Andrews, Evaluation of enzyme activity for monitoring biofiltration performance in drinking water treatment, Water Res. 205 (2021).
Y.S. Ko, Y.J. Lee, S.H. Nam, Evaluation of a pilot scale dual media biological activated carbon process for drinking water, Kor. J. Chem. Eng. 24 (2) (2007) 253–260.
S.M. Korotta-Gamage, A. Sathasivan, A review: potential and challenges of biologically activated carbon to remove natural organic matter in drinking water purification process, Chemosphere 167 (2017) 120–138.
S.W. Kramer, M.J. Sclimenti, B.M. Coffey, Testing biologically active filters for removing aldehydes formed during ozonation, J. Am. Water Works Assoc. 85 (5) (1993) 62–71.
M.J. Kirisits, M.B. Emelko, A.J. Pinto, Applying biotechnology for drinking water biofiltration: advancing science and practice, Curr. Opin. Biotechnol. 57 (2019) 197–204.
C. Lauderdale, P. Chadik, M.J. Kirisits, J. Brown, Engineered biofiltration: enhanced biofilter performance through nutrient and peroxide addition, J. Am. Water Works Assoc. 104 (5) (2012) E298–E309.
P. Laurent, M. Prévost, J. Cigana, P. Niquette, P. Servais, Biodegradable organic matter removal in biological filters: evaluation of the CHABROL model, Water Res. 33 (6) (1999) 1387–1398.
M.W. LeChevallier, W. Schulz, R.G. Lee, Bacterial nutrients in drinking water, Appl. Environ. Microbiol. 57 (3) (1991) 857–862.
M.W. LeChevallier, W.C. Becker, P. Schorr, R.G. Lee, Evaluating the performance of biologically active rapid filters, J. Am. Water Works Assoc. 84 (4) (1992) 136–146.
B. Li, A.K. Stoddart, G.A. Gagnon, Mean oxidation state of organic carbon: a novel application to evaluate the extent of oxidation of natural organic matter in drinking water biological treatment. Microbiological Sensors for the Drinking Water Industry, 2018, p. 197.
X. Liao, C. Chen, J. Zhang, Y. Dai, X. Zhang, S. Xie, Operational performance, biomass and microbial community structure: impacts of backwashing on drinking water biofilter, Environ. Sci. Pollut. Res. 22 (1) (2015) 546–554.
Y.H. Lin, H.J. Hsien, Characteristics transformation of humic acid during ozonation and biofiltration treatment processes, Water Environ. Res. 83 (5) (2011) 450–460.
X. Liu, P.M. Huck, R.M. Slawson, Factors affecting drinking water biofiltration, J. Am. Water Works Assoc. 93 (12) (2001) 90–101.

M.J. McKie, L. Taylor-Edmonds, S.A. Andrews, R.C. Andrews, Engineered biofiltration for the removal of disinfection by-product precursors and genotoxicity, Water Res. 81 (2015) 196–207.

M.J. McKie, M.C. Ziv-El, L. Taylor-Edmonds, R.C. Andrews, M.J. Kirisits, Biofilter scaling procedures for organics removal: a potential alternative to piloting, Water Res. 151 (2019) 87–97.

E. Melin, B. Eikebrokk, M. Brugger, H. Ødegaard, Treatment of humic surface water at cold temperatures by ozonation and biofiltration, Water Sci. Technol. Water Supply 2 (5–6) (2002) 451–457.

R.J. Miltner, R.S. Summers, J.Z. Wang, Biofiltration performance: part 2, effect of backwashing, J. Am. Water Works Assoc. 87 (12) (1995) 64–70.

M.A. Mirnasab, H. Hashemi, M.R. Samaei, A. Azhdarpoor, Advanced removal of water NOM by pre-ozonation, enhanced coagulation and bio-augmented granular activated carbon, Int. J. Environ. Sci. Technol. (2021) 1–10.

D.M. Moll, R.S. Summers, A.C. Fonseca, W. Matheis, Impact of temperature on drinking water biofilter performance and microbial community structure, Environ. Sci. Technol. 33 (14) (1999) 2377–2382.

N. Moona, A. Holmes, U.J. Wünsch, T.J. Pettersson, K.R. Murphy, Full-scale manipulation of the empty bed contact time to optimize dissolved organic matter removal by drinking water biofilters, ACS ES&T Water 1 (5) (2021) 1117–1126.

J. Müller, S. Levai, L. Titzschkau, N. Popović, D. Carevic, J.E. Drewes, U. Hübner, Role of reduced empty bed contact times and pre-treatment by coagulation with Fe (III) salts on the removal of trace organic compounds during sequential biofiltration, Sci. Total Environ. 685 (2019) 220–228.

V.A. Nemani, Impact of Operational Parameters and Nutrient Enhancement on Biofiltration Performance, Doctoral Dissertation, University of Toronto, Canada, 2016.

V.A. Nemani, M.J. McKie, L. Taylor-Edmonds, R.C. Andrews, Impact of biofilter operation on microbial community structure and performance, J. Water Proc. Eng. 24 (2018) 35–41.

P.D. Nguyen, T.D.Q. Le, N.H. Nguyen, K.T. Tran, M.T. Nguyen, K.A. Huynh, Reducing disinfection byproduct precursors and chlorine consuming substances by a special integration of biofiltration and ozonation: a pilot study, J. Water Proc. Eng. 37 (2020) 101419.

W. Nishijima, M. Okada, Particle separation as a pretreatment of an advanced drinking water treatment process by ozonation and biological activated carbon, Water Sci. Technol. 37 (10) (1998) 117–124.

J. Nyfennegger, C. Lauderdale, J. Brown, K. Scheitlin, Engineered biofiltration for drinking water treatment: optimizing strategies to enhance performance, Florida Water Resour. J. (2013) 12–17.

D.M. Owen, G.L. Amy, Z.K. Chowdhury, R. Paode, G. McCoy, K. Viscosil, NOM characterization and treatability, J. Am. Water Works Assoc. 87 (1) (1995) 46–63.

N.M. Peleato, M. McKie, L. Taylor-Edmonds, S.A. Andrews, R.L. Legge, R.C. Andrews, Fluorescence spectroscopy for monitoring reduction of natural organic matter and halogenated furanone precursors by biofiltration, Chemosphere 153 (2016) 155–161.

N.M. Peleato, B.S. Sidhu, R.L. Legge, R.C. Andrews, Investigation of ozone and peroxone impacts on natural organic matter character and biofiltration performance using fluorescence spectroscopy, Chemosphere 172 (2017) 225–233.

F. Persson, G. Heinicke, W. Uhl, T. Hedberg, M. Hermansson, Performance of direct biofiltration of surface water for reduction of biodegradable organic matter and biofilm formation potential, Environ. Technol. 27 (9) (2006) 1037–1045.

L. Pharand, M.I. Van Dyke, W.B. Anderson, P.M. Huck, Assessment of biomass in drinking water biofilters by adenosine triphosphate, J. Am. Water Works Assoc. 106 (10) (2014) E433–E444.

L. Pharand, M.I. Van Dyke, W.B. Anderson, Y. Yohannes, P.M. Huck, Full-scale ozone–biofiltration: seasonally related effects on NOM removal, J. Am. Water Works Assoc. 107 (8) (2015) E425–E435, https://doi.org/10.5942/jawwa.2015.107.0121.

M.L. Price, R.W. Bailey, A.K. Enos, M. Hook, S.W. Hermanowicz, Evaluation of Ozone/biological Treatment for Disinfection Byproducts Control and Biologically Stable Water, 1993.

B.E. Rittmann, D. Stilwell, J.C. Garside, G.L. Amy, C. Spangenberg, A. Kalinsky, E. Akiyoshi, Treatment of a colored groundwater by ozone biofiltration: pilot studies and modeling interpretation, Water Res. 36 (2002) 3387.

B.E. Rittmann, D. Stilwell, Modelling biological processes in water treatment: the integrated biofiltration model, J. Water Supply Res. Technol. Aqua 51 (1) (2002) 1–14.

M. Scholz, R.J. Martin, Ecological equilibrium on biological activated carbon, Water Res. 31 (12) (1997) 2959–2968.

M. Selbes, J. Brown, C. Lauderdale, T. Karanfil, Removal of selected C-and N-DBP precursors in biologically active filters, J. Am. Water Works Assoc. 109 (3) (2017) E73–E84.

B. Seredyńska-Sobecka, M. Tomaszewska, A.W. Morawski, Removal of humic acids by the ozonation–biofiltration process, Desalination 198 (1–3) (2006) 265–273.

P. Servais, A. Anzil, C. Ventresque, Simple method for determination of biodegradable dissolved organic carbon in water, Appl. Environ. Microbiol. 55 (10) (1989) 2732–2734, https://doi.org/10.1128/aem.55.10.2732-2734.1989.

P. Servais, G. Billen, P. Bouillot, Biological colonization of granular activated carbon filters in drinking-water treatment, J. Environ. Eng. 120 (4) (1994) 888–899.

D. Sharma, L. Taylor-Edmonds, R.C. Andrews, Comparative assessment of ceramic media for drinking water biofiltration, Water Res. 128 (2018) 1–9.

M. Sillanpää, A. Matilainen, T. Lahtinen, Characterization of NOM. Natural Organic Matter in Water 2, 2015, pp. 17–53.

A.K. Stoddart, G.A. Gagnon, Application of photoelectrochemical chemical oxygen demand to drinking water, J. Am. Water Works Assoc. 106 (9) (2014) E383–E390, https://doi.org/10.5942/jawwa.2014.106.0106.

A.K. Stoddart, G.A. Gagnon, Full-scale prechlorine removal: impact on filter performance and water quality, J. Am. Water Works Assoc. 107 (12) (2015) E638–E647.

A.K. Stoddart, G.A. Gagnon, Water quality and filter performance of nutrient-, oxidant-and media-enhanced drinking water biofilters, Environ. Sci. Water Res. Technol. 3 (3) (2017) 520–533.

L.G. Terry, R.S. Summers, Biodegradable organic matter and rapid-rate biofilter performance: a review, Water Res. 128 (2018) 234–245.

P. Thiel, L. Zappia, P. Franzmann, B. Warton, M. Alessandrino, A. Heitz, et al., Activated carbon vs anthracite as primary dual media filters – a pilot plant study, in: 9th Annual Water Industry Engineers and Operators Conference, Bendigo Exhibition Centre, 2006.

J. Thompson, Removal of Organic Matter by Classical Biofiltration: Mechanistic Insights Regarding "Biodegradation" (Master's Thesis), University of Waterloo, 2019.

D. Urfer, P.M. Huck, S.D. Booth, B.M. Coffey, Biological filtration for BOM and particle removal: a critical review, J. Am. Water Works Assoc. 89 (12) (1997) 83–98.

D. Van Der Kooij, A. Visser, W.A.M. Hijnen, Determining the concentration of easily assimilable organic carbon in drinking water, J. Am. Water Works Assoc. 74 (10) (1982) 540–545, https://doi.org/10.1002/j.1551-8833.1982.tb05000.x.

M. Vines, L.G. Terry, Evaluation of the biodegradability of fluorescent dissolved organic matter via biological filtration, AWWA Water Sci. 2 (5) (2020) e1201.
C.J. Volk, M.W. Lechevallier, Effects of conventional treatment on AOC and BDOC levels, J. Am. Water Works Assoc. 94 (6) (2002) 112–123, https://doi.org/10.1002/j.1551-8833.2002.tb09494.x.
J.Z. Wang, R.S. Summers, R.J. Miltner, Biofiltration performance: part 1, relationship to biomass, J. Am. Water Works Assoc. 87 (12) (1995) 55–63.
E.C. Wert, J.J. Neemann, D.J. Rexing, R.E. Zegers, Biofiltration for removal of BOM and residual ammonia following control of bromate formation, Water Res. 42 (1–2) (2008) 372–378.
A. Wong, Investigating the Enhancement of Biological Filtration with Capping Material Designs and Nutrient Amendments (Master's Thesis, University of Waterloo, 2015.
W.M. Wulfeck Jr., R.S. Summers, Control of DBP Formation using retrofitted GAC filter-adsorbers and ozonation, in: Proceedings of the 1994 AWWA Water Quality Technology Conference, American Water Works Association, San Francisco, CA, 1994.
B.M. Yang, J.K. Liu, C.C. Chien, R.Y. Surampalli, C.M. Kao, Variations in AOC and microbial diversity in an advanced water treatment plant, J. Hydrol. 409 (1–2) (2011) 225–235.
A.A. Yavich, K.H. Lee, K.C. Chen, L. Pape, S.J. Masten, Evaluation of biodegradability of NOM after ozonation, Water Res. 38 (12) (2004) 2839–2846.
S. Zhang, P.M. Huck, Removal of AOC in biological water treatment processes: a kinetic modeling approach, Water Res. 30 (5) (1996) 1195–1207.
X.H. Zhang, G.S. Jia, W.J. Liu, Z.S. Wang, L. Peyton, Synergistic combination of coagulation with biofiltration for drinking water treatment, J. Environ. Sci. Health A 33 (5) (1998) 729–747.
S. Zhang, S.W. Gitungo, L. Axe, R.F. Raczko, J.E. Dyksen, Biologically active filters—an advanced water treatment process for contaminants of emerging concern, Water Res. 114 (2017) 31–41.

CHAPTER 8

NOM removal by advanced oxidation processes

Mika Sillanpaa[1,2,3,4,5] and Anu Matilainen[6]

[1]Department of Chemical Engineering, School of Mining, Metallurgy and Chemical Engineering, University of Johannesburg, Doornfontein, South Africa; [2]Department of Applied Physics, Faculty of Science and Technology, Universiti Kebangsaan Malaysia, Bangi, Selangor, Malaysia; [3]Zhejiang Rongsheng Environmental Protection Paper Co. LTD, East Zhennan Road, Pinghu Economic Development Zone, Zhejiang, P.R. China; [4]Department of Civil Engineering, University Centre for Research & Development, Chandigarh University, Gharuan, Mohali, Punjab, India; [5]International Research Centre of Nanotechnology for Himalayan Sustainability (IRCNHS), Shoolini University, Solan, Himachal Pradesh, India; [6]Finnish Safety and Chemicals Agency, Tampere, Finland

Abstract

Efficient methods for the removal of natural organic matter (NOM) in waters have emerged. Among these are advanced oxidation processes (AOPs). These processes include O_3/H_2O_2, O_3/UV, UV/H_2O_2, TiO_2/UV, H_2O_2/catalyst, Fenton and photo-Fenton processes, and ultrasound. This chapter presents an overview of the recent research studies dealing with AOP methods for the removal of NOM and related compounds from drinking water.

Keywords: Advanced oxidation processes (AOP)s; Fenton; Natural organic matter (NOM); Ozone; Photocatalysis; Ultrasound; UV light; Water treatment.

Abbreviations

AC	Activated carbon
ALD	Atomic layer deposition
AOPs	Advanced oxidation processes
BAC	Biologically activated carbon
COD	Chemical oxygen demand
DBP	Disinfection by-product
DBPFP	Disinfection by-product formation potential
DOC	Dissolved organic carbon
FP	Formation potential
HA	Humic acids
HAAs	Haloacetic acids
HMM	High molar mass
LMM	Low molecular mass
MM	Molar mass
MWD	Molecular weight distribution
NEU	(hydrophilic) neutral

Natural Organic Matter in Water
ISBN 978-0-12-824274-2
https://doi.org/10.1016/B978-0-12-824274-2.02004-4

NOM	Natural organic matter
SUVA	Specific UV absorbance
THM	Trihalomethane
THMFP	Trihalomethane formation potential
TOC	Total organic carbon
US	Ultrasonic
VUV	Vacuum UV

8.1 Introduction

The natural organic matter (NOM) removal results of advanced oxidation processes (AOPs) are always study specific because they depend on the water characteristics, such as the amount of organic matter. Therefore, characterization of the NOM in water necessarily precludes the design and optimization of any AOP treatment. The organic characteristics of the treated water must also be determined to assess their influence on downstream processes. Furthermore, individual disinfection by-product (DBP) measurements and disinfection by-product formation potential (DBPFP) measurements are important when assessing the suitability of an AOP treatment.

AOP research is well established. Various combinations of oxidants, radiation, and catalysts have been developed for the removal of NOM and organic pollutants, e.g., O_3/H_2O_2, UV/H_2O_2, UV/O_3, UV/TiO_2, Fe^{2+}/H_2O_2, $Fe^{2+}/H_2O_2 + -hv$, vacuum ultraviolet (VUV) radiation, or ionizing radiation (Fujishima, 1971; Glaze et al., 1987, 1992; Legrini et al., 1993; Frimmel, 1994; Nagata et al., 1996; Fukushima et al., 2001; Thomson et al., 2002; Suty et al., 2004). These processes involve the generation of highly reactive radical intermediates, especially the •OH radical (Glaze et al., 1987). The appeal of AOPs is their ability to achieve complete oxidation or mineralization of organic contaminants at near ambient temperature and pressure (Matilainen and Sillanpaa, 2010).

The •OH radical is one of the most powerful known oxidants. The reaction rate of a compound in •OH radical mediated oxidation is usually several orders of magnitude higher than that achieved by molecular ozone under the same conditions. The reaction rate constants between •OH radicals and organic species are in the range of 10^8–10^{10} M^{-1} s^{-1} (Parsons and Byrne, 2004). •OH radicals are highly nonselective oxidants, enabling a very large number of reactions. Once free radical reaction has been initiated, whether by photolysis, ozone, hydrogen peroxide, or heat, a series of simple reactions will ensue. The reactions of •OH radicals with NOM proceeds in three different ways: (1) by the addition of •OH radicals to double bonds, (2) by H-atom abstraction, which yields carbon-centered radicals, and (3) by the •OH radical gaining an electron from an organic substituent. The carbon-centered radicals then react very rapidly with oxygen to form organic peroxyl radicals. The mutual reactions of peroxyl radicals can lead to the production of ketones or aldehydes and/or carbon dioxide (Kleiser and Frimmel, 2000). The rate of oxidation depends on radical, oxygen, and pollutant concentrations. Other factors affecting radical formation are pH, temperature, the presence of ions, the type of pollutant, and the presence of scavengers such as bicarbonate ions (Parsons and Byrne, 2004).

Various AOP investigations have found that NOM decreases more when measured as UV_{254} than as dissolved organic carbon (DOC), for example, in UV/H_2O_2 (Goslan et al., 2006; Sanly et al., 2007; Moncayo-Lasso et al., 2008), TiO_2/UV (Uyguner et al., 2007b; Huang et al., 2008; Liu et al., 2008a,b, 2010), and Fenton processes (Sanly et al., 2007; Moncayo-Lasso et al., 2008, 2009; Nie et al., 2010). The greater drop in UV_{254} values suggests that the chromophores in the NOM macromolecules, which mostly consist of high molar mass (HMM) aromatic rings, are rapidly broken down into lower MM by-products with no UV absorbance. These by-products are thus less susceptible to attack by •OH radicals, and therefore are not mineralized completely (Sanly et al., 2007). The intermediate fragments usually undergo a series of complicated reactions prior to mineralization. Many researchers have observed the preferential degradation of hydrophobic compounds (Zhang and Jian, 2006; Katsumata et al., 2008; Liu et al., 2010). Fractionation studies have revealed that although the fraction of NOM with very hydrophobic acids has been reduced, the corresponding increase happens in the hydrophilic charged and/or hydrophilic neutral (NEU) NOM fractions (Buchanan et al., 2006; Sanly et al., 2007; Liu et al., 2008b, 2010). However, this hydrophilic NEU fraction has not been found to be very reactive in forming, e.g., trihalomethanes (THMs) (Buchanan et al., 2006; Liu et al., 2008b, 2010).

It has been observed that HMM compounds are removed more readily, and the removal of the smaller compounds occurs only after the large molecules have been degraded. Nevertheless, a small amount of organic matter with low molecular mass (LMM) has remained after treatment in several instances (Liu et al., 2008a,b, 2010; Tercero Espinoza and Frimmel, 2009; Tercero Espinoza et al., 2009b). Also, the shift in molecular weight distribution (MWD) toward lower MM compounds during oxidation has been noted in many cases, e.g., on UV/H_2O_2 (Goslan et al., 2006; Wang et al., 2006; Sanly et al., 2007; Sarathy and Mohseni, 2007), TiO_2/UV (Liu et al., 2008a,b, 2010; Tercero Espinoza and Frimmel, 2009; Tercero Espinoza et al., 2009b), and Fenton processes (Goslan et al., 2006; Sanly et al., 2007; Katsumata et al., 2008; Nie et al., 2010). It has been noted that incomplete oxidation increases DBPFP under VUV (Buchanan et al., 2006; Bond et al., 2009), TiO_2/UV (Liu et al., 2008a,b), and UV/H_2O_2 treatments (Toor and Mohseni, 2007). Moreover, during some oxidation processes, the DBP speciation has been found to shift to more brominated DBPs, especially in VUV (Buchanan et al., 2006) and UV/H_2O_2 (Wang et al., 2006) treatments, but contradictory results have also been reported (Tercero Espinoza and Frimmel, 2008). The formation of brominated DBPs is demonstrably dependent on the characteristics of water and the design parameters, e.g., a higher concentration of TiO_2 has been observed to produce more bromoform (Tercero Espinoza et al., 2009a). In addition, UV/H_2O_2 treatment was found to increase the formation potential of nitroso-DBP (Zhao et al., 2008).

A vast amount of research has been conducted on the removal of organic and inorganic micropollutants (Hu et al., 2008; Li et al., 2008; Swaim et al., 2008), including pharmaceuticals (Benotti et al., 2009), endocrine-disrupting chemicals (Chen et al., 2007; Benotti et al., 2009) or pesticides (Rastogi et al., 2009), and the inactivation of microorganisms (Gerrity et al., 2008; Lanao et al., 2008; Navalon et al., 2009). Reviews of such AOP efficiency research include Pera-Titus et al. (2004), Ikehata and El-Din (2005), and Klavarioti et al. (2009). Yet, many of these AOP NOM studies focus on improving the removal of DBP precursors. An overview of the recent research on NOM removal by AOPs was presented by Matilainen and Sillanpaa (2010). This chapter is an update of that earlier work.

8.2 Aops in NOM removal

8.2.1 Ozone-based applications

Ozone has long been used for disinfection, taste and odor management, and color removal in water treatment facilities (Langlais et al., 1991). Ozone reacts with NOM by an electrophilic addition to double bonds. This reaction is very selective. In addition to the direct reaction of ozone with NOM, a nonselective and fast reaction occurs with the •OH radicals formed when ozone decomposes in water. The •OH radical formation potential is much lower in ozonation than in AOPs. The oxidation potentials for ozone and •OH radicals are 2.42 and 2.86 eV, respectively (Kleiser and Frimmel, 2000). There are several solutions to promote the •OH radical production in ozonation processes. The simplest and most cost-efficient way to convert a conventional ozonation process into an AOP is the addition of H_2O_2 (Duguet et al., 1985; Glaze et al., 1992; von Gunten, 2003). Another alternative is the combination of ozone with UV irradiation or the combination of all three (Glaze et al., 1982, 1987; Legrini et al., 1993).

Ozone-AOP systems have recently been tested for the disinfection and detoxification of pollutants in different types of waters (Agustina et al., 2005; Ikehata and El-Din, 2005; Lanao et al., 2008). Ozonation has been used for NOM removal in a number of cases. Molnar et al. (2013) observed recently that ozone treatment results in partial oxidation of the fulvic acid fraction, increasing the amount of hydrophilic NOM. This finding is in agreement with Yan et al. (2007) and Treguer et al. (2010), who reported that an elevated ozone dose makes the NOM more hydrophilic and lower in molecular weight, and decreases DOC content in the hydrophobic fraction. Adding H_2O_2 into the ozonation process enhances the reduction of DOC compared to the ozone alone according to Irabelli et al. (2008). This research demonstrated that a postozonation O_3/H_2O_2 system achieves better removal of DOC than a preozonation O_3/H_2O_2 system. Yet, total THM concentrations were higher in the O_3/H_2O_2 samples than in ozonated samples in both cases (Irabelli et al., 2008). Similarly, Mosteo et al. (2009) did not observe any significant gains in THM reduction after adding H_2O_2 or TiO_2 to the ozone treatment. Tubic et al. (2011) concluded that a higher ratio of H_2O_2 (2:1 H_2O_2:O_3) is less favorable for NOM removal than a lower one (1:2 H_2O_2:O_3). Yet, the simultaneous addition of H_2O_2 and TiO_2 to ozonation resulted in lower THM production. The influence of ozonation on NOM fraction properties and weight distribution has been the subject of several recent studies (Bose and Reckhow, 2007; Lamsal et al., 2011; Chen et al., 2011; Molnar et al., 2012b, 2013).

Combined O_3-UV treatment results in significant mineralization of DOC, reduction of trihalomethane formation potential (THMFP), and haloacetic acid formation potential (HAAFP), according to Chin and Berube (2005). In addition, photocatalytic ozonation, $O_3/UV/TiO_2$, has been used to remove recalcitrant organic compounds from water (Giri et al., 2008) and appears to degrade NOM much more efficiently than sole ozonation or photocatalysis (Zhang and Jian, 2006). Furthermore, O_3 alone

Table 8.1 Ozone based AOPs target compounds matrix.

Target compounds	Matrix	AOP features	Parameters monitored	Summary of results	References
O_3/H_2O_2, O_3/UV, and $O_3/H_2O_2/UV$					
THM precursors	Synthetic humic solution (HA and FA)	H_2O_2/O_3 ratios 0.088, 0.032, 0.04	TOC, COD	The best NOM removal was achieved with 0.04 H_2O_2/O_3 at 20 min, removing 78% and 50% of TOC and COD, respectively. COD removal was enhanced with time. The best result achieved was with H_2O_2/O3 0.088 at 30 min.	Alsheyab and Muñoz (2006)
Probe compound pCBA	Two raw waters and simulated groundwater	$H_2O_2:O_3$ ratios 1:2 and 1:1		Addition of peroxide to the ozone process resulted in faster formation of %OH radicals, but did not appear to significantly increase the ultimate %OH radical exposure.	Rosenfeldt et al. (2006)
THM precursors	Raw water	$H_2O_2:O_3$ ratios in the range 0.0–0.35	DOC, THMs	The removal of DOC from raw water by peroxone was more effective than by ozone alone (61% and 53%, respectively), with the postozonation peroxone system achieving the greatest removal. Yet, THM concentration was higher in the peroxone samples	Irabelli et al. (2008)
THM precursors	Synthetic HA solution (Fluka HA)	Ozone 3 mg O_3 L^{-1}, suspended Degussa TiO_2 1 mg L^{-1} with H_2O_2 1.5 mg L^{-1}	DOC, THMFP, THMs	O_3/TiO_2 did not improve THM reduction, due to the working pH 8, since at higher pH an inhibition takes place. Simultaneous addition of H_2O_2 and TiO_2 produced a synergic effect: Lower production of THMs.	Mosteo et al. (2009)
	Groundwater	O_3 and O_3/TiO_2	DOC	O_3/TiO_2 catalyzed process resulted in significantly more effective reduction. The nature of residual NOM is key to DBP formation. Ozonation leads to	Molnar et al. (2012b); Molnar

				partial oxidation of the fulvic acid fraction and increases hydrophilic NOM fractions.	
HAAs (DCAA and TCAA)	DCAA and TCAA solutions	Ozone 0.3 mg min^{-1}/ LP UV at 254 nm with addition of H_2O_2 2.5 mgL^{-1}	HAAs decomposition	Within 30 min reaction time, TCAA was decomposed >50% and DCAA >90%. Humic acids (conc. 1.2 $mgTOCL^{-1}$) decreased the decomposition rates of DCAA and TCAA by 41% and 24%, respectively. $O_3/H_2O_2/UV$ was not remarkably more efficient than O_3/UV in HAA decomposition.	Wang et al. (2009)
Bromate	Simulated natural waters	LP UV at 254 nm	Bromate formation	Bromate formation and HA removal by ozone via O_3/UV with ozone generation by UV lamp and silent electric discharge. The concentrations of bromate formed were lower with O_3/UV than with ozone alone, once HA concentrations reached a certain level.	Zhao et al. (2013)
O_3/catalyst					
DBP precursors	Natural river water after coagulation and sand filtration	O_3/several metal oxides	DOC, THMFP	Catalytic ozonation with metal oxides significantly reduced THMFP compared to ozonation.	Zhang and Jian (2006)

COD, chemical oxygen demand; *pCBA*, parachlorobenzoic acid; *DBP*, disinfection by-products; *DCAA*, dichloroacetic acid; *DOC*, dissolved organic carbon; *FA*, fulvic acid; *HAA*, haloacetic acid; *HA*, humic acid; *LP*, low pressure; *NOM*, natural organic matter; *TCAA*, trichloroacetic acid; *THM*, trihalomethane; *THMFP*, trihalomethane formation potential; *TOC*, total organic carbon.

was found to be less efficient than O_3/TiO_2 in NOM oxidation (Molnar et al., 2012a). Ozone-based AOPs are summarized in Table 8.1.

8.3 UV light-based applications

UV photolysis is widely used for disinfection purposes. It involves the interaction of artificial or natural light with the target molecules and compounds, which then induce a series of photochemical reactions (Gjessing and Kallqvist, 1991; Frimmel, 1994, 1998). Direct photolysis is dependent on the ability of the compounds to absorb the emitted light.

Its efficiency is enhanced when irradiation is combined with hydrogen peroxide, whose photocatalytic dissociation yields •OH radicals (Legrini et al., 1993; Rosenfeldt et al., 2006), thus facilitating the degradation processes of different contaminants in water and water disinfection (Alkan et al., 2007; Chen et al., 2007; Kruithof et al., 2007; Hu et al., 2008; Li et al., 2008; Swaim et al., 2008). Under VUV radiation, water molecules dissociate into hydrogen atoms, hydroxyl radicals, and hydrated electrons (Kutschera et al., 2009). The degradation of target compounds with UV/ H_2O_2 and VUV occurs several times more efficiently than with direct UV photolysis, as observed by Dobrovic et al. (2007), Bond et al. (2009), and Imoberdorf and Mohseni (2011).

Although it has been proved that UV photolysis alone changes the chemical and biological properties of NOM, e.g., MWD (Frimmel, 1998), NOM reduction has been observed only with rather high UV doses (Goslan et al., 2006; Huang et al., 2008). The combination of UV irradiation at 254 nm and H_2O_2 treatment (with an oxidation potential of 1.8 eV) has been shown to enhance NOM reduction and promote •OH radical formation (Goslan et al., 2006; Toor and Mohseni, 2007).

UV/H_2O_2 treatment combines direct photolysis and advanced oxidation through indirect photolysis (Pereira et al., 2007; Sindelar et al., 2014). Differences in the spectrum of medium-pressure and low-pressure lamps can affect chemical bonds in various ways, generating different degradation products (Metz et al., 2011). It has been concluded, however (Sarathy and Mohseni, 2007), that NOM is not removed under the fluence and H_2O_2 concentrations typically applied in drinking water treatment applications (the standard UV dose for disinfection is about 400 mJ cm^{-2}; Chin and Berube, 2005), although a recent study observed changes in NOM structure with economically feasible UV fluences and H_2O_2 concentrations (Sarathy et al., 2011). Yet, other research demonstrated >60% reduction of

NOM with a UV dose of 540 mJ cm^{-2} and a H_2O_2 dose of 6 mg L^{-1} (Kruithof et al., 2007). UV/H_2O_2 treatment has been found to decrease molecular size (Sarathy and Mohseni, 2009), create more biodegradable decomposition products (Thomson et al., 2004; Sarathy et al., 2011), reduce NOM aromaticity (Thomson et al., 2004; Kleiser and Frimmel, 2000; Sarathy et al., 2011), and decrease hydrophobicity (Sarathy and Mohseni, 2010; Lin and Wang, 2011). Moreover, significant changes in NOM structural characteristics have been noted even at lower fluences (Wang et al., 2006; Sanly et al., 2007; Sarathy and Mohseni, 2007). DOC degradation has been rather low, however (Sanly et al., 2007). Taken together, these results indicate that a combination of high UV fluence and high peroxide concentration is required to generate enough •OH radicals to significantly reduce NOM and DBPFP (Tuhkanen, 2004; Sarathy and Mohseni, 2007; Toor and Mohseni, 2007). A pilot investigation into seasonal variations concluded that some NOM mineralization (2%–7%) occurred. It is important to note that proper pretreatment before a UV/H_2O_2 process increases the NOM mineralization rate (Sarathy and Mohseni, 2010). Suitable pretreatment methods include coagulation, sand filtration, and membrane technology, which increase UV transmittance and remove OH scavengers.

The UV fluence and oxidant dose needed to reduce UV_{254} and DOC in water containing high levels of NOM was tested for three oxidants, namely hydrogen peroxide, percarbonate, and perborate. Sindelar et al. (2014) concluded that although H_2O_2 is the best chemical oxidant to use in combination with H_2O_2, percarbonate and perborate are rather comparable in efficiency and offer other potential benefits over H_2O_2. Advances in UV light efficiency may make UV/H_2O_2 AOPs feasible for natural water applications in the future, however (Sindelar et al., 2014).

It is noteworthy that excess H_2O_2 can play the role of •OH radical scavenger, thus making the process less effective (Tuhkanen, 2004; Wang et al., 2006; Rosenfeldt and Linden, 2007; Song et al., 2008). Wang et al. (2006) observed that H_2O_2 at a concentration of 0.0082 M could absorb most of the UVC energy (200–280 nm), and dissociate into •OH radicals. They suggested that the optimal H_2O_2 concentration for HA degradation would be between 0.0032 and 0.0163 M. For surface waters with high DOC, even higher concentrations (~0.032 M) would be necessary.

AOPs can also be used as a pretreatment to enhance the biodegradability of NOM if the resulting intermediates are easily degradable in a further biological treatment. The more biodegradable compounds are readily utilized as substrates by microorganisms in biologically activated carbon (BAC) filtration (Toor and Mohseni, 2007). In combined UV/H_2O_2 AOP-BAC treatment, significant reductions of DBPs, TOC, and UV_{254} have been observed compared to AOP treatment without BAC (Toor and Mohseni, 2007). Similarly, photocatalysis combined with activated carbon (AC) filtration has been found to greatly improve removal efficiencies (measured as $color_{436}$ and UV_{254}) compared to photocatalysis results without AC. A combination of ozonation, photocatalytic oxidation, and AC adsorption obtained complete decolorization (Uyguner et al., 2007b).

In VUV treatment, highly oxidizing hydroxyl radicals are generated by the photolysis of water at wavelengths lower than 190 nm (Gonzalez et al., 2004).

VUV oxidation has been found highly efficient in NOM reduction (Thomson et al., 2002; Imoberdorf and Mohseni, 2011). NOM surrogates were degraded very efficiently when measured as DOC with high fluent dose (47–48J cm^{-2}, Bond et al., 2009). At lower doses (21J cm^{-2}); however, incomplete mineralization led to an increase in HAA levels, which was particularly dramatic in the case of two amino acids (Bond et al., 2009). These findings are in accordance with results obtained by others concerning THMFP (Buchanan et al., 2006) and UV/H_2O_2 treatment (Toor and Mohseni, 2007). Raw water NOM was found to decline with increased VUV dosage (0–128J cm^{-2}), resulting in total THMFP reduction (Buchanan et al., 2006). Despite this, and although bromoform accounted for only a small fraction of the total THMFP, the speciation of DBPs shifted to more brominated compounds, and the formation of bromoform was observed in the hydrophilic fraction of NOM (Buchanan et al., 2006). The UV/H_2O_2 treatment has also been noted to increase the formation potential of nitroso-DBPs (Zhao et al., 2008). Furthermore, inorganic compounds also play a role in the radiation propagation during VUV process.

The formation of undesirable by-products presents a challenge to the application of UV in general and VUV in particular. The main by-product is nitrite, which is subject to European Union regulation (EU Drinking Water Standard, maximum concentration 0.5 mg L^{-1}). The formation of nitrite can be avoided by the addition of ozone when using UV/VUV (Kutschera et al., 2009). Nitrite formation during the irradiation of natural water has not been observed in all cases. One investigation found, however, that rapid formation of nitrite occurs in nitrate-spiked water. The concentration significantly decreases after VUV radiation of 25 J cm^{-2}

Table 8.2 UV light based AOPs target.

Target compounds	Matrix	AOP features	Parameters monitored	Summary of results	References
UV/H_2O_2					
NOM	Raw water	LP UV at 254 nm	DOC, UV_{254}, HPSEC	At a UVC dose of 22 J cm^{-1}, reduction of NOM as UV_{254} and DOC were 94% and 78%, respectively. DOC removal was increased with increasing UVC dosage up to 37 J cm^{-1} while UV_{254} reduction did not change significantly at doses higher than 22 J cm^{-1}.	Goslan et al. (2006)
Probe compound pCBA	Two raw waters and simulated groundwater	LP UV and MP UV/ H_2O_2 2, 10, and 50 mg L^{-1}	UV_{254}, specific compound	H_2O_2 was the most energy-intensive portion of the process. The MP UV systems were consistently less efficient for •OH radical production than LP UV systems.	Rosenfeldt et al. (2006)
NOM	Raw water and Synthetic HA (Aldrich HA)	HP UV/ H_2O_2 0–32.4 mM	NPDOC, UV_{254}, IC, FTIR, THMFP	At an irradiation time of 30 mm, 90% of NPDOC was mineralized (0.01% H_2O_2), THMFP was reduced, although a shift from chlorine to bromine THMs was noted.	Wang et al. (2006)

Continued

Table 8.2 UV light based AOPs target.—cont'd

Target compounds	Matrix	AOP features	Parameters monitored	Summary of results	References
HA	Synthetic HA solution (Fluka HA)	UVA with maxi mum centered at 365 nm	DOC, UV_{254}, HPSEC, fractionation	UVA/H_2O_2 treatment exhibited preferential degradation of organic compounds with HMM (1.5 −10 kDa). There was a significant shift of average MWD to the lower range, although only 33% of DOC was removed after 1 h irradiation.	Sanly et al. (2007)
NOM	Raw water	LP UV	TOC, UV_{254}, HPSEC	Under AOP conditions typically applied in drinking water treatment applications (500 mJ cm^{-2}, 4 mgL^{-1} H_2O_2), NOM was not mineralized but was partially oxidized, resulting in significant reduction in aromaticity.	Sarathy and Mohseni (2007)
DBP precursors		LP UV at 254 nm	TOC, UV_{254}, UV_{203}, ADOC and BDOC, AOC, DBPFP	AOP was only effective at reducing DBPs at UV fluences >1000 mJ cm^{-1} and initial H_2O_2 concentrations of about or greater than 23 mgL^{-1}. Combined AOP-BAC treatment showed significant reductions.	Toor and Mohseni (2007)
	Groundwater and NOM-free water	LP UV at 254 nm	UV_{254} and UV_{310}, TOC,	NOM decomposition kinetics was substantially enhanced with	Song et al. (2008)

NOM and synthetic organic compounds			color, specific compounds	oxidation over time. Progressive decrease in•OH radical scavengers and UV adsorbing species may increase NOM decomposition rates.	
DBP precursors (especially nitrosamines)	Source waters from several locations	MP UV	TOC, UV_{254}, color	AOP oxidation treatment may be a source of nitroso-DBP precursors.	Zhao et al. (2008)
Model compounds (analogues of LMM and hydrophilic NOM)		N-lamp 254 nm	DOC, BDOC, HAAFP, specific compounds	DOC removal was 91% with a high fluence of 47–48 J cm^{-2} and 78% with UVC dose of 21 Jcm^{-2}. Oxidation degraded both hydrophilic and hydrophobic surrogates. Partial biodegradation increased HAAFP, especially DCAAFP.	Bond et al. (2009)
NOM and synthetic organic compounds	Surface water	LP UV at 254 nm	BDOC,TOC, SUVA	Partial mineralization of NOM occurred. This led to a reduction in aromatic structures and the degree of substitution of aromatic species, and to BDOC formation, including formaldehyde and acetaldehyde	Sarathy and Mohseni (2010, 2011)
DBP precursors		LP UV and MP UV	SUVA, TOC, THM	Minor mineralization of NOM.	Metz et al. (2011)

Continued

Table 8.2 UV light based AOPs target.—cont'd

Target compounds	Matrix	AOP features	Parameters monitored	Summary of results	References
VUV					
NOM	Raw surface water	VUV at 254 and 185 nm	DOC, BDOC, fractionation, THMFP, HAAFP	THMFP increased at VUV doses <40J cm^{-2} but reduced higher fluences. HAAFP reduced by VUV irradiation. Concentrations of individual THM and HAA decreased with increasing radiation, but bromoform increased linearly.	Buchanan et al. (2006)
		LP UV at 254 and 185 nm	UV254	NOM was reduced by 42% −46%.VUV photodegradation was over 10 times faster than basic UV.	Dobrovic et al. (2007)
LMM and hydrophilic NOM surrogates	Model compounds in solution referring to LMM and hydrophilic NOM	H-lamp 254 and 185 nm	DOC, BDOC, HAAFP, specific compounds	DOC removal was 97% with a UVC dose of 1″ IS Ian and 58% with 21 J cm^{-2}. Oxidation degraded both hydrophilic and hydrophobic surrogates. Partial oxidation and biodegradation increased HAAFP, especially DCAAFP.	Bond et al. (2009)
Taste and odor compounds (geosmin and 2-	Ultrapure and raw water	LP UV at 254 and 185 nm	DOC, UV_{254},	In raw water, DOC and SUVA decreased by 6% and 18%, respectively. NOM acted as a	Kutschera et al. (2009)

methyliso-borneol (MIB))			specific compound	scavenger for geosmin and MIB reduction, although they were oxidized quite well. Nitrite formed as a by-product.	
NOM	Surface water	VUV Hg lamp	TOC, HPSEC	VUV was much more efficient than UV/H_2O_2. Carbonates and bicarbonates were important in radiation propagation.	Imoberdorf and Mohseni (2011)

ADOC, adsorbable dissolved organic carbon; *AOC*, assimilable organic carbon; *BDOC*, biodegradable dissolved organic carbon; *pCBA*, parachlorobenzoic acid; *DCAA*, dichloroacetic acid; *DOC*, dissolved organic carbon; *FTIR*, Fourier transform infrared spectroscopy; *HA*, humic acid; *HAAFP*, haloacetic acid formation potential; *HPSEC*, high pressure size exclusion chromatography; *IC*, ion chromatography; *LMM*, low molecular mass; *LP*, low pressure; *MP*, medium pressure; *MWD*, molecular weight distribution; *NOM*, natural organic matter; *NPDOC*, nonpurgeable dissolved organic carbon; *THM*, trihalomethane; *THMFP*, trihalomethane formation potential; *TOC*, total organic carbon.

(Buchanan et al., 2006). UV light based AOPs are summarized in Table 8.2.

8.4 Fenton processes

The conventional Fenton process involves the use of an oxidizing agent (usually H_2O_2) and a catalyst (metal salt or oxide, usually iron). Fenton's reagent (H_2O_2 with Fe(II)/Fe(III) ions) in water produces •OH radicals and peroxy radicals (Voelker and Sulzberger, 1996; Park and Yoon, 2007). The photo-Fenton process also involves irradiation with sunlight or an artificial light source, which leads to the production of additional •OH radicals, thus increasing the rate of organic compound or contaminant degradation (Zepp et al., 1992; Fukushima et al., 2001; Moncayo-Lasso et al., 2009). A considerable body of research has been conducted on the application of electro-Fenton and photoelectro-Fenton processes, especially in wastewater and contaminated soil treatment (Jiang and Zhang, 2007). Compared to the conventional methods, the electro-Fenton process produces H_2O_2 electrochemically via oxygen reduction on the cathode surface. Further reactions occur between H_2O_2 and Fe^{2+}, as in the conventional technique.

With UV/Fe(III) treatment, HMM compounds were formed as an agglomeration of the organic molecules by Fe, with a significant level of LMM compounds remaining even after a long irradiation time (Sanly et al., 2007). Thus, the UV/Fe(III) process was ineffective unless H_2O_2 was added, resulting in more efficient NOM removal across the whole range of MM. This, in turn, suggests that the UV/Fe(III)/H_2O_2 system has the potential to remove organics with a broad range of MMs (Sanly et al., 2007; Moncayo-Lasso et al., 2009). Even though high NOM removal rates were detected, the remaining organic compounds appear fairly reactive to chlorine, thus resulting in low reduction of THMFP. The presence of iron may have been responsible for the elevated THMFP values (Sanly et al., 2007). Nevertheless, efficient DBPFP reduction has been detected with an irradiation time of over 90 min (Nie et al., 2010). While the optimal pH for Fenton processes has been confirmed at about pH 4 in several cases (Park and Yoon, 2007; Sanly et al., 2007; Moncayo-Lasso et al., 2008), 254 nm absorption decay and TOC reduction is also most effective in the pH range of 3–7 (Goslan et al., 2006). This is more feasible and cost effective for full-scale treatment applications, since no pH adjustment would be necessary.

Solar light irradiation has gained advocates for environmental and economic reasons (Malato et al., 2007, 2009; Moncayo-Lasso et al., 2008).

Solar light has been used in photo-Fenton processes, e.g., to inactivate bacteria, by Moncayo-Lasso et al. (2009), who demonstrated that NOM can be partially degraded at natural water pH by solar photo-Fenton. A significant limitation to the application of Fenton and photo-Fenton processes has been the low optimal pH (<3), which most often requires acidification and neutralization steps. Fenton based AOPs are summarized in Table 8.3.

8.5 Heterogeneous photocatalysis and catalytic oxidation

Heterogeneous photocatalysis employs several different semiconductors such as TiO_2, ZnO, ZnS, WO_3, and $SrTiO_3$ (Hoffmann et al., 1995; Pirkanniemi and Sillanpaa, 2002).

TiO_2 particles are the most commonly used photocatalysts, due to their high photocatalytic activity and environmentally benign properties, though several others have been investigated. TiO_2 in anatase form is stable, efficient, inexpensive, and nontoxic. UV irradiation of the TiO_2 induces the excitement of electrons from the valence band to the conduction band, leading to the creation of highly oxidative holes on the valence band and the formation of radicals, in particular •OH radicals. Organic compounds are degraded by holes in the TiO_2 surface and by radicals in the bulk solution (Hoffmann et al., 1995; Carp et al., 2004). The parameters affecting the oxidation are catalyst concentration, light wavelength and intensity, water matrix, solution pH, and the addition of H_2O_2 as an extra oxidant to promote reactions (Hand et al., 1995; Bekbolet and Ozkosemen, 1996; Eggins et al., 1997). Uyguner-Demirel and Bekbolet (2011) have recently reviewed the literature on the photocatalytic degradation of NOM. The major advantage of TiO_2 photocatalysis is optical adsorption in the near UV region in comparison to UVC-driven AOPs (Uyguner-Demirel and Bekbolet, 2011).

Photocatalytic oxidation has been proposed as a good candidate for NOM degradation, especially for the removal of DBPs and their precursors (Valencia et al., 2014). This is due to the three-way removal of compounds: oxidation by •OH radicals, reductive dechlorination by superoxide radicals, and physical adsorption by TiO_2 (Hand et al., 1995; Bekbolet and Ozkosemen, 1996; Eggins et al., 1997; Wiszniowski et al., 2002; Gerrity et al., 2009). Adsorption has been found to play a prominent role in the photocatalytic degradation of NOM, especially under acidic pH conditions

Table 8.3 Fenton based AOPs.

Target compounds	Matrix	AOP features	Parameters monitored	Summary of results	References
Fenton andphoto-Fenton					
NOM	Raw water	Fenton's reagent/ LP UV at 254 nm	DOC, UV_{254}, HPSEC	Efficient NOM removal at pH 3–7. DOC reduction after 20 min irradiation 82 (UV_{254}) and 96% (DOC) by Fenton reaction (FR), and after 1 min at 31 (UV254) and 88% (DOC) by photo-FR (PFR), respectively. The DOC reduction was much faster for PFR. UV_{254} reduction was faster for FR. H_2O_2 and Fe^2+ interfered with the UV254 measurement.	Goslan et al. (2006)
HA	Synthetic HA solution (Aldrich HA)	Fenton's reagent/LP UV, mineral particles	DOC, UV-HA, THMFP, resin fractionation, oxidation by-products	Modified-Fe-ions improved the H_2O_2 degradation rate in Fenton and photo-Fenton processes. Adding glass improved THM removal and decomposition rate of H_2O_2.	Park and Yoon (2007)
	Synthetic HA solution (Fluka HA)	Fenton's reagent/UVA with maximum centered at 365 nm	DOC, UV_{254}, HPSEC, fractionation	More than 80% and 90% DOC and UV_{254} removal, respectively, over 2.5 h. Addition of H_2O_2 enhanced removal kinetics. The CHA and NEU fractions increased and were hardest to remove.	Sanly et al. (2007)
NOM	Natural source water	Fenton's reagent	DOC, UV_{254}, fractionation, HPSEC	Potential method for combined oxidation of any formed DBPs, poor removal of recalcitrant NOM.	Jarvis et al. (2008)

HA	Synthetic HA solution (Aldrich HA)	Fenton's reagent/UV	TOC, UV_{254}, HPSEC, HA concentration	Absorbance of HA disappeared after 8 h irradiation and TOC reduced 80% after 15 h. HMM compounds were preferentially removed.	Katsumata et al. (2008)
NOM, and especially DBP precursors	Natural river water before and after sand filtration, river water spiked with NOM model compounds	Fe^{3+} (1.0 mgL^{-1}) H_2O_2/ sunlight	TOC, specific compounds, UV_{254}	Light/H_2O_2: UV_{254} reduction higher (c. 68%, 80%) than TOC reduction (c. 55%, 62%), before and after filtration, respectively. Addition of $^{-}$1 mgL^{-1} Fe^{3+} significantly increased UV_{254} and TOC removal (70%–90%). Prechlorination filtration led to almost total NOM removal, and remaining TOC was suitable to prevent THM formation.	Moncayo Lasso et al. (2008)
Fenton and photo-Fenton					
Organic probes (benzoate and methanol)	Synthetic solutions	Fenton's reagent/ $PW_{12}O_{40}^{3-}$	Specific compound	Novel homogenous Fentonlike system capable of oxidizing recalcitrant organic compounds at neutral pH.	Lee and Sedlak (2009)
NOM and *Escherichia coli*	Natural river water	Fenton's reagent/ sunlight	TOC, *E. coli* inactivation	Addition of low conc. of Fe^{3+} in the presence of H_2O_2 at near-neutral pH using CPC solar reactor led to simultaneous	Moncayo Lasso et al. (2009)

Continued

Table 8.3 Fenton based AOPs.—cont'd

Target compounds	Matrix	AOP features	Parameters monitored	Summary of results	References
	after sand filtration			decrease of TOC (55%) and total *E. coliE. coli* inactivation (10^5–10^6 CFUmL^{-1}).	
HA	Synthetic HA solution (Aldrich HA)	Iron oxides/ Fe^0/H_2O_2/ HP-UV	TOC, FTIR, DBPFP, fractionation, MWD by UF	The core–shell structure of nanoparticles of iron oxides coated on Fe^0 (nIOCl) produced a highly efficient photo-Fenton catalyst for HA degradation. After 90 min irradiation, about 62% of DOC was removed and DBPFP almost disappeared.	Nie et al. (2010)
NOM	Surface water	Solar photo-Fenton	TOC, THM	Solar photo-Fenton was effective for the transformation and partial mineralization of NOM.	Moncayo-Lasso et al. (2012)

CFU, colony forming units; *CHA*, hydrophilic charged; *DBP*, disinfection by-products; *DOC*, dissolved organic carbon; *FTIR*, Fourier transform infrared spectroscopy; *HA*, humic acid; *HPSEC*, high pressure size exclusion chromatography; *HMM*, high molecular mass; *MWD*, molecular weight distribution; *NOM*, natural organic matter; *PEC*, photoelectrocatalysis; *THM*, trihalomethane; *THMFP*, trihalomethane formation potential; *TOC*, total organic carbon; *UF*, ultrafiltration.

when carboxylate surface groups of NOM are adsorbed onto TiO_2 (Wiszniowski et al., 2002; Liu et al., 2008a). This adsorption of NOM onto the TiO_2 surface seems to occur rapidly (Huang et al., 2008) leading to the conclusion that hydrophobic NOM is preferably adsorbed onto TiO_2 surface, and more hydrophilic intermediate oxidation products are then released. In photocatalytic NOM removal, pH plays a central role, determining the mechanism but not the sequence of photocatalytic degradation (Valencia et al., 2012a,b). Copper greatly inhibits the larger molecular weight fraction of NOM in particular and reduces the formation of formic and oxalic acids (Tercero Espinoza et al., 2011). Several anions lessened photocatalytic NOM removal at elevated concentrations (Uyguner-Demirel et al., 2013). Reductive dechlorination of, e.g., THMs can also be a beneficial use of photocatalytic oxidation in full-scale water treatment at targeted locations (Gerrity et al., 2009). Though TiO_2-based photocatalytic oxidation has significant potential for NOM removal, this process will only become economically viable if its energy requirements are significantly reduced (Philippe et al., 2010). Uyguner-Demirel and Bekbolet (2011) concluded that TiO_2 based photocatalytic oxidation of NOM diminishes DBP formation capacity.

Photocatalytic processes can be conducted by using a slurry of fine particles of the semiconductor. Nevertheless, the use of catalyst in slurry form requires an additional treatment step to remove it from the treated effluent, which can make the process expensive (Murray and Parsons, 2006; Klavarioti et al., 2009). TiO_2 slurry can be separated through a microfiltration (Fu et al., 2006b; Choo et al., 2008; Benotti et al., 2009) or ultrafiltration membrane (Fu et al., 2006a). The powder catalyst may penetrate into the pores of the membrane, however, which in turn reduces the TiO_2 concentration and blocks membrane pores. TiO_2/Al_2O_3 microspheres consisting of nanosized catalyst crystals could present one solution to this problem (Xu et al., 2009).

Another alternative to filtration is to immobilize the catalyst on a suitable support matrix. This, however, can reduce the surface area available for reactions compared to suspended systems. To make the process viable, the catalyst therefore needs to be immobilized on a high surface area material. Different coating matrices have been studied, including nonwoven webs (Gumy et al., 2006), nonwoven porous fabrics such as silica-supported titania (Navalon et al., 2009), and magnetic core matrices with silica layer (Chen et al., 2009). Durable supported films can be obtained in various ways, for example by the sol-gel process (Murray and Parsons, 2006; Chang et al., 2009) and the atomic layer deposition (ALD) technique (Vilhunen et al., 2009). TiO_2 was synthesized by the sol-gel method coupled with a solvothermal technique for photocatalytical NOM degradation (Valencia et al., 2012a,b). The resulting differences in crystallographic and morphological properties did not affect the selectivity of NOM photodegradation, but the synthesized material had higher efficiency than Degussa P-25.

The aromaticity of the organic matter is reduced after limited photocatalysis, as indicated, e.g., by lowered specific UV absorbance (SUVA) values. The applied energy values, however, are insufficient to demonstrate organic mineralization (Gerrity et al., 2009; Liu et al., 2010). Higher TiO_2 concentrations have achieved greater DOC removal rates, but also a possible increase in THMFP (Fu et al., 2006a; Gerrity et al., 2009). Beyond a certain level, additional TiO_2 particles cannot be involved because of increased turbidity and thus decreased UV light penetration. Extended photocatalysis (>10 kWh m^{-3}) has yielded high DOC removal and mineralization, and nearly complete reduction of UV_{254} and SUVA values (Gerrity et al., 2009). Even further extended illumination (160 kWh m^{-3}) significantly reduced THMFP, although the energy requirements of this experiment make it infeasible for treatment plant application (Gerrity et al., 2009).

Renewable energy sources such as sunlight can aid photocatalysis. Solar irradiation has been observed to change the characteristics on NOM (Frimmel, 1998; Chow et al., 2004), resulting in several studies of the use of natural sunlight irradiation for water treatment (Malato et al., 2007, 2009). Sunlight can provide a low-cost, sustainable, and simple method of treating drinking waters (Ljubas, 2005), especially in developing countries with consistently sunny climates. Solar photocatalysis has been proposed mainly for water disinfection and detoxification (Malato et al., 2007, 2009; Rincon and Pulgarin, 2007). Tercero Espinoza et al. (2009a, 2009b) have used a solar UV simulator for photocatalytic research. The disadvantage of solar light is that only UVA irradiation can be used for catalyst photoactivation, and the solar irradiation reaching the surface of the earth contains only 3%–5% UVA (Ljubas, 2005). Thus, there is a great incentive to extend the absorbance wavelength range of TiO_2 to the visible region without losing photocatalytic activity. This has been done by doping TiO_2 with nonmetallic elements such as nitrogen, sulfur, boron, carbon, and selenium (Comninellis et al., 2008; Malato et al., 2009) or by noble metals like Ag, Pt, Ni, Cu, and Pd (Chen et al., 2009; Kumar et al., 2009; Malato et al., 2009).

Target compounds	Matrix	AOP features	Parameters monitored	Summary of results	References
UV/TiO$_2$					
FA	Synthetic FA solution (Beijing BioChem)	Suspended Degussa TiO_2/LP UV	TOC, UV_{254}	TOC removal improved with low pH, increased additive concentration ($K_2S_2O_8$) and relatively high light intensity.	Fu et al. (2006a)
NOM	Natural raw water and Suwannee river NOM (SRNOM)	LP-UV/suspended Degussa TiO_2	TOC, HPSEC	Operating values are feedwater dependent. With low-NOM water requires a pH 4–5 and low TiO_2 concentration (0.1 g L−1).	Le-Clech et al. (2006)
DBP precursors	Synthetic HA solution (Aldrich HA)	UVA at 365 nm/ thin-film- coated TiO_2 slides	DOC, UV_{254}, SUVA, HPSEC	UV_{254} reduced c. 48%. Immobilized catalysts need a larger surface for removal in a slurry reactor.	Murray and Parsons (2006)
NOM	Raw water	Suspended Degussa TiO_2/UV	DOC, UV_{254}, UV_{280}, THMFP, fluorescence	Photocatalytic oxidation after enhanced coagulation increased NOM degradation by $<$ 5%–32% measured as DOC. UV_{254} and SUVA decreased c.10%–33% and 15%–47%, respectively. Differences depended on the related properties of NOM.	Uyguner et al. (2007a)
NOM	Natural lake water and synthetic HA water (Aldrich HA)	Degussa TiO2/8-W UV-lamp/MF membrane/ ferrihydrite particles	TOC, UV_{254}	Addition of ferrihydrite enhanced HA removal in synthetic water, but lake water NOM with low SUVA and high pH values was adversely affected. Photocatalytic degradation controlled membrane fouling well.	Choo et al. (2008)

Continued

Table 8.4 Catalytic studies.—cont'd

Target compounds	Matrix	AOP features	Parameters monitored	Summary of results	References
	Suwannee river NOM (SRNOM)	Suspended Degussa TiO_2/LP UV at 254 nm	TOC, UV_{254}, HPSEC	At 40 min reaction time, DOC and UV_{254} were reduced by 50% and 85%, respectively. At 120 min reaction time, more than 80% of NOM as TOC was removed.	Huang et al. (2008)
HA	Synthetic HA solution (Fluka HA)	Suspended Degussa TiO_2/UVA with maximum centered at 365 nm	DOC, UV254, HPSEC, resin fractionation, THMFP, zeta potential analysis	After 150 min, UV_{254} reduced by over 90%, DOC reduced by only c. 80%. THMFP also lowered, and H_2O_2 addition enhanced the reaction rate. HPSEC showed that AMW shifted toward LMM compounds. Hydrophobic fraction removed efficiently, NEU fraction most difficult to remove.	Liu et al. (2008a)
NOM	Surface raw water	Suspended Degussa TiO_2/UVA with maximum centered at 365 nm	DOC, UV_{254}, HPSEC, resin fractionation, THMFP, HAAs, aldehydes, and ketones formation	Higher UV254 reduction than DOC after 120 min (over 90 and c. 70%, respectively). H_2O_2 addition enhanced initial kinetics of the degradation reactions, but residual concentrations of DOC remained similar. LMM (200–1000 Da) matter remained (hydrophilic charged or neutral molecules). THMFP and HAAFP reduced significantly.	Liu et al. (2008b)
NOM, model	Lake water and model			No formation of BrO_3^- and THMs in DOC containing water	Tercero Espinoza

compounds, DHBA, salicylic acid and bromide	compound solution	Suspended Degussa TiO_2/LP UV	DOC, IC, AOX, THM, DHBA concentration	was detected with up to 45 min of irradiation. With higher initial bromide concentrations, adsorbable organic Br increased.	and Frimmel (2008)
HA	Synthetic HA solution (Aldrich HA)	Suspended Degussa TiO_2/UV/MF	TOC	Photocatalytic oxidation and MF process effectively degraded HA, membrane fouling was well controlled.	Bai et al. (2009)
DBP precursors	Surface raw water	Degussa TiO_2 slurry/Photo-Cat Lab® reactor with LP UV bulbs	DOC, UV254, SUVA, TTHM, THMFP	Limited photocatalysis (<5 kWh m^{-3}) achieved reductions in DOC and UV254 (nearly 50%) but led to increased chlorine demand and THMFP. THMFP also started to decrease when energy consumption exceeded 20 kWh m^{-3}.	Gerrity et al. (2009)
NOM		Suspended Degussa TiO_2/UVA with maximum centered at 365 nm	DOC, UV_{254}, HPSEC, resin fractionation, THMFP	UV_{254} reduced over 95% after 150 min. DOC reduced only 80%–85%. HPSEC showed that AMW shifted toward LMM compounds. The hydrophilic fraction dominated the organic distribution, NEU fraction was the most difficult to remove.	Liu et al. (2010)
NOM model compound, DHBA, and bromide	Model compound and bromide solution	Suspended Degussa TiO_2/solar UV simulator	DOC, THMs, DHBA concentration	Considerable variability in the extent and time scale of brominated by-product formation depending on the raw water and design parameters.	Tercero Espinoza et al. (2009a)

Continued

Table 8.4 Catalytic studies.—cont'd

Target compounds	Matrix	AOP features	Parameters monitored	Summary of results	References
NOM, especially HA	Humic-rich lake water	Suspended Degussa TiO_2/ solar UV simulator	SEC-DOC/UV, IC, bacterial incubation	Sequential photocatalytic NOM degradation from the larger to LMM fractions. Bioavailable products formed, e.g., LMM organic acids.	Tercero Espinoza et al. (2009b)
HA	Synthetic HA solution	TiO2/Al2O3 microsphere/LP-UV/ultrafiltration	HA concentration, TMP	TiO_2/Al_2O_3 microspheres showed good photocatalytic activity, mechanical strength, and fouling control capacity.	Xu et al. (2009)
NOM	Natural river water and synthetic HA solution (Aldrich HA)	Degussa TiO_2/8-W UV-lamp/MF membrane/iron oxide particles (IOP)	TOC, UV254, FE-SEM, TMP	Additional IOP precoat layers on the membrane surface did not aggravate membrane fouling, adsorptive removal of photocatalyzed NOM enhanced DOC removal efficiently, especially for high DOC waters.	Yao et al. (2009)
HA	Synthetic HA solution (Aldrich HA)	TiO2 nanotube/ LP-UV/membrane filtration	TOC, UV436	Near complete mineralization (93%–98%) of HA. Nanotubes were totally separated and recovered with membrane filtration, and reused efficiently.	Zhang et al. (2009)
NOM	Surface water	Degussa TiO_2	SEC, DOC	Cu and some other cations inhibited photocatalytic degradation of NOM.	Tercero Espinoza et al. (2011)
		Solar UV-light simulator	SEC, AOX, THM, DOC	pH determines the mechanism, but not the sequence of photocatalytic degradation.	Valencia et al. (2012a,b)

HA	Prepared HA solution	Degussa TiO_2	DOC, UV_{254}, UV_{365}	Several anions at elevated concentrations inhibited photocatalytic NOM degradation.	Uyguner-Demirel et al. (2013)
O_3/UV/TiO_2					
NOM	Natural water, pretreated with coagulation —sand filtration	LP-UV/TiO_2 coated titanium sheet	TOC, UV_{254}, BDOC, SUVA	Combined process was much more effective than ozonation or photocatalysis. HMM compounds decreased faster than LMM compounds. Ozone dosage and retention time were the main parameters affecting removal.	Zhang and Jian (2006)
Photoelectrocatalysis					
FA	Synthetic FA solution (Bajrng BioChem)	Ti/TiO_2 electrode/LP-UV	TOC, SEM	Under optimal process parameters, 57% removal of FA was obtained.	Fu et al. (2007)
HA	Synthetic HA solution (Aldrich HA)	Thin-film TiO_2 photoelectrodes/ Xe–Hg arc lamp	TOC, UV_{254}	PEC improved photo-catalytic degradationof HA 2- to 4-fold, depending on system conditions.	Selcuk and Bekbolet (2008)

AC, activated carbon; *AOP*, advanced oxidation processes; *DOC*, dissolved organic carbon; *FA*, fulvic acid; *LP*, low pressure; *HA*, humic acid; *HP*, high pressure; *HPSEC*, high pressure size exclusion chromatography; *IC*, ion chromatography; *BDOC*, biodegradable dissolved organic carbon; *DHBA*, 2,4-dihydroxybenzoic acid; *z:NEU*, neutral; *TMP*, transmembrane pressure; *FE-SEM*, field emission scanning electron microscope; *PEC*, photoelectrocatalysis; *TOC*, total organic carbon; *NOM*, natural organic matter; *THM*, trihalomethane; *THMFP*, trihalomethane formation potential.

Aqueous H_2O_2 decomposes over heterogeneous catalysts including metals such as Fe, Cu, Pt, Ti, and Ni and metal oxides immobilized on various support materials including sand, silica, zeolites, and alumina (Pirkanniemi and Sillanpaa, 2002). As in Fenton reactions, the decomposition of peroxide through interactions with the surface sites of such catalysts results in the formation of strong oxidants including •OH radicals (Kitis and Kaplan, 2007). Contaminated soil and groundwater remediation has been attempted with various oxidants, such as sulfate radical-based AOPs, which use peroxymonosulfate and persulfate as precursor oxidants (Anopsitakis and Dionysiou, 2004). The application of a chelating agent noticeably improved the effectiveness of Fentonlike reactions (Lee and Sedlak, 2009; Rastogi et al., 2009).

Natural pumice has been coated with iron oxides and sand for catalytic oxidation (Kitis and Kaplan, 2007). An investigation into the influence of mineral particles added to the Fenton processes concluded that although bentonite enhanced NOM removal, THMFP also increased (Park and Yoon, 2007). The addition of glass improved both THM removal and the of H_2O_2 decomposition rate.

Catalytic ozonation can double the NOM removal efficiency achieved by ozonation alone (Legube and Vel Leitner, 1999; Kasprzyk-Hordern et al., 2005; Zhang and Jian, 2006), Catalytic ozonation with MnO_2 improved the elimination of NOM measured as TOC and chemical oxygen demand (COD) (Alsheyab and Munoz, 2007). Furthermore, catalytic wet peroxide oxidation, developed with Al/Fe-pillared clays, has proved to be an efficient Fentonlike method, achieving over 95% DOC removal in 4 h (Galeano et al., 2012). Zhou et al. (2014) also synthesized efficient TiO_2-graphene composites with different graphene weight addition ratios by a solvothermal process for the removal of FA under UV light. Catalytic studies are summarized in Table 8.4.

8.6 Ultrasound irradiation and e-beam irradiation

Ultrasonic (US) irradiation can also be deployed for NOM removal (Nagata et al., 1996). It is attractive because of its application simplicity. It has been shown that high-power ultrasound produces strong cavitation in aqueous solution, and the violent collapse of the cavitation bubble causes shock waves and reactive free radicals. These effects should contribute to both the physical disruption and inactivation of microbial structures and the decomposition of organic compounds (Naddeo et al., 2007). US also

Table 8.5 Ultrasonic studies.

Target compounds	Matrix	AOP features	Parameters monitored	Summary of results	References
NOM	Synthetic HA solution (Aldrich HA)	20 kHz ultrasonic probe/ membrane filtration	TOC, UV, FTIR	Ultrasound significantly reduced membrane fouling.	Chen et al. (2006)
		Horn sonotrode at 20 kHz	TOC, UV_{254}	20 min sonication at 42W cm^{-2} provided the best TOC removal.	Naddeo et al. (2007)
	Natural surface water	Horn sonotrode at 20 kHz, sonication at 800, 1400, or 1500W/MF membrane	Fouling	The shorter sonication produced smaller molecular NOM, and longer sonication may encourage aggregation or readier adsorption.	Tran et al. (2007)
HA	Synthetic HA solution (Aldrich HA)	Ultrasonic genera-tor of 24 kHz	TOC, DOC, SUVA, UV_{254}	US/O3 enhanced HA mineralization	Stepniak et al. (2012)

DOC, dissolved organic carbon; *FTIR*, Fourier transform infrared spectroscopy; *HA*, humic acid; *TOC*, total organic carbon.

appears to influence the chemical degradation of humic matter in the cavitation bubbles by oxidation and the physical aggregation of HA fragments (Naddeo et al., 2007).

The effect of NOM fouling on membrane filtration may be mitigated by ultrasound (Chen et al., 2006; Tran et al., 2007). A shorter sonication time appears to reduce flux decline and degrade NOM to a less adsorbable form, but this beneficial effect is diminished as the sonication time is extended.

To combat the low rate of sonochemical decomposition of aromatic compounds in aqueous solutions, the combined treatments, such as US/UV and US/O_3, have also been applied. The combination of US with other oxidation methods has resulted in higher rates of NOM oxidation and mineralization (Goncharuk et al., 2008). US and O_3 have proved more effective and energy efficient for HA mineralization in combination than alone (Stepniak et al., 2012). Higher US intensities improved the efficiency. US/O_3 is one of the more highly efficient O_3-based NOM removal methods (Ziylan and Ince, 2013).

HAs were removed by electron beam irradiation with the addition of persulfate (Roshani and Leitner, 2011).

The advantages of persulfate include high solubility, low cost, environmentally friendly end products, and high stability. Persulfate anions can be activated to generate sulfate radicals, one of the strongest oxidizing species (Roshani and Leitner, 2011). Yet aromatic structures limit the positive impact of persulfate on e-beam irradiation. US studies are summarized in Table 8.5.

8.7 Conclusions

Different raw waters contain variable amounts of NOM. The characteristics of NOM also vary with geological conditions and the surrounding watershed. An increase in the amount of NOM has been observed worldwide in the last few years. Seasonal variability in the same location and natural disasters such as floods or droughts present even more challenges to water purification facilities. New and improved methods are thus urgently needed to reduce NOM, hence DBPs, in drinking water.

Hydrophobic and HMM NOM compounds appear more amenable to oxidation with AOPs such as photocatalysis or UV/H_2O_2.The photo-Fenton process also removes LMM organics (Murray and Parsons, 2004), which are the most difficult to remove in traditional treatment. The efficiencies of AOPs for DOC removal are as follows: UV/ H_2O_2 < Fenton < photo-Fenton (Goslan et al., 2006; Jarvis et al., 2008). The addition of chelating agents further improves the efficiency of Fenton-like processes (Lee and Sedlak, 2009), as do various catalyst solutions such as iron oxides coated on Fe^0 core shell nanospheres (Nie et al., 2010).

Photocatalysis has been the subject of a great deal of research (Malato et al., 2009; Choi et al., 2010). Catalyst separation, recovery, and reuse are significant obstacles to the practical applications of TiO_2 powder heterogeneous photocatalysis, mainly due to the small particle size of the TiO_2. More advanced catalyst separation techniques have been developed, such as membrane filtration, although membrane fouling is a problem. Another alternative is to immobilize the catalyst on suitable support matrices. Different matrices and TiO_2 coating techniques such as ALD have emerged recently (Vilhunen et al., 2009; Chen et al., 2009).

Current advances in photocatalytic oxidation include nanotechnology (Choi et al., 2010), which uses, for example, titanium-derived 1D nanomaterials such as titanate nanotubes and nanowires (Doong and Chiang, 2008; Zhang et al., 2009) or high-strength TiO_2 fiber (Giri et al., 2008). The latter technique looks promising for overcoming the reduced efficiency of an immobilized TiO_2 catalyst.

TiO_2 doped with noble metals such as platinum significantly improves photocatalytic activity (Chen et al., 2009). Alternative semiconductors to TiO_2, such as ZnO (Akyol and Bayramoglu, 2010), have also been considered, although these are not generally considered to be as attractive for application as TiO_2.

Due to the high energy demand of conventional UV lamps, other sources of UV light appear increasingly attractive, and the VUV process has proved its efficiency. Sunlight is environmentally friendly and cost effective but less efficient (Ljubas, 2005; Malato et al., 2007, 2009). To increase efficiency, the absorbance wavelength range of TiO_2 can be extended by doping it with nonmetallic elements such as nitrogen, sulfur, boron, carbon, and selenium (Comninellis et al., 2008; Chang et al., 2009; Chen et al., 2009; Yang et al., 2010). Some new catalyst alternatives more efficient than TiO_2, which are more applicable on the solar spectrum (Shifu et al., 2009; Song et al., 2009), and new photolysis oxidants could advance the process (Anopsitakis and Dionysiou, 2004; Criquet and Vel Leitner, 2009).

Photoelectrocatalysis uses TiO_2 coated surfaces in the electrochemical process to enhance the removal of NOM significantly further than photocatalysis (Selcuk et al., 2006; Selcuk and Bekbolet, 2008). In bromide-containing waters, however, bromate formation was observed during oxidation.

As an alternative to oxidative degradation, enzyme-mediated oxidative coupling processes are a potentially cost effective. These processes reconfigure the NOM molecules to HMM compounds, thus enhancing NOM removal, for example by coagulation, thus reducing DBPFP (Lu and Huang, 2009).

To enable better control of the treatment processes, further research is needed on the oxidation product formation and reaction mechanisms of AOPs in conjunction with other treatments. AOPs in combination with other treatments may enhance the efficiency of NOM removal (Buchanan et al., 2006; Toor and Mohseni, 2007; Uyguner et al., 2007a,b; Huang et al., 2008), and often have better NOM removal efficiency than individual treatments (Lamsal et al., 2011). Coagulation prior to oxidation improves AOP treatment by removing most of the HMM and hydrophilic organic matter present (Uyguner et al., 2007a). AOP treatment may reduce harmful DBPs more effectively when combined with BAC (Toor and Mohseni, 2007). The more biodegradable compounds formed during an AOP can readily be utilized as substrates by microorganisms in the BAC. Moreover, in full-scale application, UV-AOP treatment has proved to be very robust and a reliable barrier against pathogenic microorganisms and organic micropollutants with postfiltration, which was found to efficiently remove residual H_2O_2, biodegradable compounds, and any formed nitrite (Kruithof et al., 2007). AOP combined with membrane filtration has also been found to improve membrane performance efficiently.

AOPs are among the most studied and promising technologies for drinking water purification and disinfection. Although total reduction of NOM has not been achieved with AOPs, several studies have shown efficient NOM reduction and mitigation of DBP formation. Besides AOP process efficiency, the nature of residual NOM after AOP processes also needs consideration to identify the reactive precursors of DBPs (Molnar et al., 2012a,b).

Due to their nonselective nature, AOPs may be suitable processes across the range of NOM (Bond et al., 2011). It should be emphasized that AOPs have site-specific effects, so pilot-scale and full-scale tests must be conducted to obtain the greatest removal efficiency (Rahman et al., 2010). Studies are often conducted under conditions that are not economically feasible in commercial applications (Tuhkanen, 2004; Toor and Mohseni, 2007; Jarvis et al., 2008). The full-scale application of AOPs in drinking water treatment plants is still limited because of high cost, high level of pretreatment required, a lack of experience, and operational difficulties. Nevertheless, the interest attracted by AOPs implies that their application to full-scale treatment will increase (Parsons and Byrne, 2004; Comninellis et al., 2008). The objective of the treatment needs to be defined, whether it is to enhance biodegradability or mineralize organic compounds, and the optimal conditions for each case must be determined according to the type and amount of organic compounds present in the water.

References

T. Agustina, H. Ang, V. Vareek, A review of synergistic effect of photocatalysis and ozonation on wastewater treatment, J. Photochem. Photobiol. Photochem. Rev. 6 (2005) 264–273.

A. Akyol, M. Bayramoglu, Preparation and characterization of supported ZnO photocatalyst by zincate method, J. Hazard Mater. 175 (2010) 484–491.

U. Alkan, A. Teksoy, A. Ateply, H. Bapkaya, Efficiency of the UV/H_2O_2 process for the disinfection of humic surface waters, J. Environ. Sci. Health A Tox. Hazard. Subst. Environ. Eng. 42 (2007) 497–506.

M. Alsheyab, A. Munoz, Reducing the formation of trihalomethanes (THMs) by ozone combined with hydrogen peroxide (H_2O_2/O_3), Desalination 194 (2006) 121–126.

M. Alsheyab, A. Munoz, Comparative study of ozone and MnO_2/O_3 effects on the elimination of TOC and COD of raw water at Valmayor station, Desalination 207 (2007) 179–183.

G. Anopsitakis, D. Dionysiou, Transition metal/UV-based advanced oxidation technologies for water decontamination, Appl. Catal. B Environ. 54 (2004) 155–163.

H. Bai, X. Zhang, J. Pan, D. Sun, J. Shao, Combination of nano TiO_2 photocatalytic oxidation with microfiltration (MF) for natural organic matter removal, Water Sci. Tech- nol. Water Supply 9 (2009) 31–37.

M. Bekbolet, G. Ozkosemen, A preliminary investigation on the photocatalytic degradation of a model humic acid, Water Sci. Technol. 33 (1996) 189–194.

M. Benotti, B. Stanford, E. Wert, S. Snyder, Evaluation of a photocatalytic reactor membrane pilot system for the removal of pharmaceuticals and endocrine disrupting compounds from water, Water Res. 43 (2009) 1513–1522.

T. Bond, E. Goslan, B. Jefferson, F. Roddick, L. Fan, S. Parsons, Chemical and biological oxidation of NOM surrogates and effect on HAA formation, Water Res. 43 (2009) 2615–2622.

T. Bond, E.H. Goslan, S.A. Parsons, B. Jefferson, Treatment of disinfection by— product precursors, Environ. Technol. 32 (2011) 1–25.

P. Bose, D. Reckhow, The effect of ozonation on natural organic matter removal by alum coagulation, Water Res. 41 (2007) 1516–1524.

W. Buchanan, F. Roddick, N. Porter, Formation of hazardous by-products resulting from irradiation of natural organic matter: comparison between UV and VUV irradiation, Chemosphere 63 (2006) 1130–1141.

O. Carp, C. Huisman, A. Reller, Photoinduced reactivity of tianium dioxide, Prog. Solid State Chem. 32 (2004) 33–177.

P.-Y. Chang, C.-H. Hsin, R. Doong, Characterization and photocatalytic activity of vanadium-doped titanium dioxide nanocatalyst, Water Sci. Technol. 59 (2009) 523–530.

D. Chen, L. Weavers, H. Walker, J. Lenhart, Ultrasonic control of ceramic membrane fouling caused by natural organic matter and silica particles, J. Membr. Sci. 276 (2006) 135–144.

P. Chen, E. Rosenfeldt, S. Kullman, D. Hinton, K. Linden, Biological assessments of a mixture of endocrine disruptors at environmentally relevant concentrations in water following UV/H_2O_2 oxidation, Sci. Total Environ. 376 (2007) 18–26.

Y.-H. Chen, L.-L. Chen, N.-C. Shang, Photocatalytic degradation of dimethyl phthalate in an aqueous solution with Pt-doped TiO_2-coated magnetic PMMA micro- pheres, J. Hazard Mater. 172 (2009) 20–29.

C. Chen, X. Zhang, L. Zhu, W. He, H. Han, Changes in different organic matter fractions during conventional treatment and advanced treatment, J. Environ. Sci. 23 (2011) 582–586.

A. Chin, P. Berube, Removal of disinfection by-product precursors with ozone-UV advanced oxidation process, Water Res. 39 (2005) 2136–2144.

H. Choi, S.R. Al-Abed, D.D. Dionysiou, E. Stathatos, P. Lianos, TiO_2-based advanced oxidation nanotechnologies for water purification and reuse, in: M.A. Abraham, R. Sheldon (Eds.), Sustainability Science and Engineering, Book Series, in: I.C. Escobar, A.I. Schafer (Eds.), Sustainable Water for the Future: Water Recycling versus Desalination, vol. 2, Elsevier, 2010, pp. 229–254.

K.-W. Choo, R. Tao, M.-J. Kim, Use of photocatalytic membrane reactor for the removal of natural organic matter in water: effect of photoinduced desorption and fer- rihydrite adsorption, J. Membr. Sci. 322 (2008) 368–374.

C. Chow, R. Fabris, M. Drikas, A rapid fractionation technique to characterize natural organic matter for the optimization of water treatment processes, J. Water Supply Res. Technol. - Aqua 53 (2004) 85–92.

C. Comninellis, A. Kapalka, S. Malato, S. Parsons, I. Poulios, D. Mantzavinos, Advanced oxidation processes for water treatment: advances and trends for R&D, J. Chem. Technol. Biotechnol. 83 (2008) 769–776.

J. Criquet, K. Vel Leitner, Degradation of acetic acid with sulfate radical generated by persulfate ions photolysis, Chemosphere 77 (2009) 194–200.

S. Dobrovic, H. Juretic, N. Ruzinski, Photodegradation of natural organic matter in water with UV irradation at 185 and 254 nm: importance of hydrodynamic conditions on the decomposition rate, Separ. Sci. Technol. 42 (2007) 1421–1432.

R.-A. Doong, L.-F. Chiang, Coupled removal of organic compounds and heavy metals by titanate/carbon nanotube composites, Water Sci. Technol. 58 (2008) 1958–1992.

J. Duguet, E. Brodard, B. Dussert, J. Mallevialle, Improvement in the effectiveness of ozonation of drinking water through the use of hydrogen peroxide, Ozone Sci. Eng. 7 (1985) 241–253.

B. Eggins, F. Palme, J. Byrne, Photocatalytic treatment of humic substances in drinking water, Water Res. 31 (1997) 1223–1226.

F. Frimmel, Photochemical aspects related to humic substances, Environ. Int. 20 (1994) 373–385.

F. Frimmel, Impact of light on the properties of aquatic natural organic matter, Environ. Int. 24 (1998) 559–571.

J. Fu, M. Ji, Y. Zhao, L. Wang, Kinetics of aqueous photocatalytic oxidation of fulvic acids in a photocatalysis-ultrafiltration reactor (PUR), Separ. Purif. Technol. 50 (2006a) 107–113.

J. Fu, J. Min, Z. Wang, L. Jin, D. An, A new submerged membrane photocatalysis reactor (SMPR) for fulvic acid removal using nano-structured photocatalyst, J. Hazard Mater. 131 (2006b) 238–242.

J. Fu, Y. Zhao, Q. Wu, Optimising photoelectrocatalytic oxidation of fulvic acid response surface methodology, J. Hazard Mater. 144 (2007) 499–505.

A. Fujishima, Electrochemical photolysis of water at a semiconductor electrode, Nature 238 (1971) 37–38.

M. Fukushima, K. Tatsumi, S. Nagao, Degradation characteristics of humic acid during photo-Fenton processes, Environ. Sci. Technol. 35 (2001) 3683–3690.

L. Galeano, P. Bravo, C. Luna, M. Vicente, A. Gil, Removal of natural organic matter for drinking water production by Al/Fe-PILC-catalyzed wet peroxide oxidation: effect of the catalyst preparation from concentrated precursors, Appl. Catal. B Environ. 111–112 (2012) 527–535.

D. Gerrity, H. Ryu, J. Crittenden, M. Abbaszadegan, Photocatalytic inactivation of viruses using titanium dioxide and low-pressure UV light, J. Environ. Sci. Health A Tox. Hazard. Subst. Environ. Eng. 43 (2008) 1261–1270.

D. Gerrity, B. Mayer, H. Ryu, J. Crittenden, M. Abbaszadegan, A comparison of pilot-scale photocatalysis and enhanced coagulation for disinfection byproduct mitigation, Water Res. 43 (2009) 1597–1610.

R. Giri, H. Ozaki, R. Takanami, S. Taniguchi, Heterogenous photocatalytic ozonation of 2,4-D in dilute aqeuous solution with TiO_2 fiber, Water Sci. Technol. 58 (2008) 207–216.

E. Gjessing, T. Kallqvist, Algicidal and chemical effect of u.v.-radiation of water containing humic substances, Water Res. 25 (1991) 491–494.

W.H. Glaze, F. Beltran, T. Tuhkanen, J.-W. Kang, Chemical models of advanced oxidation processes, Water Pollut. Res. J. Can. 27 (1992) 23–42.

W. Glaze, G. Peyton, S. Lin, R. Huang, J. Burleson, Destruction of pollutants in water with ozone in combination with ultraviolet radiation. 2. Natural trihalomethane precursors, Environ. Sci. Technol. 16 (1982) 454–458.

W. Glaze, J. Kang, D. Chapin, The chemistry of water treatment processes involving ozone, hydrogen peroxide and ultraviolet radiation, Ozone Sci. Eng. 9 (1987) 335–352.

V. Goncharuk, V. Malyarenko, V. Yaremenko, Use of ultrasound in water treatment, J. Water Chem. Technol. 30 (2008) 137–150.

M. Gonzalez, E. Oliveros, M. Worner, A. Braun, Vacuum-ultraviolet photolysis of aqueous reaction systems, J. Photochem. Photobiol. C Photochem. Rev. 5 (2004) 225–246.

E. Goslan, F. Gurses, J. Banks, S. Parsons, An investigation into reservoir NOM reduction by UV photolysis and advanced oxidation processes, Chemosphere 65 (2006) 1113–1119.

A. Gumy, A.G. Rincon, R. Hajdu, C. Pulgarin, Solar photocatalysis for detoxification and disinfection of water: different types of suspended and fixed TiO_2 catalysts study, Sol. Energy 80 (2006) 1376–1381.

U. von Gunten, Ozonation of drinking water: part I. Oxidation kinetics and product formation, Water Res. 37 (2003) 1443–1467.

D. Hand, D. Perram, J. Crittenden, Destruction of DBP precursors with catalytic oxidation, J. Am. Water Works Assoc. 87 (1995) 84–96.

M. Hoffmann, S. Martin, W. Choi, D. Bahnemann, Environmental applications of semiconductor photocatalysis, Chem. Rev. 95 (1995) 69–96.

Q. Hu, C. Zhang, Z. Wang, Y. Chen, K. Mao, X. Zhang, Y. Xiong, M. Zhu, Photodegradation of methyl *tert-butyl* ether (MTBE) by UV/H_2O_2 and UV/TiO_2, J. Hazard Mater. 154 (2008) 795–803.

X. Huang, M. Leal, Q. Li, Degradation of natural organic matter by TiO_2 photo- catalytic oxidation and its effect on fouling of low-pressure membranes, Water Res. 42 (2008) 1142–1150.

K. Ikehata, M. El-Din, Aqeuous pesticide degradation by ozonation and ozone-based advanced oxidation processes: a review (part II), Ozone Sci. Eng. 27 (2005) 173–202.

G. Imoberdorf, M. Mohseni, Degradation of natural organic matter in surface water using vacuum-UV irradiation, J. Hazard Mater. 186 (2011) 240–246.

A. Irabelli, S. Jasim, N. Biswas, Pilot-scale evaluation of ozone vs. peroxone for trihalomethane formation, Ozone Sci. Eng. 30 (2008) 356–366.

P. Jarvis, J. Banks, R. Molinder, T. Stephenson, S. Parsons, B. Jefferson, Processes for enhanced NOM removal: beyond Fe and Al coagulation, Water Sci. Technol. Water Supply 8 (2008) 709–716.

C. Jiang, J. Zhang, Progress and prospect in electro-Fenton process for wastewater treatment, J. Zhejiang Univ. - Sci. 8 (2007) 1118–1125.

B. Kasprzyk-Hordern, U. Raczyk-Stanislawiak, J. Swietlik, J. Nawrocki, Catalytic ozonation of natural organic matter on alumina, Appl. Catal. B Environ. 62 (2005) 345–358.

H. Katsumata, M. Sada, S. Kaneco, T. Suzuki, K. Ohta, Y. Yobiko, Humic acid degradation in aqeuous solution by the photo-Fenton process, Chem. Eng. J. 137 (2008) 225–230.

M. Kitis, S. Kaplan, Advanced oxidation of natural organic matter using hydrogen peroxide and iron-coated pumice particles, Chemosphere 68 (2007) 1846–1853.

M. Klavarioti, D. Mantzavinos, D. Kassinos, Removal of residual pharmaceuticals from aqueous systems by advanced oxidation processes, Environ. Int. 35 (2009) 402–417.

G. Kleiser, F. Frimmel, Removal of precursors for disinfection by-products (DBPs) - differences between ozone- and OH-radical-induced oxidation, Sci. Total Environ. 256 (2000) 1–9.

J. Kruithof, P. Kamp, B. Martijn, UV/H_2O_2 treatment: a practical solution for organic contaminant control and primary disinfection, Ozone Sci. Eng. 29 (2007) 273–280.

P. Kumar, A. Manivel, S. Anandan, Synthesis of Ag-ZnO nanoparticles for enhanced photocatalytic degradation of acid red 88 in aqeuous environment, Water Sci. Technol. 59 (2009) 1423–1430.

K. Kutschera, H. Bornick, E. Worch, Photoionitiated oxidation of geosmin and 2-methylisoborneol by irradiation with 254 nm and 185 nm UV light, Water Res. 43 (2009) 2224–2232.

R. Lamsal, M. Walsh, G. Gagnon, Comparison of advanced oxidation processes for the removal of natural organic matter, Water Res. 45 (2011) 3263–3269.

M. Lanao, M. Ormad, C. Ibarz, N. Miguel, J. Ovelleiro, Bactericidal effectiveness of O_3, O_3/H_2O and O_3/TiO_2 on, Clostridium perfringens. Ozone Sci. Eng. 30 (2008) 431–438.

B. Langlais, D.A. Reckhow, D.R. Brink, Ozone in Water Treatment: Application and Engineering, Lewis Publishers, Chelsea, Michigan, USA, 1991.

P. Le-Clech, E.-K. Lee, V. Chen, Hybrid photocatalysis/membrane treatment for surface waters containing low concentrations of natural organic matters, Water Res. 40 (2006) 323–330.

C. Lee, D. Sedlak, A novel homogenous Fenton-like system with Fe(III)- phosphotungstate for oxidation of organic compounds at neutral pH values, J. Mol. Catal. Chem. 311 (2009) 1–6.

O. Legrini, E. Oliveros, A. Braun, Photochemical processes for water treatment, Chem. Rev. 83 (1993) 671–698.

B. Legube, N.K. Vel Leitner, Catalytic ozonation: a promising advanced oxidation technology for water treatment, Catal. Today 53 (1999) 61–72.

K. Li, D. Hokanson, J. Crittenden, R. Trussell, D. Minakata, Evaluating UV/H_2O_2 processes for methyl tert-butyl ether and tertiary butyl alcohol removal: effect of pretreatment options and light sources, Water Res. 42 (2008) 5045–5053.

H.-C. Lin, G.-S. Wang, Effects of UV/H_2O_2 on NOM fractionation and corresponding DBPs formation, Desalination 270 (2011) 221–226.

S. Liu, M. Lim, R. Fabris, C. Chow, K. Chiang, M. Drikas, R. Amal, Removal of humic acid using TiO_2 photocatalytic process - fractionation and molecular weight characterization studies, Chemosphere 72 (2008a) 263–271.

S. Liu, M. Lim, R. Fabris, C. Chow, M. Drikas, R. Amal, TiO_2 photocatalysis of natural organic matter in surface water: impact on trihalomethane and haloacetic acid formation potential, Environ. Sci. Technol. 42 (2008b) 6218–6223.

S. Liu, M. Lim, R. Fabris, C. Chow, M. Drikas, R. Amal, Comparison of photo- catalytic degradation of natural organic matter in two Australian surface waters using multiple analytical techniques, Org. Geochem. 41 (2010) 124–129.

D. Ljubas, Solar photocatalysis-a possible step in drinking water treatment, Energy 30 (2005) 1699–1710.

J. Lu, Q. Huang, Removal of acetaminophen using enzyme-mediated oxidative coupling processes: II. Cross-coupling with natural organic matter, Environ. Sci. Technol. 43 (2009) 7068–7073.

S. Malato, J. Blanco, D. Alarcon, M. Maldonado, P. Fernandez-Ibanez, W. Gernjak, Photocatalytic decontamination and disinfection of water with solar collectors, Catal. Today 122 (2007) 137–149.

S. Malato, P. Fernandez-Ibanez, M. Maldonado, J. Blanco, W. Gernjak, Decontamination and disinfection of water by solar photocatalysis: recent overview and trends, Catal. Today 147 (2009) 1–59.

A. Matilainen, M. Sillanpaa, Removal of natural organic matter from drinking water by advanced oxidation processes: a review, Chemosphere 80 (2010) 351–365.

D.M. Metz, M. Meyer, A. Dotson, E. Beerendonk, D. Dionysiou, The effect of UV/H_2O_2 treatment on disinfection by-product formation potential under simulated distribution system conditions, Water Res. 45 (2011) 3969–3980.

J. Molnar, J. Agbaba, B. Dalmacija, M. Klasnja, M. Dalmacija, M. Kragulj, A comparative study of the effects of ozonation and TiO_2-catalyzed ozonation on the selected chlorine disinfection by-product precursor content and structure, Sci. Total Environ. 425 (2012a) 169–175.

J. Molnar, J. Agbaba, B. Dalmacija, S. Roncevic, M. Prica, A. Tubic, Influence of pH and ozone dose on the content and structure of haloacetic acid precursors in groundwater, Environ. Sci. Pollut. Res. 19 (2012b) 3079–3086.

J. Molnar, J. Agbaba, B. Dalmacija, A. Tubic, D. KrCmar, S. Maletic, D. Tomasevic, The effects of matrices and ozone dose on changes in the characteristics of natural organic matter, Chem. Eng. J. 222 (2013) 435–443.

A. Moncayo-Lasso, C. Pulgarin, N. Benitez, Degradation of DBPs' precursors in river water before and after slow sand filtration by photo-Fenton process at pH 5 in solar CPC reactor, Water Res. 42 (2008) 4125–4132.

A. Moncayo-Lasso, J. Sanabria, C. Pulgarin, N. Benitez, Simultaneous *E. coli* inactiva- tion and NOM degradation in river water via photo-Fenton process at natural pH in solar CPC reactor: a new way for enhancing solar disinfection of natural water, Chemosphere 77 (2009) 296–300.

A. Moncayo-Lasso, A.-G. Rincon, C. Pulgarin, N. Benitez, Significant decrease of THMs generated during chlorination of river water by previous photo-Fenton treatment at near neutral pH, J. Photochem. Photobiol. Chem. 229 (2012) 46–52.

R. Mosteo, N. Miguel, S. Martin-Muniesa, M. Ormad, J. Ovelleiro, Evaluation of trihalomethane formation potential in function of oxidation processes used during the drinking water production process, J. Hazard Mater. 172 (2009) 661–666.

C. Murray, S. Parsons, Removal of NOM from drinking water: fenton's and photo-Fenton's processes, Chemosphere 54 (2004) 1017–1023.

C. Murray, S. Parsons, Preliminary laboratory investigation of disinfection by-product precursor removal using an advanced oxidation process, Water Environ. J. 20 (2006) 123–129.

V. Naddeo, V. Belgiorno, R. Napoli, Behaviour of natural organic matter during ultrasonic irradiation, Desalination 210 (2007) 175–182.

Y. Nagata, K. Hirai, H. Bandow, Y. Maeda, Decomposition of hydrobencoic and humic acids in water by ultrasonic irradiation, Environ. Sci. Technol. 30 (1996) 1133–1138.
S. Navalon, M. Alvaro, H. Garcia, D. Escrig, V. Costa, Photocatalytic water disinfection of *Cryptosporidium parvum* and *Giardia lamblia* using a fibrous ceramic TiO_2 photocatalysis, Water Sci. Technol. 59 (2009) 639–645.
Y. Nie, C. Hu, L. Zhou, J. Qu, Q. Wei, D. Wang, Degradation characteristics of humic acid over iron oxides/Fe^0 core-shell nanoparticles with UVA/H_2O_2, J. Hazard Mater. 173 (2010) 474–479.
S. Park, T. Yoon, The effects of iron species and mineral particles on advanced oxidation processes for the removal of humic acids, Desalination 208 (2007) 181–191.
S. Parsons, A. Byrne, Water treatment applications, in: S. Parsons (Ed.), Advanced Oxidation Processes for Water and Wastewater Treatment, IWA Publishing, London, 2004, pp. 329–346.
M. Pera-Titus, V. Garcia-Molina, M. Banos, J. Gimenes, S. Esplugas, Degradation of chlorophenols by means of advanced oxidation processes: a general review, Appl. Catal. B Environ. 47 (2004) 219–256.
V. Pereira, H. Weinberg, K. Linden, P. Singer, UV degradation kinetics and modeling of pharmaceutical compounds in laboratory grade and surface water via direct and indirect photolysis at 254 nm, Environ. Sci. Technol. 41 (2007) 1682–1688.
K. Philippe, C. Hans, J. MacAdam, B. Jefferson, J. Hart, S. Parsons, Photocatalytic oxidation of natural organic matter surrogates and the impact on trihalomethane formation potential, Chemosphere 81 (2010) 1509–1516.
K. Pirkanniemi, M. Sillanpaa, Heterogeneous water phase catalysis as an environmental application: a review, Chemosphere 48 (2002) 1047–1060.
M. Rahman, S. Jasim, E. Yanful, S. Ndiongue, D. Borikar, Advanced oxidation treatment of drinking water: part II. Turbidity, particles and organics removal from Lake Huron water, Ozone Sci. Eng. 32 (2010) 295–304.
A. Rastogi, S. Al-Abed, D. Dionysiou, Effect of inorganic, synthetic and naturally occurring chelating agents on Fe(II) mediated advanced oxidation of chlorophenols, Water Res. 43 (2009) 684–694.
A.-G. Rincon, C. Pulgarin, Fe^3+ and TiO_2 solar-light-assisted inactivation of *E.coli* at field scale. Implications in solar disinfection at low temperature of large quantities of water, Catal. Today 122 (2007) 128–136.
E. Rosenfeldt, K. Linden, S. Canonica, U. von Gunten, Comparison of the efficiency of OH radical formation during ozonation and advanced oxidation processes O_3/H_2O_2 and UV/H_2O_2, Water Res. 40 (2006) 3695–3704.
E.J. Rosenfeldt, K.G. Linden, The ROH, UV concept to characterize and the model UV/H_2O_2 process in natural waters, Environ. Sci. Technol. 41 (2007) 2548–2551.
B. Roshani, N. Leitner, Effect of persulfate on the oxidation of benzotriazole and humic acid by e-beam irradiation, J. Hazard Mater. 190 (2011) 403–408.
Sanly, M. Lim, K. Chiang, R. Amal, R. Fabris, C. Chow, M. Drikas, A study on the removal of humic acid using advanced oxidation process, Separ. Sci. Technol. 42 (2007) 13911404.
S. Sarathy, M. Mohseni, The impact of UV/H_2O_2 advanced oxidation on molecular size distribution of chromophoric natural organic matter, Environ. Sci. Technol. 41 (2007) 8315–8320.
S.R. Sarathy, M. Mohseni, The fate of natural organic matter during UV/H_2O_2 advanced oxidation of drinking water, Can. J. Civ. Eng. 36 (2009) 160–169.
S. Sarathy, M. Mohseni, Effects of UV/H_2O advanced oxidation on chemical characteristics and chlorine reactivity of surface water natural organic matter, Water Res. 44 (2010) 4087–4096.

S. Sarathy, M. Stefan, A. Royce, M. Mohseni, Pilot-scale UV/H_2O_2 advanced oxidation process for surface water treatment and downstream biological treatment: effects on natural organic matter characteristics and DBP formation potential, Environ. Technol. 32 (2011) 1709–1718.

H. Selcuk, M. Bekbolet, Photocatalytic and photoelectrocatalytic humic acid removal and selectivity of TiO_2 coated photoanode, Chemosphere 73 (2008) 854–858.

H. Selcuk, H. Sarikaya, M. Bekbolet, M. Anderson, Bromate formation on the non- porous TiO_2 photoanode in the photoelectrocatalytic system, Chemosphere 62 (2006) 715–721.

C. Shifu, Z. Wei, L. Wei, Z. Huaye, Y. Xiaoling, C. Yinghao, Preparation, characterization and activity evaluation of p-n junction photocatalyst p-$CaFe_2O_4$/n-Ag_3VO_4 under visible light irradiation, J. Hazard Mater. 172 (2009) 1415–1423.

H. Sindelar, M. Brown, T. Boyer, Evaluating UV/H2O2, UV/percarbonate, and UV/ perborate for natural organic matter reduction from alternative water sources, Chemosphere 105 (2014) 112–118.

W. Song, V. Ravindran, M. Pirbazari, Process optimization using a kinetic model for the ultraviolet radiation-hydrogen peroxide decomposition of natural and synthetic organic compounds in groundwater, Chem. Eng. Sci. 63 (2008) 3249–3270.

H. Song, O. Orr, Y. Hong, T. Karanfil, Isolation and fractionation of natural organic matter: evaluation of reverse osmosis performance and impact of fractionation parameters, Environ. Monit. Assess. 153 (2009) 307–321.

L. Stepniak, E. Stanczyk-Mazanek, M. Kusiak, Ultrasounds and ozone in removal of humic substances from water, Inzynieria Ochrona Srodowiska 15 (2012) 143–154.

H. Suty, C. De Traversay, M. Cost, Applications of advanced oxidation processes: present and future, Water Sci. Technol. 49 (2004) 227–233.

P. Swaim, A. Royce, T. Smith, T. Maloney, D. Ehlen, B. Carter, Effectiveness of UV advanced oxidation for destruction of mirco-pollutants, Ozone Sci. Eng. 30 (2008) 34–42.

L. Tercero Espinoza, F. Frimmel, Formation of brominated products in irradiated titanium dioxide suspensions containing bromide and dissolved organic carbon, Water Res. 42 (2008) 1778–1784.

L. Tercero Espinoza, F. Frimmel, A simple simulation of the degradation of natural organic matter in homogenous and heterogenous advanced oxidation processes, Water Res. 43 (2009) 3902–3909.

L. Tercero Espinoza, M. Rembor, C. Matesanz, A. Heidt, F. Frimmel, Formation of bromoform in irradiated titanium dioxide suspensions with varying photocatalyst, dissolved organic carbon and bromide concentrations, Water Res. 43 (2009a) 4143–4148.

L. Tercero Espinoza, E. ter Hazeborg, M. Weber, F. Frimmel, Investigation of the photocatalytic degradation of brown water natural organic matter by size exclusion chromatography, Appl. Catal. B Environ. 87 (2009b) 56–62.

L. Tercero Espinoza, E. ter Haseborg, M. Weber, E. Karle, R. Peschke, F. Frimmel, Effect of selected metal ions on the photocatalytic degradation of bog lake water natural organic matter, Water Res. 45 (2011) 1039–1048.

J. Thomson, F. Roddick, M. Drikas, Natural organic matter removal by enhanced photo-oxidation using low pressure mercury vapour lamps, Water Sci. Technol. Water Supply 2 (2002) 435–443.

J. Thomson, F.A. Roddick, M. Drikas, Vacuum ultraviolet irradiation for natural organic matter removal, J. Water Supply Res. Technol. - Aqua 53 (2004) 193–206.

R. Toor, M. Mohseni, UV-H_2O_2 based AOP and its integration with biological activated carbon treatment for DBP reduction in drinking water, Chemosphere 66 (2007) 2087–2095.

T. Tran, S. Gray, B. Bolto, T. Farmer, T. Collings, Ultrasound enhancement of microfiltration performance for natural organic matter removal, Org. Geochem. 38 (2007) 1091–1096.
R. Treguer, R. Tatin, A. Couvert, D. Wolbert, A. Tazi-Pain, Ozonation effect on natural organic matter adsorption and biodegradation - application to a membrane bioreactor containing activated carbon for drinking water production, Water Res. 44 (2010) 781–788.
A. Tubic, J. Agbaba, B. Dalmacija, S. Perovic, M. Klasnja, S. Roncevic, I. IvanCev-Tumbas, Removal of natural organic matter from groundwater using advanced oxidation processes at a pilot scale drinking water treatment plant in the Central Banat Region (Serbia), Ozone Sci. Eng. 33 (2011) 267–278.
T.A. Tuhkanen, UV/H_2O_2 processes, in: S. Parsons (Ed.), Advanced Oxidation Processes for Water and Wastewater Treatment, IWA Publishing, London, 2004, pp. 86–110.
C. Uyguner, M. Bekbolet, H. Selcuk, A comparative approach to the application of a physico-chemical and advanced oxidation combined system to natural water samples, Separ. Sci. Technol. 42 (2007a) 1405–1419.
C. Uyguner, S. Suphandag, A. Kerc, M. Bekbolet, Evaluation of adsorption and coagulation characteristics of humic acids preceded by alternative advanced oxidation techniques, Desalination 210 (2007b) 183–193.
C. Uyguner-Demirel, M. Bekbolet, Significance of analytical parameters for the understanding of natural organic matter in relation to photocatalytic oxidation, Chemosphere 84 (2011) 1009–1031.
C. Uyguner-Demirel, C. Birben, M. Bekbolet, Key role of common anions on the photocatalytic degradation profiles of the molecular size fractions of humic acid, Catal. Today 209 (2013) 122–126.
S.H. Valencia, J.A. Marin, G.M. Restrepo, Evolution of natural organic matter by size exclusion chromatography during photocatalytic degradation by solvothermal- synthesized titanium dioxide, J. Hazard Mater. 213–214 (2012a) 318–324.
S. Valencia, J. Marina, J. Velasqueza, G. Restrepoa, F. Frimmel, Study of pH effects on the evolution of properties of brown-water natural organic matter as revealed by size-exclusion chromatography during photocatalytic degradation, Water Res. 46 (2012b) 1198–1206.
S. Valencia, J.M. Marin, G. Restrepo, F.H. Frimmel, Evaluation of natural organic matter changes from Lake Hohloh by three-dimensional excitation-emission matrix fluorescence spectroscopy during TiO_2/UV process, Water Res. 51 (2014) 124–133.
S. Vilhunen, M. Bosund, M.-L. Kaariainen, D. Cameron, M. Sillanpaa, Atomic layer deposited TiO_2 films in photodegradation of aqueous salicylic acid, Separ. Purif. Technol. 66 (2009) 130–134.
B. Voelker, B. Sulzberger, Effects of fulvic acid on Fe(II) oxidation by hydrogen peroxide, Environ. Sci. Technol. 30 (1996) 1106–1114.
G.-S. Wang, C.-H. Liao, H.-W. Chen, H.-C. Yang, Characteristics of natural organic matter degradation in water by UV/H_2O_2 treatment, Environ. Technol. 27 (2006) 277–287.
K. Wang, J. Guo, M. Yang, H. Junji, R. Deng, Decomposition of two haloacetic acids in water using UV radiation, ozone and advanced oxidation processes, J. Hazard Mater. 162 (2009) 1243–1248.
J. Wiszniowski, D. Robert, J. Surmacz-Gorska, K. Miksch, J.-V. Weber, Photocatalytic decomposition of humic acids on TiO_2 part I: discussion of adsorption and mechanism, J. Photochem. Photobiol. Chem. 152 (2002) 267–273.
S. Xu, X. Zhang, J. Ng, D. Sun, Preparation and application of TiO_2/Al_2O_3 micros- perical photocatalyst for water treatment, Water Sci. Technol. Water Supply 9 (2009) 39–44.

M. Yan, D. Wang, B. Shi, M. Wang, Y. Yan, Effect of pre-ozonation on optimized coagulation of a typical North-China source water, Chemosphere 69 (2007) 1695–1702.

S. Yang, C. Sun, X. Li, Z. Gong, X. Quan, Enhanced photocatalytic activity for titanium dioxide by co-modifying with silica and fluorine, J. Hazard Mater. 175 (2010) 258–266.

P. Yao, K. Choo, M. Kim, A hydridized photocatalysis-microfiltration system with iron oxide-coated membranes for the removal of natural organic matter in water treatment: effects of iron oxide layers and colloids, Water Res. 43 (2009) 4238–4248.

R. Zepp, B. Faust, J. Hoigne, Hydroxyl radical formation in aqueous reactions (pH 3-8) of iron(II) with hydrogen peroxide: the photo-Fenton reaction, Environ. Sci. Technol. 26 (1992) 313–319.

P. Zhang, L. Jian, Ozone-enhanced photocatalytic degradation of natural organic matter in water, Water Sci. Technol. Water Supply 6 (2006) 53–61.

X. Zhang, J.H. Pan, W. Fu, A.J. Du, D.D. Sun, TiO_2 nanotube photocatalytic oxidation for water treatment, Water Sci. Technol. Water Supply 9 (2009) 45–49.

Y.-Y. Zhao, J. Boyd, M. Woodbeck, R. Andrews, F. Qin, S. Hrudey, X.-F. Li, Formation of N-nitrosamines from eleven disinfection treatments of seven different surface waters, Environ. Sci. Technol. 42 (2008) 4857–4862.

G. Zhao, X. Lu, Y. Zhou, Q. Gu, Simultaneous humic acid removal and bromate control by O_3 and UV/O_3 processes, Chem. Eng. J. 232 (2013) 74–80.

Q. Zhou, Y.-H. Zhong, X. Chen, J.-H. Liu, X.-J. Huang, Y.-C. Wu, Adsorption and photocatalysis removal of fulvic acid by TiO_2-graphene composites, J. Mater. Sci. 49 (2014) 1066–1075.

A. Ziylan, N. Ince, Ozonation-based advanced oxidation for pre-treatment of water with residuals of anti-inflammatory medication, Chem. Eng. J. 220 (2013) 151–160.

CHAPTER 9

NOM removal by adsorption

Mohamed Chaker Necibi[1], Almotasembellah Abushaban[1], Yuri Park[2] and Mika Sillanpää[3,4,5]

[1]Mohammed VI Polytechnic University (UM6P), International Water Research Institute (IWRI), Green City, Benguerir, Morocco; [2]Institute of Environmental Technology, Department of Environmental Engineering, Seoul National University of Science and Technology, Seoul, South Korea; [3]Department of Chemical Engineering, School of Mining, Metallurgy and Chemical Engineering, University of Johannesburg, South Africa; [4]Chemistry Department, College of Science, King Saud University, Riyadh, Saudi Arabia; [5]Department of Applied Physics, Faculty of Science and Technology, Universiti Kebangsaan Malaysia, Selangor, Malaysia

Abstract

Natural organic matter (NOM) is a complex matrix of organic substances present in various aquatic ecosystems through intricate geological, hydrological and biological cycles. Such variability, along with the important increase of its concentration in fresh waters and municipal wastewater effluents, makes the removal of NOM and its constituents from water sources a challenging issue, especially in the treated drinking water supplies. Hence, there is a need to develop robust and highly efficient water treatment processes targeting NOM removal.

This chapter focuses on the removal of NOM and its constituents from water using adsorption technology, which was confirmed to be one of the best technologies for the removal of various kinds of organic pollutants in freshwaters and wastewaters. Recent NOM removal results obtained with different adsorbents have been summarized, as well as studies focusing on enhancing the adsorption-based process by optimizing the operating conditions and coupling it with other water treatment technologies, mainly coagulation and membrane separation.

Keywords: Adsorbing materials; Adsorption; Integrated processes; Natural organic matter; Removal efficiency; Water treatment.

Abbreviations

AC	Activated carbon
BSA	Barberry stem ash
BSP	Barberry stem powder
CF	Coconut fiber
CGFS	Coal gasification fine slag
CNF	Cellulose nanofiber
CNT	Carbon nanotube
CTF	covalent triazine-based framework
DAAO	N,N-dimethyl dehydroabietylamine oxide

Natural Organic Matter in Water
ISBN 978-0-12-824274-2
https://doi.org/10.1016/B978-0-12-824274-2.00005-3

DBP	Disinfection by-product
DOC	Dissolved organic carbon
HA	Humic acid
HAAs	Haloacetic acids
HANs	Haloacetonitriles
HAOPs	Heated aluminum oxide particles
LDH	Layered double hydroxide
MAR	mesoporous adsorbent resin
MIPs	Molecularly imprinted polymers
MPR	Magnetic attapulgite-resin
nACP	nano-amorphous calcium phosphate
NOM	Natural organic matter
PAC	Powdered activated carbon
POF	Palm oil fiber
SMNP	silica-coated magnetite nanoparticle
SMZ	surfactant-modified zeolite
THM	Trihalomethane
UF	ultrafiltration
XPS	X-ray photoelectron spectroscopy
μGAF	Microgranular adsorptive filtration

9.1 Introduction

Natural organic matter (NOM) is one of the common and widely distributed organic constituents found in soil, sediment, and natural water sources. Complex molecules of NOM serve as a substrate for the growth of heterotrophic bacteria in water distribution system and thus influences drinking water quality and the efficiency of any applied water treatment technologies (Sillanpää et al., 2018).

For instance, NOM can initiate the inactivation of pathogenic microorganisms by consuming chemical disinfectants, such as chlorine, as well as triggering serious issues of interference with water treatment technologies, mainly fouling in membrane-based separation or increasing the dosage of coagulants in coagulation processes (Crafton et al., 2014; Yu et al., 2018).

Intrinsically, if NOM is present in freshwater sources, it does not pose a direct threat to living organisms. The issue emerges when NOM and its constituents are not removed prior to chemical disinfection processes in water treatment. In this case, the reaction of NOM with disinfectant results in the formation of disinfection by-products (DBPs) (Hua et al., 2015), which have been linked to elevated cancer risk, low birth weights and birth defects (Mian et al., 2018; Yin et al., 2018).

Adsorption is among the most applied processes for water and wastewater treatment. The application of various types of adsorbents to remove NOM for drinking water supplies has received a great deal of attention for researchers worldwide because it was efficient and helped avoiding the formation of DBPs (Gora and Andrews, 2019; Yang et al., 2021). Nonetheless, with the increased fluctuation of NOM in water (concentration and composition), the efficiency of commercially available adsorbents was substantially reduced, hence the need to develop enhanced adsorption processes by (i) developing more robust and efficient organic or inorganic adsorbents, (ii) optimizing the operating conditions and related operational and maintenance costs, and (iii) integrating adsorption with other water treatment technologies.

In the present chapter, recent research studies dealing with the application of adsorption for NOM removal from drinking water supplies are presented and analyzed. In addition, integration schemes combining coagulation and other water treatment processes are presented, along with promising integration processes coupling adsorption with coagulation and membrane separation.

9.2 Removal of NOM and its constituents from water using adsorbents

Research studies on the applicability and efficiency of the adsorption technique to remediate the serious issue of NOM in water gained momentum in the last couple of decades for two main reasons: First, the increasing occurrence on NOM in water bodies and its highly heterogonous character complicated the remediation measures, especially with the issue of DBPs and the negative interference in the removal of other pollutants. The second reason is the specificities in the various contamination scenarios with NOM (i.e., its origin, composition and interaction schemes with the other pollutants present in the water to be treated).

Thus, multiple recent research investigations were conducted with the main objective of optimizing the efficiency of the developed and/or modified (chemically and/or thermally) absorbents to remove NOM while trying to control the various possible interferences with the adsorption process.

In the following sections, several adsorbing materials from various feedstocks, used as such or subjected to various modification treatment, will be presented and studied.

9.2.1 Raw and modified carbon-based adsorbents

9.2.1.1 Biochars

In a recent study, Yazdani et al. (2019) reported the effective use of tailored mesoporous biochars from pinecone biomass for the adsorptive removal of NOM from lake water. For this, several biochars were developed via two procedures: pre-pyrolysis/activation/post-pyrolysis and activation/post-pyrolysis processes, using NaOH and $ZnCl_2$ as activating agents. The objective of such combined chemically and thermally induced modifications of the pine biomass was to improve the surface chemistry and mesoporous structure for higher NOM adsorption. The related results showed that the best performing tailored biochar was fabricated by NaOH activation via pre- and post-pyrolysis processes. Under optimum operating conditions, about 97% of the NOM was removed from the lake water. The proposed reaction mechanisms involved during the adsorption of NOM using the pine-derived biochars are depicted in Fig. 9.1.

In another study, biochar and fly ash were activated and surface-modified with Polyethyleneimine (PEI) for enhanced adsorptive removal of NOM from water (Truong et al., 2020). For the case of biochar, the PEI-functionalized derivate was found to have higher surface areas ($294\ m^2\ g^{-1}$ compared to $230\ m^2\ g^{-1}$), and more abundant functional groups (especially amine groups) compared to the raw biochar, which translated in an improved adsorption capacity ($80.6\ mg\ g^{-1}$ for the modified biochar and $24.6\ mg\ g^{-1}$ for the raw one). Furthermore, the authors evaluated the adsorptive behaviors of the both biochars (raw and

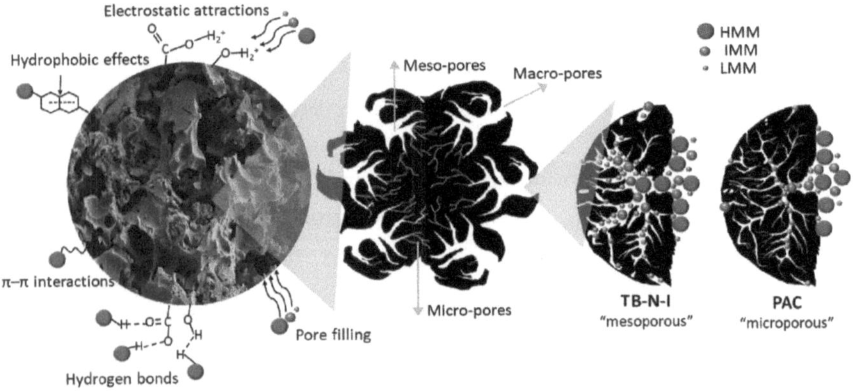

Figure 9.1 Schematic illustration of key mechanisms involved in NOM adsorption from lake water by tailored mesoporous biochars (Yazdani et al., 2019). *(With reprint permission from Elsevier.)*

functionalized) using analytical techniques for dissolved organic carbon (DOC), UV-absorbing and fluorescent moieties. The proposed interaction mechanisms for the tested adsorption system involved electrostatic attraction, hydrogen bonding, and π-π interaction.

9.2.1.2 Activated carbons

Activated carbon (AC) has been widely used as an adsorbent in water and wastewater treatment to remove a wide array of inorganic and organic pollutants including several investigations to eliminate NOM and its constituents (mainly humic and fulvic acids) from water.

Recently, Tafvizi et al. (2021) investigated the use of tailored ACs from waste materials of a pulp and paper company to remove NOM synthetic and natural water, after a chemical activation step using by nitric acid, followed by physical activation and metal coating. First, it was observed that the nitric acid activation increased the surface area and promoted the development of mesoporous structure. Optimizing the operating conditions revealed that the highest adsorption capacity toward NOM (18.25 mg g^{-1}) was attained using a dosage of 0.15 g L^{-1} of AC for 24 h. Metal coating on the tested waste-derived and chemically activated AC using three valent ions of Fe and Al (derived from ferric chloride and aluminum sulfate) enhanced the adsorptive removal. Thus, the ratio (metal:AC) of 1.00% for Al and 0.75% for Fe respectively improved the adsorption capacities by 44% and 35%.

Naghizadeh et al. (2018) assessed the adsorption capacity of a commercial AC to remove humic acid (HA) from aqueous media, followed by AC regeneration suing ultrasound. The main results revealed that HA adsorption was rapid (around 60 min), with a maximum adsorption of around 30 mg g^{-1} occurring at pH 3.0. The related thermodynamic study showed that the adsorption of HA into the tested AC was endothermic with a positive entropy of adsorption. Desorption wise, the ultrasonic regeneration of the spent AC was most effective at 37 kHz, for pH 11 and during 60 min. In addition, the adsorption capacity decreased by only 2% after five regeneration cycles.

In another study, the adsorptive removal of HA on various coconut-based activated carbons (raw, HNO_3-modified, H_2SO_4-modified, and base-modified ACs) was investigated (Yang and Fox, 2018). Overall, the acid treatment tended to increase the isoelectric point of the carbon samples to about pH 3.5, and to increase the size of pores while removing particulate contaminants. Such combined structural and chemical modification led to enhanced adsorption capacities of HA for acid-modified ACs (50 mg g^{-1} and 70 mg g^{-1}, respectively for nitric acid and sulfuric activation), in comparison with 29 mg g^{-1} for the untreated ACs and only 7.6 mg g^{-1} for the base-modified one.

Algal blooms constitute a serious issue in water treatment as it intensifies operational difficulties related to increases of NOM content and taste- and odor-causing compounds. To remediate this problem, a study investigated the use of powdered activated carbon (PAC) during algal blooms and its impact on NOM removal and the formation potential of trihalomethanes (THMs), haloacetic acids (HAAs), and haloacetonitriles (HANs) (Park et al., 2019). Hence, it was reported that the PAC addition was effective in removing NOM, especially low molecular weights NOM. Indeed, a PAC dose of 200 mg L^{-1} led to the removal rates of 46% and 84% for DOC and UV254, respectively. As well, the PAC addition showed consistent reduction of formation potentials with increasing PAC dosage for THMs, HAAs, and HANs, with the former showing the highest reduction (from 182 μg L^{-1} without PAC to 78.8 μg L^{-1} for a PAC dose of 200 mg L^{-1}). According to the authors, such tendency implies that removal of organic nitrogen such as aliphatic amines and free amino acids in algogenic organic matter (autochthonous organic derivative of algae) was substantial by PAC adsorption, as previously highlighted by Ma et al. (2017).

9.2.2 Metal oxides and hydroxides-based adsorbents

9.2.2.1 Metal oxides

Titanium dioxide (TiO_2) is the most commonly used metal oxide in water treatment applications for the photocatalytic degradation of organic contaminants though its ability to mineralize organic matter (Al-Mamun et al., 2019). In several other related research studies, TiO_2 was proposed as an absorbent with interesting NOM capabilities, especially after some modifications (i.e., coatings, immobilization, in composites, etc.).

Along with a quicker removal ability, such trend was mainly justified with the ability to use TiO_2 for the adsorption of NOM, thus preventing the formation potentialities of oxidation by-product (compared with the application photocatalysis). Furthermore, after the adsorption process, the spent TiO_2-based adsorbents can be regenerated using their own photocatalytic property.

In this context, an Australian research team developed and tested a magnetic titanium dioxide ($Fe_3O_4@SiO_2@TiO_2$) for the adsorptive removal of NOM from synthetic and natural water, with the subsequent use of ultraviolet light or simulated solar light to regenerate the adsorbents via photocatalysis, as shown in Fig. 9.2 (Ng et al., 2014).

The main results shed that within the first 5 min, at a neutral pH range, and with a magnetic particle loading of 3 g L^{-1}, around 65% and 39% of NOM was removed from synthetic water and natural water (NW), respectively. Further, the values of formation potentials of THMs in both synthetic water and natural water samples were suppressed to a similar range (205 ± 2 μg L^{-1}) in the first adsorption cycle, despite the initial difference in NOM composition in both waters.

In another study, the adsorption and desorption of a key NOM constituent, HA, was investigated using magnetite (Fe_3O_4) (Fan et al., 2015). It was found that HA adsorption was highly pH dependent, with an optimum removal capacity of 11 mg g^{-1} at pH 3. Desorption wise, 65% of HA was leached after 6 min of contact time in a solution set around pH 10, up until 80% of desorption after 24 h.

For a better understanding of the adsorption mechanism, Zhang et al. (2016) analyzed the hydrophobic interaction in the Fe_2O_3-HA complexes, along with the functional groups of HA under various operating conditions. Their finding revealed that under different ionic strengths (0, 0.005, 0.01, and 0.05 mol/kg), the thermal weight loss of the complex, which indicated its hydrophilicity and hydrophobicity, decreased at first and then increased as the pH value increased from 7 to 12. Overall, it was reported that the

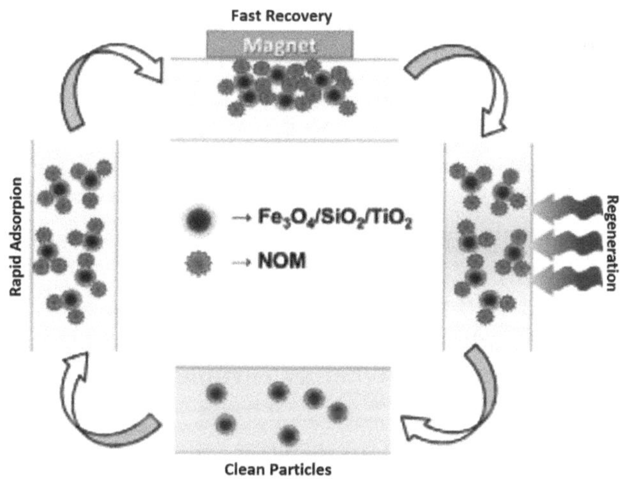

Figure 9.2 Illustration of adsorptive and regenerative magnetic TiO_2 for NOM removal in water (Ng et al., 2014). *(With reprint permission from Elsevier.)*

change trend of HA adsorption amount with solution conditions coincided with that of weight loss of HA-Fe_2O_3 complex. Thus, hydrophobic interactions did affect the adsorption of HA on the metal oxides.

In a more recent investigation, Ma et al. (2020) employed Molecular dynamics simulation to understand the mechanisms through which heated aluminum oxide particles (HAOPs) can remove NOM for water and thus mitigate NOM-fouling more effectively than a conventional adsorbent, namely, powdered ACs. Thus, six model foulants were selected and studied as representative of both high molecular weight (HMW) constituents of NOM (i.e., polysaccharide, humic acid and fulvic acid) and Low MW (LMW) of NOM (i.e., oxalic acid, pyruvic acid and DBP precursor). The main results revealed that, based on the analysis of the trends from the radial distribution function (RDF) plots, HAOPs had higher adsorption capacities toward HMW NOM than powdered AC. According to the authors, such tendency can be explained by the following mechanisms:

- Higher foulant-HAOPs interaction energy
- Increased hydration of the HMW NOM, thus increasing the affinity to the more hydrophilic HAOPs
- Diminished mobility of the foulant in both horizontal and vertical directions once adsorbed, which impedes desorption
- And, higher peak intensity in the RDF between multiple functional groups on the HMW NOM foulants and HAOPs.

9.2.2.2 Metal hydroxides

The most investigated group of metal hydroxides in water treatment are layered double hydroxides (LDHs). Indeed, various studies have demonstrated the efficiency of LDHs to adsorb different organic and inorganic pollutants (Zubair et al., 2017; Khandaker et al., 2021). For the case of NOM remove from water, several research studies were conducted to assess the applicability and efficacy of using LDHs. For instance, Fang et al. (2018) investigated the removal efficiency of HA and FA by raw and calcinated LDHs. The related results showed that calcinating LDH at 500°C enhanced NOM removal with adsorption capacities around 99 $mg\ g^{-1}$ for HA and 98 $mg\ g^{-1}$ for FA at pH 9.5. From an environmental perspective, assays revealed that the release of Al from LDH and its calcinated derivate was negligible and safe to aquatic organisms at $pH > 6.5$. Furthermore, regeneration experiments showed that up to five adsorption/desorption cycles, calcinated LDH showed a good reusability for NOM removal in water treatment with a slight decrease in the first run followed with a steady performance thereafter at a range of 74–83 $mg\ g^{-1}$ for HA and 82–89 $mg\ g^{-1}$ for FA.

More recently, the related research effort focused on the use of LDHs in association with other martials to develop more efficient composites (i.e., higher NOM removal, better reusability, etc.). In this regard, Hu et al. (2021) developed a composite using magnetic biochar (MB) from Enteromorpha marine algae and Mg/Al LDH (MB@MgAL-LDH), and tested it for the adsorptive HA from water. The impacts of varying different operating conditions were studied using the Box-Behnken Design model. The results showed that the highest adsorption capacity of HA by the composite was 5.8 mgg^{-1} at 25°C. Using a thermal regeneration approach, the removal rate remained above 90% after five reusability cycles.

In another related study, a composite of magnetite and Zn/Al LDH was a magnetically separable adsorbent for effective removal of HA (Santosa et al., 2021). The adsorption experiments revealed that the removal of HA using the composite Fe_3O_4−Zn/Al LDH was optimum at pH 4.0, with a maximum capacity estimated according to the Langmuir isotherm model of about 35.28 mg g^{-1} . Furthermore, the authors used alum, the commonly used coagulant, to compare its NOM removal capability with the developed composite. As illustrated in Fig. 9.3, the related tests conducted on real raw water sample showed that Fe_3O_4−Zn/Al LDH was more effective in removing dissolved organic carbons.

9.2.3 Nano-adsorbents

Several R&D studies were conducted to test the applicability and efficiency of various nanoparticulate materials for the removal of NOM and its constituents from water.

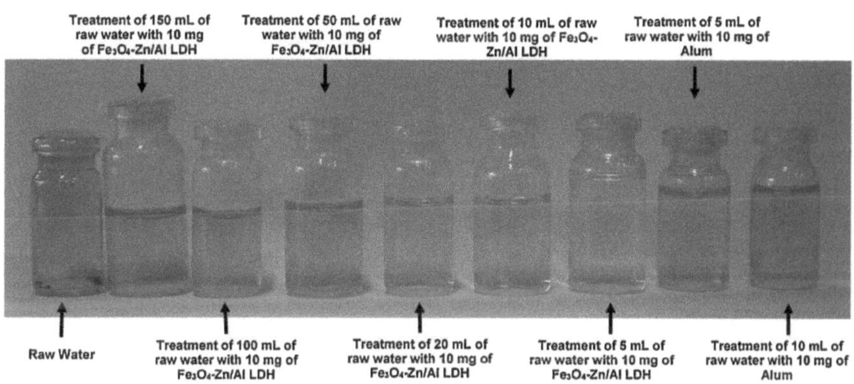

Figure 9.3 Alum versus Fe_3O_4−Zn/Al LDH for the removal of humic substances in real raw water (Santosa et al., 2021). *(With reprint permission from Elsevier.)*

In a related recent study, the efficiency of cerium oxide nanoparticles (CONPs) to remove NOM from water with a special focus on its various UV-Vis spectral indices (Mahato and Gupta, 2021). The developed CONPs were synthesized via two methods, the efficient microwave combustion (CONP-I) and hydroxide mediated approach (CONP-II), and were extensively characterized. For instance, the X-ray powder diffraction analysis showed a cubic structure with an average crystal size of 20.16 nm (CONP-I) and 6.75 nm (CONP-II). The BET analysis revealed respective surface areas of $85.43\ m^2\ g^{-1}$ and $78.59\ m^2\ g^{-1}$, and pore volumes of $0.0073\ cm^3\ g^{-1}$ and $0.0067\ cm^3\ g^{-1}$. Under optimum operating conditions, the maximum NOM adsorption capacity of $239\ mg\ g^{-1}$ was reported at neutral pH for CONP-I. Overall, the authors confirmed that the developed CONPs possess promising efficiency for the simultaneous removal of DOC (94%), UV254 (93%), adsorption slop index (95%), phenolic content (88%), and carboxylic content (73%).

Carbons nanotubes (CNTs) are among the widely experimented nanoparticles in water treatment (Ncibi and Sillanpää, 2015; Minh et al., 2019). For the case for NOM removal, a research study tested modified CNTs for this application. First, the CNTs were oxidized at 140°C for 5 h using nitric acid, and then were bound to the surface of silica sand and used for HA removal (Skandari et al., 2016). The main results showed that the CNTs were able to remove HA with a maximum adsorption capacity of around $82\ mg\ g^{-1}$. Regeneration wise, the experiments revealed that up to five cycles, the adsorption capabilities of the modified CNTs remained roughly stable. Afterward, the adsorption capacity was educed to around 50% of the initial adsorption capacity.

Zulfikar et al. (2016) investigated the synthesis and application of magnetic Fe_3O_4 nanoparticles for the adsorption of HA from aqueous solutions. The maximum adsorption capacity was reported as $43.32\ mg\ g^{-1}$ in this study. In a related study, Fe_3O_4 nanoparticles (MNPs) and silica-coated magnetite nanoparticles (SMNPs) were synthesized and tested as adsorbents for the removal of HA from water (Pasandideh et al., 2016). The main reported results showed that the highest adsorption capacities of HA by MNPs and SMNPs occurred at acidic conditions around pH 3. The kinetic modeling revealed that the main mechanism in the investigated adsorption process involved physisorption (i.e., hydrogen bonding and electrostatic interaction). As for the isotherm modeling analysis, the best fitting model was the Langmuir isotherm with maximum adsorption capacities clearly showing the higher efficacy of SMNPs to remove HA from water (196.07 mg g^{-1}), in comparison with MNPs (96 mg g^{-1}).

In another study, Jiang et al. (2018) investigated the adsorption of HA on nano-amorphous calcium phosphate (nACP). Under optimum operating conditions, the tested nano-absorbent showed the highest HA removal capacity (248 mg g^{-1}) at 30°C. The X-ray photoelectron spectroscopy (XPS) results showed that the adsorption mechanic was involving surface complexation reaction between oxygen-containing group and calcium of HA and nACP, respectively. Furthermore, the regeneration assays, conducted using 0.1 M NaOH, revealed that HA could be effectively desorbed from the absorbent's surface with a slight decrease in the initial adsorption capacity after four adsorption/desorption cycles.

9.2.4 Modified zeolites

Zeolites are highly microporous crystalline aluminosilicate minerals with tetrahedrally connected three-dimensional (3D) frameworks, generally containing either Si or Al as the central cation, and extra framework charge complementary cations such as K^+, Na^+, Ca^{2+}, and Mg^{2+} ions (Islam et al., 2020), thus resulting in valuable physicochemical properties, such as cation exchange, molecular sieving, catalysis, and adsorption (Wang and Peng, 2010).

It is also known that zeolites have little affinity toward organic pollutants due to permanent negative charges in their crystal structures. This same property enables these aluminosilicate minerals to be modified by cationic surfactants for an enhanced adsorption capacity toward organic pollutants (Li et al., 2011).

Thus, several research studies were conducted to investigate the efficacy of various modified zeolites to remove NOM and its constituents from water. For instance, Wang et al. (2016) proposed an innovative zinc oxide-coated zeolite as adsorbent for the removal of HA. The developed materials were extensively characterized (porosimetry analysis, scanning electron microscopy, XRD analysis, and high-resolution transmission electron microscope). The maximum adsorption capacity registered at 21°C was about 60 mg g^{-1}. The authors also confirmed that the positive charge density of ZnO-coated zeolites was proportional to the amount of ZnO coated on zeolite, leading to an enhanced affinity of the ZnO-coated zeolite adsorbents toward negatively charged ions. Also, the tested adsorbent exhibited enhanced adsorptive properties for HA in a wide pH range from 4 to 9.5. Furthermore, as sketched in Fig. 9.4, the authors proposed a 2-stage

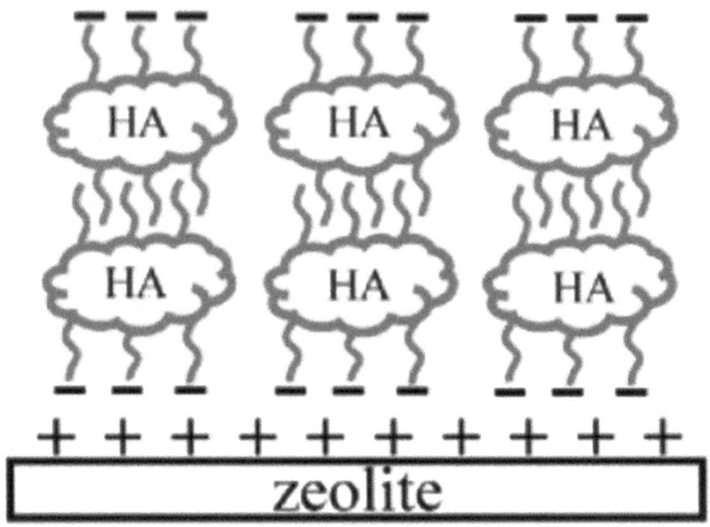

Figure 9.4 Illustration of the 2-stage adsorption mechanism of HA on ZnO-coated zeolite (Wang et al., 2016). *(With reprint permission from Elsevier.)*

adsorption process starting with HA adsorption on the modified zeolite surface, followed by the second step involving HA self-aggregation by π-π stacking interactions due to the abundance of aromatic rings on HA.

In another related work, Liu et al. (2018) studied the adsorption capabilities, in batch and fixed bed column modes, of a surfactant-modified zeolite (SMZ) using the N,N-dimethyl dehydroabietylamine oxide (DAAO). The main results showed that the adsorptive removal ability of the tested adsorbent increased with increasing HA concentrations and contact time, but decreased with increasing pH, temperature, and ionic strength. In the presence of other ionic species, the adsorption of HA on SMZ was increased by the presence of the cations Ca^{2+} and Mg^{2+}, whereas anions such as NO^{3-}, HCO^{3-}, SO_4^{2-}, and PO_4^{3-} showed the opposite trend, and competed with HA adsorption. In addition, it was reported that the maximum adsorption capacity in batch experiments was around 126 mg g^{-1}, with a possible recovery rate of 94% through desorption using 0.1 M NaOH.

In a recent research study, Wang et al. (2021) investigated the mutual influence of HA and Cr(VI) in a water treatment system using ZnO−30N-zeolite (20wt% loading of ZnO on zeolite) as adsorbent. The removal of HA by the modified zeolite reached 90% in 5 min of reaction for both pollutants. Interestingly, the adsorption reaction with HA alone took more than 1 day to reach the same removal rate. Thus, synergetic interaction between HA and Cr(VI) was confirmed for the case of adsorptive removal using ZnO−30N-zeolite.

However, the presence of HA showed no obvious effect on the removal capacity of Cr(VI) by the same absorbent. Overall, the results revealed that both HA and ZnO−30N-zeolite could effectively reduce Cr(VI) to Cr(III), thus reducing chromium's potential toxicity to living organisms.

9.2.5 Polymers

Several polymeric substances were investigated for the adsorptive removal of NOM from water. In this context, molecularly imprinted polymers (MIPs) were proposed as promising adsorbents for the removal of HA from peat water (Zulfikar et al., 2016b). The MIPs were synthesized using methyl methacrylate as a monomer, HA as templates, ethylene glycoldimethacrylate as a cross-linker, and dimethyl formamide as a solvent. The experiments conducted in batch mode showed that the equilibrium data fitted well the Langmuir adsorption isotherms, with a maximum adsorption capacity of around 45 mg g^{-1}.

In another study, Tavengwa et al. (2016) assessed the HA adsorption of the biopolymer cellulose, in comparison with PAC. For the case of cellulose, the optimum operating conditions for HA removal were 2−3 pH, 50 mg of sorbent dose, and a contact time of 30 min. Furthermore, according to the Langmuir isotherm modeling, the maximum adsorption capacity of cellulose was 89 mg g^{-1}, compared with 30.5 mg g^{-1} for PAC.

Liu and Zhou (2018) studied the removal of HA from aqueous solution using polyacrylamide/chitosan semi-interpenetrating polymer network (semi-IPN) hydrogel. The polymeric hydrogel was synthesized via a simple and fast in-situ free radical cross-linking polymerization of acrylamide and N,N′-methylenebisacrylamide, in the presence of chitosan by thermal initiation without nitrogen protection. The effects of several operating conditions on the adsorption capacity of semi-IPN hydrogel toward HA were assessed, including pH, ionic strength, contact time, initial HA concentration, and temperature. The related results showed that the developed semi-IPN hydrogel can be effectively applied in a wide pH range (3−9) and at low ionic strength. The maximum adsorption capacity was registered at around 166 mg g^{-1} at 25°C, attributed mainly to electrostatic interactions and hydrogen bonding.

In a recent investigation, a quaternary ammonium-functionalized β-cyclodextrin polymer (β-CDP) was synthesized and applied for the removal of various organic micropollutants from water, including NOM constituents' HA and FA (Hu et al., 2020). The characterization of the functionalized polymer revealed a structure containing mainly ultra-micropores and a surface area of 89 $m^2 g^{-1}$. Removal wise, β-CDP showed maximum adsorption capacities of 40 $mg g^{-1}$ and 166 $mg g^{-1}$ for HA and FA, respectively.

9.2.6 Miscellaneous adsorbents

Numerous other adsorbents were studied for the removal of NOM and its constituents from water. This selectively includes the following:

- Micro- and mesoporous covalent triazine-based framework (CTF) which were synthesized and applied in batch experiments to assess its adsorption capability toward HA (Liu et al., 2015). The structural, porous, and surface characterization of the tested adsorbents revealed that mesoporous CTF showed faster adsorption kinetics of HA, mainly attributed to the open and wide pore size distribution. It was also reported that, as the pH increased from 3 to 11, the adsorption removal capacity decreased from 48 $mg g^{-1}$ to 5.6 $mg g^{-1}$, which implies that the pH-mediated protonation or deprotonation of the surface functional groups has a significant effect on adsorption of HA on CTF.
- Biowastes such coconut fiber (CF) and palm oil fiber (POF) were examined for the adsorptive removal of NOM (Din et al., 2016). Studies on isotherms and kinetic models for NOM adsorption showed that the equilibrium data were fitted to Langmuir isotherm model for both biowastes with adsorption capacities estimated around 16 $mg g^{-1}$ and 31 $mg g^{-1}$ by the CF and POF, respectively. For the kinetics analysis, the most suitable adsorption model was the pseudo-first-order kinetic model for POF and pseudo-second-order kinetic model for CF.
- Another wastes, coal gasification fine slag (CGFS) was tested for the adsorption and desorption of HA from water. This amorphous silica-alumina solid waste, with large surface area and broad pore-size distribution, showed promising adsorption performance with a Langmuir's maximum adsorption capacity of 61 $mg g^{-1}$, at pH 7 and 20°C. The desorption rates were registered at 75.5% ± 0.5%, and electrostatic interaction and hydrogen bonding were reported to be the main adsorption mechanisms during HA adsorption on CGFS.
- Nanocomposites of magnetic attapulgite-resin (MPR) were developed by ultrasonic dispersion and tested for HA adsorption from water (Xi-Jun et al., 2018). The experimental results showed that, with a

specific surface area of about 674 $m^2\ g^{-1}$, MPR composite exhibited a higher adsorption capacity than its separate components of resin and attapulgite. It was also reported that the adsorption capacity increased with the contact time, and reached equilibrium after 233 mi with an adsorption capacity of 13 $mg\ g^{-1}$.

- Two nanoclays, bentonite and montmorillonite nanoparticles, were investigated to their adsorption capabilities to remove HA form was in batch mode (Derakhshani and Naghizadeh, 2018). The impacts of different operating conditions such as pH, contact time, HA initial concentration, and adsorbent dosage were assessed. Under optimum conditions (pH 3 for an HA concentration of 40 $mg\ L^{-1}$), the highest adsorption capacities reached 58 $mg\ g^{-1}$ and 48 $mg\ g^{-1}$, respectively for bentonite and montmorillonite nanoparticles.
- Waste pulp residues were used for the synthesis of an adsorbent applicable for HA removal from water (Sehaqui et al., 2015). Cellulose pulp was etherified with a quaternary ammonium salt in water to introduce positive charges onto the surface of the pulp fibers, and subsequently mechanically disintegrated into cellulose nanofibers (CNF) with substantially higher surface area. The main results showed that the adsorption capacity of CNTs and the related kinetics in HA removal were found to be promoted by both acidic pH conditions and the surface charge content of the pulp derived nano-adsorbent. At pH 6, the maximum adsorption capacity of the functionalized CNF was estimated around 310 $mg\ g^{-1}$.
- Physically and chemically treated clay/biopolymer composite based on bentonite and chitosan was prepared and tested for adsorptive removal of HA from water (Dehghani et al., 2018). After assessing the effects of several parameters, it was found that the optimum pH to remove HA was pH 4. Further, based on the Langmuir isotherm model, the maximum monolayer adsorption capacity for HA by the bentonite/chitosan composite was estimated at 91 $mg\ g^{-1}$ (under pH 4 and 25°C). Kinetic studies were also carried out and the analysis revealed that the adsorption of HA followed the pseudo-second order model, thus suggesting a predominately chemisorption.
- Barberry stem powder (BSP) and barberry stem ash (BSA) were investigated for HA removal from water via adsorption (Khodadadi et al., 2019). Under optimum conditions (pH 7, reaction time 10 min, 15°C, initial HA concentration = 40 $mg\ L^{-1}$, and adsorbent amount = 1 $g\ L^{-1}$), the results showed that the adsorption process is

favorable for both adsorbents. The equilibrium data were in good fit with the Langmuir model, with maximum removal capacities of 20 mg g^{-1} and 17 mg g^{-1} for BSP and BSA, respectively.

9.2.7 Integrated processes

Despite the huge numbers of adsorbents from various origins and subjected to different chemical and thermal treatment produces, along with high NOM removal capabilities, several issues were reported for large scale applications of the adsorption process to remove NOM, and other pollutants for that matter. Such challenges include the pollution transfer issue when desorption and regeneration studies are not conducted, which is the case in most lab-scale adsorption research studies. Fluctuating parameters during large scale application such pH, pollutants varying composition and concentration, temperature, etc. also substantially impact the efficiency and scalability of adsorption systems.

Overall, continuing the search for highly efficient adsorbents, either for a versatile or specific application is very important, but to ensure a highly efficient water treatment processes (which is the ultimate goal), integrating two or more of the technologies with proven NOM removal capabilities is a viable strategy. Thus, in the following section, two key adsorption-comprising integration schemes for NOM removal from water are presented, namely adsorption-coagulation with and adsorption-membrane separation.

9.2.7.1 Adsorption and coagulation

This integration involving adsorption and coagulation is an interesting scenario considering the simple fact that both techniques with widely used in water treatment processes around the world. Research wise, numerous studies were, and are being, conduced to assess the applicability and efficiency of such hybrid water treatment system for the removal of NOM and its constituents.

Among those studies, Zhou et al. (2020) investigated the use of nano-MgO for the removal of NOM from water (cf. Fig. 9.5). Interestingly, it was proven that MgO acted as a two-in-one coagulant and adsorbent with the very high HA removal capacity of 1260 mg g^{-1}.

The MgO treatment process was also found to be pH sensitive and weak acidic condition was enhanced NOM removal efficiency. Regeneration wise, MgO can be reused in more than 10 cycles in a chemicals-free method by annealing $Mg(OH)_2$/Mg-NOM composite at 500°C. More

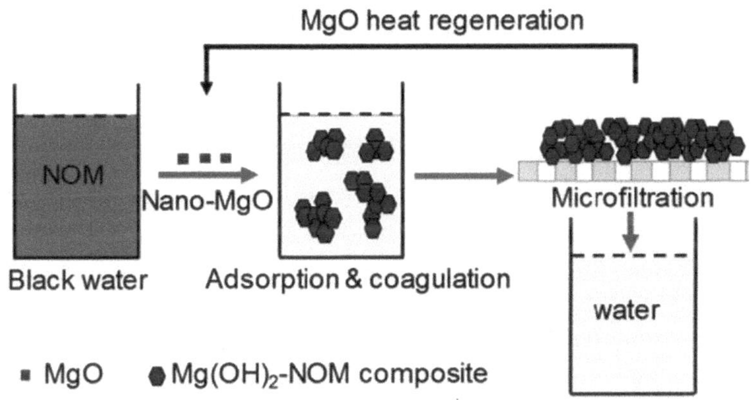

Figure 9.5 Recycling-based process for NOM removal using MgO (Zhou et al., 2020). *(With reprint permission from Elsevier.)*

importantly, no solid waste is generated in this novel process. This MgO-recycling NOM-removal process is simple, efficient, and sustainable for water NOM removal and will be significant in promoting novel sustainable technologies for NOM- or HA-related water remediation and treatment while minimizing the generation of solid waste.

In another recent study, a novel process of adsorption combined with enhanced coagulation-flocculation (i.e., spiked with magnetic nano-adsorbents) was developed and tested for the removal of NOM from drinking water (Kumari and Gupta, 2020). The simultaneous adsorption and enhanced coagulation-flocculation process for NOM and turbidity removal was conducted for the first stage using magnetic nano-adsorbents, followed in the second stage by coagulation-flocculation experiments using conventional coagulants like alum and ferric chloride. Overall, the results showed that a combined process using magnetic nano-adsorbents and alum was highly efficient in removing dissolved organic matter (98.7%), ultraviolet absorbance at 254 nm (99.7%), and trihalomethane formation potential (91%).

Wongcharee et al. (2020) investigated separate and combined processes of enhanced coagulation and adsorption for NOM removal from dam water. After assessing the separate efficiencies of coagulation (using alum) and adsorption (using activated carbon-zeolite nano-adsorbents), the simultaneous application of enhanced coagulation and complementary adsorption using 80 mg L^{-1} of alum and 150 mg L^{-1} of the nano-adsorbents showed high removal performances with 76% ammonia and

90% DOC reduction in 20 min time. Application wise, the authors suggested that such concurrent treatment can be simply implemented utilizing readily available rapid mixers and flocculators.

9.2.7.2 *Adsorption and membrane separation*

The main goal of such integration scheme is to applied adsorption as a pretreatment to mitigate the serious issue of fouling during the subsequent membrane separation process, which is highly relevant for the case of treating NOM-rich waters.

Several interesting R&D studies were carried out in this topic including a study focusing on the application of different adsorbents, and the impact of their use on ultrafiltration (UF) membrane fouling caused by NOM (Li et al., 2020). The integrated adsorption−UF system was made of a continuous-flow hollow fiber UF set-up and two types of adsorbents, commercial PAC and a mesoporous adsorbent resin (MAR). For the synthetic water composed of algal organic matter, membrane fouling of PAC/UF and MAR/UF systems was reduced by 40.7% and 83.3%, respectively. Using NOM-rich natural water from the Songhua River, it was reported that membrane fouling of the PAC/UF system was increased by 39.5% compared with that of UF alone, whereas the addition of MAR mitigated membrane fouling by 66.4%.

In another investigation, the efficiency of two pretreatment methods based on heated aluminum oxide particles (HAOPs) as adsorbents for NOM removal was assessed, along with the impacts on membrane fouling and THM formation potential (Wang and Benjamin, 2016).

- The first method was referred to as pre-adsorption in which HAOPS were added to the water and thoroughly mixed. Afterward, the particles were separated from the water either upstream of or by the membrane.
- In the second method, pre-deposition, mode, a thin layer of adsorbent particles was deposited on a support media or the membrane) itself, prior to passing the water feed through the layer and the membrane.

The main results showed that DOC removal was similar for the two methods, increasing steadily with increasing HAOPs dose up to 50%−55% removal for an adsorbent dosage of at a dose of 20 mg L^{-1}. Then, it was revealed that passing the water through a pre-deposited layer of HAOPs removed more NOM from the solution than was achieved by pre-adsorption.

Another research team also studied the impacts of HAOPs to remove NOM and alleviate membrane fouling, but in association with PAC

(Modarresi and Benjamin, 2019). First, the authors demonstrated that, for any given adsorbents dosage, slightly more NOM removal occurred in microgranular adsorptive filtration (μGAF) reactor than in the batch one. Also, it was found that HAOPs and PAC selectively adsorbed different fractions of NOM, with HAOPs being more effective at accumulating larger molecules (humic substances and building blocks), and PAC more effective for smaller molecules (LMW acids and neutrals).

Finally, an interesting R&D study investigated the impact of four different pre-treatment methods on membrane fouling and NOM removal efficiency, namely adsorption, coagulation, coagulation followed by adsorption (C-A), and simultaneous coagulation and adsorption (C + A) (Bu et al., 2019). The results showed that the adsorption pretreatment required a large amount of adsorbent, which led to the formation of a dense cake layer on the membrane surface, and thus a severe membrane fouling. Compared to adsorption alone, the coagulation and C-A processes decreased the transmembrane pressure by 4.9 kPa due to a reduced accumulation of particles on the membrane surface. Overall, among the various tested processes, the integrated C-A/ultrafiltration system achieved the highest NOM removal efficiency.

9.3 Conclusion and outlook

In the present chapter, we aimed at summarizing recent R&D effort around the globe on the adsorption of NOM and its constituents (mainly HA and to a lesser extent FA) from synthetic and natural waters using a wide range of adsorbents, including both inexpensive and locally available materials as well as advanced composites and nanomaterials. The far from extensive list of the selected and studied adsorbents included biochars, granular and powdered activated carbons, carbon nanotubes, iron oxide particles (e.g., goethite and hematite), iron oxide-coated materials, chitosan and zeolite composite, molecularly imprinted polymers, semi-interpenetrating polymer network, several nanoclays, and some industrial wastes (e.g., coal gasification fine slag) and agricultural residues (e.g., coconut fiber, palm oil fiber, and barberry stem).

In most studies, it was noticed that researchers methodically opted for a pretreatment stage to modify the structure and/or surface chemistry of potential adsorbents because of the limited removal capacities of unmodified adsorbents to remove NOM from water. Thus, several modification methods were applied, including carbonization, chemical activation, coating, and hybridization. With such modifications, and despite the enhanced NOM removal, some issues were noticed mainly the risks of leaching of the activating chemical agents in the treated water, and the possible leak of nanomaterials into the environment with the potential adverse impact on living organisms in the receiving water bodies, hence the need for more attention to those issues in NOM removal studies using adsorption.

Overall, R&D on the adsorption technique for the removal of NOM from water bodies, the development of novel adsorbing materials, and especially their application in integrated water treatment processes, need to be further developed and optimized. The main objective is to innovate a sustainable solution to control the presence of NOM and its constituents in water and alleviate their interfering negative impacts during the treatment of other toxic pollutants. Lately, with the emergence and gradual expansion worldwide of the two emerging sustainable concepts of circular economy and bioeconomy, the application of adsorption is that water treatment needs to be aligned with the principles of those paradigms. Within circular economy, the research directions to be promoted include enhancing the adsorption process without increasing the adsorbents dosage but with more highly efficient hybrid ones, as well as cost-effective schemes to regenerate and reuse those adsorbing materials. In the context of the bioeconomy concept, more R&D efforts need to be put in the production and application of low-cost bio-based carbonaceous, polymerics and nanostructured particles from bioresources, and especially from biowastes. Such research direction will enable the development of cost-effectives and eco-friendly solutions toward a sustainable water treatment process targeting the removal of NOM and other pollutants.

References

M.R. Al-Mamun, S. Kader, M.S. Islam, M.Z.H. Khan, Photocatalytic activity improvement and application of UV-TiO2 photocatalysis in textile wastewater treatment: a review, J. Environ. Chem. Eng. 7 (5) (2019) 103248.

F. Bu, B. Gao, Q. Yue, et al., The combination of coagulation and adsorption for controlling ultra-filtration membrane fouling in water treatment, Water 11 (1) (2019) 90.

E.A. Crafton, D. Wu, S.E. Duirk, Comparison of anion exchange resins and aluminum-based coagulants for natural organic matter (NOM) removal and disinfection by-product (DBP) formation, Water Sci. Technol. Water Supply 14 (4) (2014) 577–584.

M.H. Dehghani, A. Zarei, A. Mesdaghinia, et al., Production and application of a treated bentonite–chitosan composite for the efficient removal of humic acid from aqueous solution, Chem. Eng. Res. Des. 140 (2018) 102–115.

E. Derakhshani, A. Naghizadeh, Optimization of humic acid removal by adsorption onto bentonite and montmorillonite nanoparticles, J. Mol. Liq. 259 (2018) 76–81.
M.F.M. Din, M. Ponraj, W.P. Low, et al., Removal rate of organic matter using natural cellulose via adsorption isotherm and kinetic studies, Water Environ. Res. 88 (2) (2016) 118–130.
J. Fan, F. Liu, Y. Hu, J. Chen, Effects of pH and ionic composition on sorption/desorption of natural organic matter on zero-valent iron and magnetite nanoparticles, Water Sci. Technol. 72 (2) (2015) 303–310.
L. Fang, J. Hou, C. Xu, et al., Enhanced removal of natural organic matters by calcined Mg/Al layered double hydroxide nanocrystalline particles: adsorption, reusability and mechanism studies, Appl. Surf. Sci. 442 (2018) 45–53.
S.L. Gora, S.A. Andrews, Removal of natural organic matter and disinfection byproduct precursors from drinking water using photocatalytically regenerable nanoscale adsorbents, Chemosphere 218 (2019) 52–63.
X. Hu, G. Xu, H. Zhang, et al., Multifunctional β-cyclodextrin polymer for simultaneous removal of natural organic matter and organic micropollutants and detrimental microorganisms from water, ACS Appl. Mater. Interfaces 12 (10) (2020) 12165–12175.
Y. Hu, X. Zheng, S. Liu, et al., Prediction and optimization of adsorption performance of MC@ MgAl-LDH for the removal of humic acid from aqueous solution: BBD model and mechanism, J. Solid State Chem. 302 (2021) 122377.
G. Hua, D.A. Reckhow, I. Abusallout, Correlation between SUVA and DBP formation during chlorination and chloramination of NOM fractions from different sources, Chemosphere 130 (2015) 82–89.
M.A. Islam, D.W. Morton, B.B. Johnson, M.J. Angove, Adsorption of humic and fulvic acids onto a range of adsorbents in aqueous systems, and their effect on the adsorption of other species: a review, Separ. Purif. Technol. 247 (2020) 116949.
L. Jiang, Y. Li, Y. Shao, Y. Zhang, et al., Enhanced removal of humic acid from aqueous solution by novel stabilized nano-amorphous calcium phosphate: behaviors and mechanisms, Appl. Surf. Sci. 427 (2018) 965–975.
S. Khandaker, M.T. Hossain, P.K. Saha, et al., Functionalized layered double hydroxides composite bio-adsorbent for efficient copper (II) ion encapsulation from wastewater, J. Environ. Manag. 300 (2021) 113782.
M. Khodadadi, T.J. Al-Musawi, M. Kamranifar, et al., A comparative study of using barberry stem powder and ash as adsorbents for adsorption of humic acid, Environ. Sci. Pollut. Control Ser. 26 (25) (2019) 26159–26169.
M. Kumari, S.K. Gupta, A novel process of adsorption cum enhanced coagulation-flocculation spiked with magnetic nanoadsorbents for the removal of aromatic and hydrophobic fraction of natural organic matter along with turbidity from drinking water, J. Clean. Prod. 244 (2020) 118899.
C. Li, Y. Dong, D. Wu, et al., Surfactant modified zeolite as adsorbent for removal of humic acid from water, Appl. Clay Sci. 52 (4) (2011) 353–357.
K. Li, S. Li, C. Sun, et al., Membrane fouling in an integrated adsorption–UF system: effects of NOM and adsorbent properties, Environ. Sci. J. Integr. Environ. Res. 6 (1) (2020) 78–86.
J. Liu, J. Cao, H. Chen, D. Zhou, Adsorptive removal of humic acid from aqueous solution by micro-and mesoporous covalent triazine-based framework, Colloids Surf. A Physicochem. Eng. Asp. 481 (2015) 276–282.
S. Liu, W. Zhang, X. Tan, et al., Performance of a zeolite modified with N, N-dimethyl dehydroabietylamine oxide (DAAO) for adsorption of humic acid assessed in batch and fixed bed columns, RSC Adv. 8 (16) (2018) 9006–9016.
Z. Liu, S. Zhou, Removal of humic acid from aqueous solution using polyacrylamide/chitosan semi-IPN hydrogel, Water Sci. Technol. 2017 (1) (2018) 16–26.

C. Ma, H. Pei, W. Hu, et al., The enhanced reduction of C-and N-DBP formation in treatment of source water containing Microcystis aeruginosa using a novel CTSAC composite coagulant, Sci. Total Environ. 579 (2017) 1170–1178.

Y. Ma, S. Velioğlu, M.B. Tanis-Kanbur, et al., Mechanistic understanding of the adsorption of natural organic matter by heated aluminum oxide particles (HAOPs) via molecular dynamics simulation, J. Membr. Sci. 598 (2020) 117651.

J.K. Mahato, S.K. Gupta, Exceptional adsorption of different spectral indices of natural organic matter (NOM) by using cerium oxide nanoparticles (CONPs), Environ. Sci. Pollut. Control Ser. (2021) 1–10.

H.R. Mian, G. Hu, K. Hewage, M.J. Rodriguez, R. Sadiq, Prioritization of unregulated disinfection by-products in drinking water distribution systems for human health risk mitigation: a critical review, Water Res. 147 (2018) 112–131.

T.D. Minh, M.C. Ncibi, V. Srivastava, et al., Gingerbread ingredient-derived carbons-assembled CNT foam for the efficient peroxymonosulfate-mediated degradation of emerging pharmaceutical contaminants, Appl. Catal. B Environ. 244 (2019) 367–384.

S. Modarresi, M.M. Benjamin, Using adsorbent mixtures to mitigate membrane fouling and remove NOM with microgranular adsorptive filtration, J. Membr. Sci. 573 (2019) 528–533.

A. Naghizadeh, F. Momeni, H. Kamani, Study of ultrasonic regeneration and adsorption of humic acid on activated carbon, Health Scope 7 (2) (2018).

M.C. Ncibi, M. Sillanpää, Optimized removal of antibiotic drugs from aqueous solutions using single, double and multi-walled carbon nanotubes, J. Hazard Mater. 298 (2015) 102–110.

M. Ng, E.T. Kho, S. Liu, et al., Highly adsorptive and regenerative magnetic TiO2 for natural organic matter (NOM) removal in water, Chem. Eng. J. 246 (2014) 196–203.

K.Y. Park, Y.J. Yu, S.J. Yun, J.H. Kweon, Natural organic matter removal from algal-rich water and disinfection by-products formation potential reduction by powdered activated carbon adsorption, J. Environ. Manag. 235 (2019) 310–318.

E.K. Pasandideh, B. Kakavandi, S. Nasseri, et al., Silica-coated magnetite nanoparticles core-shell spheres (Fe 3 O 4@ SiO 2) for natural organic matter removal, Journal of Environmental Health Science and Engineering 14 (1) (2016) 1–13.

S.J. Santosa, P.A. Krisbiantoro, T.T.M. Ha, et al., Composite of magnetite and Zn/Al layered double hydroxide as a magnetically separable adsorbent for effective removal of humic acid, Colloids Surf. A Physicochem. Eng. Asp. 614 (2021) 126159.

H. Sehaqui, U.P. de Larraya, P. Tingaut, T. Zimmermann, Humic acid adsorption onto cationic cellulose nanofibers for bioinspired removal of copper (II) and a positively charged dye, Soft Matter 11 (26) (2015) 5294–5300.

M. Sillanpää, M.C. Ncibi, A. Matilainen, Advanced oxidation processes for the removal of natural organic matter from drinking water sources: a comprehensive review, J. Environ. Manag. 208 (2018) 56–76.

S. Skandari, A. Torabian, G. Nabi Bidhendi, et al., Preparation of engineered carbon nanotube materials and its application in water treatment for removal of hydrophobic natural organic matter (NOM), Desalination Water Treat. 57 (52) (2016) 24855–24866.

H. Tafvizi, S. Chowdhury, T. Husain, Low cost activated carbon for removal of NOM and DBPs: optimization and comparison, Water 13 (16) (2021) 2244.

N.T. Tavengwa, L. Chimuka, L. Tichagwa, Equilibrium and kinetic studies on the adsorption of humic acid onto cellulose and powdered activated carbon, Desalination Water Treat 57 (36) (2016) 16843–16854.

H.B. Truong, I.A. Ike, Y.S. Ok, J. Hur, Polyethyleneimine modification of activated fly ash and biochar for enhanced removal of natural organic matter from water via adsorption, Chemosphere 243 (2020) 125454.

L.F. Wang, M.M. Benjamin, A multi-spectral approach to differentiate the effects of adsorbent pretreatments on the characteristics of NOM and membrane fouling, Water Res. 98 (2016) 56–63.

L. Wang, D.D. Dionysiou, J. Lin, et al., Removal of humic acid and Cr (VI) from water using ZnO–30N-zeolite, Chemosphere 279 (2021) 130491.

L. Wang, C. Han, M.N. Nadagouda, D.D. Dionysiou, An innovative zinc oxide-coated zeolite adsorbent for removal of humic acid, J. Hazard Mater. 313 (2016) 283–290.

S. Wang, Y. Peng, Natural zeolites as effective adsorbents in water and wastewater treatment, Chem. Eng. J. 156 (1) (2010) 11–24.

S. Wongcharee, V. Aravinthan, L. Erdei, Removal of natural organic matter and ammonia from dam water by enhanced coagulation combined with adsorption on powdered composite nano-adsorbent, Environ. Technol. Innovat. 17 (2020) 100557.

M. Xi-Jun, X. Xing-Yong, J. Xiao-Wu, et al., Adsorption Property of Humic Acid in Micro-polluted Water by Organic-Inorganic-Magnetic Resin, 2018.

K. Yang, J.T. Fox, Adsorption of humic acid by acid-modified granular activated carbon and powder activated carbon, J. Environ. Eng. 144 (10) (2018) 04018104.

H. Yang, S. Ye, J. Wang, et al., The approaches and prospects for natural organic matter-derived disinfection byproducts control by carbon-based materials in water disinfection progresses, J. Clean. Prod. (2021) 127799.

M.R. Yazdani, N. Duimovich, A. Tiraferri, P. Laurell, M. Borghei, J.B. Zimmerman, R. Vahala, Tailored mesoporous biochar sorbents from pinecone biomass for the adsorption of natural organic matter from lake water, J. Mol. Liq. 291 (2019) 111248.

N. Yin, S. Liang, S. Liang, et al., DEP and DBP induce cytotoxicity in mouse embryonic stem cells and abnormally enhance neural ectoderm development, Environ. Pollut. 236 (2018) 21–32.

W. Yu, T. Liu, J. Crawshaw, T. Liu, N. Graham, Ultrafiltration and nanofiltration membrane fouling by natural organic matter: mechanisms and mitigation by pre-ozonation and pH, Water Res. 139 (2018) 353–362.

Y. Zhang, M. Li, S. Wu, Q. Zhang, Research on hydrophilicity and hydrophobicity of adsorption of NOM on metal oxide/water interface, Desalination and Water Treat. 57 (5) (2016) 1940–1948.

J. Zhou, Y. Xia, Y. Gong, et al., Efficient natural organic matter removal from water using nano-MgO coupled with microfiltration membrane separation, Sci. Total Environ. 711 (2020) 135120.

M. Zubair, M. Daud, G. McKay, et al., Recent progress in layered double hydroxides (LDH)-containing hybrids as adsorbents for water remediation, Appl. Clay Sci. 143 (2017) 279–292.

M.A. Zulfikar, F.I. Suri, H. Rusnadi Setiyanto, et al., Fe3O4 nano-particles prepared by co-precipitation method using local sands as a raw material and their application for humic acid removal, Int. J. Environ. Stud. 73 (1) (2016a) 79–94.

M.A. Zulfikar, D. Wahyuningrum, R.R. Mukti, H. Setiyanto, Molecularly imprinted polymers (MIPs): a functional material for removal of humic acid from peat water, Desalination Water Treat. 57 (2016b) 15164–15175.

CHAPTER 10

Ion exchange

Mika Sillanpää[1,2,3], Irina Levchuk[4] and Yuri Park[5]
[1]Department of Chemical Engineering, School of Mining, Metallurgy and Chemical Engineering, University of Johannesburg, Doornfontein, South Africa; [2]Chemistry Department, College of Science, King Saud University, Riyadh, Saudi Arabia; [3]Department of Biological and Chemical Engineering, Aarhus University, Nørrebrogade, Aarhus C, Denmark; [4]Fine Particle and Aerosol Technology Laboratory, Department of Environmental and Biological Sciences, University of Eastern Finland, Kuopio, Finland; [5]Institute of Environmental Technology, Department of Environmental Engineering, Seoul National University of Science and Technology, Seoul, South Korea

Abstract

The presence of natural organic matter (NOM) in drinking water causes difficulties during conventional water treatment processes such as coagulation. Problems also arise when alternative treatment techniques for NOM removal, such as nanofiltration (NF), are applied. The most common problem with NF is membrane fouling. Ion exchange (IE) processes proved to be an efficient NOM removal technology, and are recommended for use at the beginning of the treatment process. This approach not only significantly decreases the concentration of NOM but also prevents the formation of disinfection by-products, such as trihalomethanes. This chapter reviews recent studies conducted on NOM removal from water by IE.

Keywords: Ion exchange; Natural organic matter (NOM); Water treatment.

Abbreviations

BDOC:	Biodegradable dissolved organic carbon
DBP:	Disinfection by-product
DOC:	Dissolved organic carbon
DOM:	Dissolved organic matter
ERD:	Effective resin dose
FIX:	Fluidized ion exchange
FP:	Formation potential
HAAs:	Haloacetic acids
HRT:	Hydraulic residence time
IE:	Ion exchange
MIEX:	Magnetic ion exchange resin
NF:	Nanofiltration
NOM:	Natural organic matter
PAC:	Powdered activated carbon
SHA:	Slightly hydrophobic acid
SUVA:	Specific UV absorbance
THM:	Trihalomethane

Natural Organic Matter in Water
ISBN 978-0-12-824274-2
https://doi.org/10.1016/B978-0-12-824274-2.00001-6

THMFP: Trihalomethane formation potential
TOC: Total organic carbon
VHA: Very hydrophobic acid

10.1 Introduction

This chapter introduces the basics of ion exchange (IE) and briefly describes types of ion exchangers and recent findings on NOM removal from water by the IE process.

An *ion exchange process* is commonly defined as the reversible exchange of ions between the solid and liquid phases (Harland, 1994; Wachinski and Etzel, 1997). The solid phase is also known as *ion exchanger* or *ion exchange resin* and, being insoluble in the liquid phase, it carries exchangeable ions and does not undergo substantive structural changes during the reaction (Helfferich, 1995). Depending on exchangeable ions, ion exchangers can be divided into cation and anion exchangers. Simultaneous cation and anion exchange can also be attributed to certain types of materials, known as *amphoteric ion exchangers*. Typical cation and anion exchange reactions are shown in Eqs. (10.1) and (10.2), respectively, where X is the structural unit of the ion exchanger.

$$2NaX + CaCl_2(aq) \leftrightarrow CaX_2 + 2NaCl(aq) \tag{10.1}$$

$$2XCl + Na_2SO_4(aq) \leftrightarrow X_2SO_4 + 2Nacl(aq) \tag{10.2}$$

Reaction (10.1) is attributed to the softening of water. During the typical water softening process, hard water is pumped through an IE column (NaX) and cations such as calcium are removed from the water and replaced by an equivalent amount of sodium. When all sodium ions in the cation exchanger are replaced, it is considered to be "exhausted"; however, "regeneration" of IE resin is easily achievable by pumping in a sodium salt solution (Helfferich, 1995). Schematic presentation of this process is shown in Fig. 10.1. The left part of Fig. 10.1 demonstrates the initial step of the process, when contact between ion exchanger and water containing calcium cations starts. On the right side of Fig. 10.1, achieved equilibrium is shown.

The unique properties of ion exchangers can be explained by their structure, which, according to the common definition, is a framework carrying a positive or negative surplus charge, compensated by opposite ions also known as *counter-ions*. Since the counter-ions move freely, they can

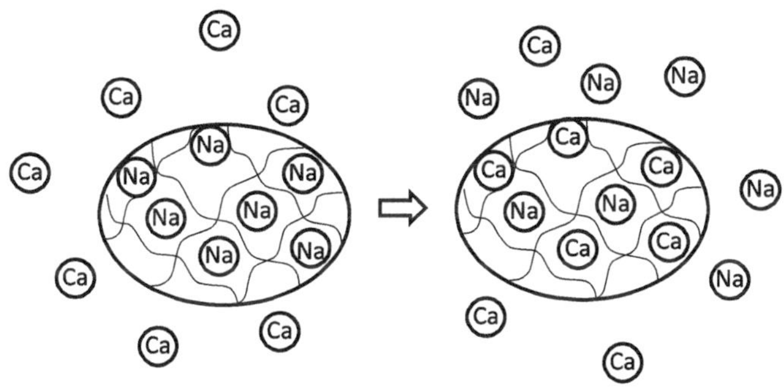

Figure 10.1 Schematic representation of NA^{+}—CA^{2+} ion exchange (IE) process. *(Adapted from Helfferich (1995).)*

easily be replaced by other ions that have the same sign. The content of counter-ions in an ion exchanger is known as *ion-exchange capacity*. Another important characteristic of ion exchangers is selectivity, which is commonly defined as the ability of an ion exchanger to distinguish between different species of counter-ions (Helfferich, 1995).

Thus, the ideal ion exchanger should meet following requirements (Harland, 1994):

- hydrophilicity;
- chemical and physical stability;
- relatively high speed of IE;
- sufficient IE capacity;
- particle size and effective surface area that are adequate to the application.

An IE process can be conducted in a batch or column method. In the batch process, IE resin is mixed with solution in a tank, after which it is separated. The main drawback of this method is the inefficiency of batchwise resin regeneration. Hence, the number of applications for the batch IE process is limited, and the column method is widely used on an industrial scale (Wachinski and Etzel, 1997).

IE properties can be attributed to a wide range of natural and synthetic materials. Among the natural materials, aluminosilicates are well known mainly as cation exchangers; however, some of them exhibit anion exchanging properties (apatite and hydroxylapatite; Wachinski and Etzel, 1997). For instance, recently it was demonstrated that natural zeolite, consisting mainly of clinoptilolite obtained from the Semnan province in Iran, is a suitable ion exchanger for NH_4^+ ion removal from water. Thus, it has potential for application in water treatment and the agricultural industry, since it is an environmentally friendly and relatively inexpensive ion exchanger (Malekian et al., 2011). Other examples of natural ion exchangers are different types of coal, alumina, alginic acid, collodine, and keratin. Some natural materials gain IE properties after certain chemical treatment, for instance, sulfonation and phosphorylation is successfully applied to wood, tar, nut shells, and cotton (Wachinski and Etzel, 1997).

Most of the ion exchangers used in water treatment and other industrial applications are synthetic resins. Different types of IE resins can be distinguished

1. strong-acid cation resins;
2. weak-acid cation resins;
3. strong-base anion resins;
4. weak-base anion resins;
5. metal selective chelating resins.

The chemical behavior of strong-acid cation resins resembles the behavior of strong acids, which explains the origin of their name. The chemical matrix of this type of resin consists of styrene and divinylbenzene and the functional groups are sulfonic acid radicals. The behavior of weak-acid cation resin is similar to that of weak organic acids that are weakly dissociated. The functional group is carboxylic acid.

Strong-base anion resins can be divided into two main types. Resins of the first type contain a functional group of three methyl groups, while in resins of the second type an ethanol group replaces one of the methyl groups. Strong-base exchangers of the first type are more stable, whereas exchangers of the second type have a higher capacity and regeneration efficiency (Wachinski and Etzel, 1997).

The weak-base anion exchangers are normally based on phenol—formaldehyde or epoxy matrices with a functional group of secondary or tertiary amines (Wachinski and Etzel, 1997). Metal selective chelating resins, such as ethylenediaminetetraacetic acid, behave in a similar way to weak-acid cation resins, and have a high selectivity for metals Eckenfelder (2000).

10.2 Removal of NOM from water by ion exchange

What follows is a brief overview of recent developments in NOM removal from water by IE. The presence of NOM in treated water implies difficulties during the purification process and problems with water quality after treatment. For instance, NOM causes membrane fouling and decreases activated carbon efficiency (Kennedy et al., 2008; Kim and Dempsey, 2013; Lee et al., 2004; Teixeira and Sousa, 2013; Zularisam et al., 2007; Gibert et al., 2013a,b). NOM can also react with chemicals used during the water treatment process, leading to the formation of disinfection by-products (DBPs), for instance, potentially carcinogenic chlorinated hydrocarbons can be formed after water chlorination (Yang et al., 2013; Tian et al., 2013; Bolto et al., 2004).

Anion exchange water treatment is well known as an alternative to conventional water treatment in cases of high NOM concentration in raw water (Hongve et al., 1999). It was reported that NOM concentration in drinking water is usually in the range of 2–10 ppm, of which only 10%–30% are defined species (Bolto et al., 2002). According to some estimation, from 10% to 40% of NOM is not removed after IE, which probably correlates with the amount of uncharged species in NOM (Bolto et al., 2004).

Three hydrophobic and three hydrophilic fractions of NOM can be distinguished (Nkambule et al., 2009). The main constituents of hydrophobic fractions are humic acids (HAs) and fulvic acids (FAs), while low molecular weight (MW) carbohydrates, proteins, and amino acids are often found in the hydrophilic fraction of NOM. The distribution of organic compounds for average river water can be presented in following way: 40% FAs, 10% HAs, 30% hydrophilic acids, 10% carbohydrates, 6% carboxylic acids, and 4% amino acids (Bolto et al., 2004). Depending on the fraction of NOM, deviations in DBP formation can be observed. One example is trihalomethane (THM) formation, which is caused by all six fractions of NOM. The hydrophilic basic fraction of NOM is the most reactive of all six fractions, and it was reported to result in higher THM formation (Nkambule et al., 2009).

Humic substances are known as "problematic" compounds because they can interact with nonpolar contaminants, such as pesticides or polychlorinated biphenyls (Bolto et al., 2004). Hence, to choose the appropriate water treatment method for NOM removal, it is critical to define the species contained in NOM in each specific case. For instance, it was demonstrated that for the removal of charged organic compounds from water, IE is more efficient than coagulation (Bolto et al., 2002). A comparative study of humic substance removal using adsorption onto activated carbon, anion exchange resins, carbonaceous resins, and metal oxides claimed that the IE process was the most efficient (Fettig, 1999).

Strongly basic anion exchange resins have been investigated as a means of NOM removal from drinking water (Bolto et al., 2004; Brattebø et al., 1987). Widely studied resins include quaternary ammonium resins in chlorine form. NOM removal from water can be presented with the following reaction, where R-is charged dissolved organic carbon (DOC) (Bolto et al., 2004):

$$\text{Resin} - \text{NMe}_3^+\text{Cl}^- + \text{R}^- \leftrightarrow \text{Resin} - \text{NMe}_3^+\text{R}^- + \text{Cl}^- \quad (10.3)$$

Regeneration of this type of resin is conducted with an excess of brine or caustic brine. The effect of physical and chemical structure on NOM elimination was studied as well, and it was concluded that the highest NOM removal is attributed to resins with a high water content (Bolto et al., 2002). It was also reported that polystyrene resins have a higher selectivity than acrylic. Studies conducted with strongly basic anion exchangers prior 2004 are comprehensively reviewed by Bolto et al. (2004). They suggested that NOM removal is not as efficient with weakly basic resins as with strongly basic resins or their acrylic analogues.

Commercially available anionic exchange resins were tested for NOM acid removal from water, and the best performance was observed for the smallest resins and/or those with the highest water content. The resins were not named in this work, however, for reasons of confidentiality (Cornelissen et al., 2008).

One of the main problems of IE implementation for NOM removal from water on an industrial scale is the need to use a packed bed of IE resin, which is possible in the later stages of the water treatment process. As an example of DOC removal from water on a large scale, a 2.5 ML day^{-1} plant in Germany can be considered (Bolto et al., 2002). Macroporous (MP) styrene resin successfully removed DOC from water (initial concentration 6.5 mg L^{-1}). Within a contact time of 1.2 min, 50% of DOC was removed. Nevertheless, the plant stopped this process due to difficulties in disposing of the regeneration solution.

A magnetic ion exchange resin (MIEX) and MIEX-DOC process developed by the Australian Water Quality Center, in collaboration with Orica Australia Pty Ltd., used MIEX in a slurry reactor prior to coagulation (Cornelissen et al., 2008; Drikas et al., 2002). MIEX is a strong base anion exchange resin with an MP polyacrylic matrix in chloride form, containing magnetic iron oxide particles (IOP) within its core (Boyer and Singer, 2006; Drikas et al., 2011). Incorporated into resin, IOPs simplify separation and recycling of the resin in a continuous process. The MIEX particle size is known to be 2–5 times smaller in diameter than that of traditional resins (Hsu and Singer, 2010), which leads to faster IE kinetics. A flow chart of the MIEX process is shown in Fig. 10.2.

The MIEX process consists of two main steps, the first being contact of raw water with MIEX slurry within 10–30 min, after which the resin is separated in a clarifier and the supernatant is sent to subsequent treatment, usually coagulation. Collected MIEX is regenerated and reused (Singer and Bilyk, 2002). MIEX can be used without pretreatment and is very stable, making it a highly beneficial option for water treatment (Kitis et al., 2007).

A comparison of NOM removal efficiency by MIEX and by conventional methods was conducted on both laboratory and pilot scales. Laboratory-scale studies demonstrated that alum in combination with MIEX has a higher efficiency of DOC removal from water than alum alone. It was also observed that a broader range of compounds was removed and the formation of THMs was lower when MIEX was included in the treatment scheme (Drikas et al., 2003). Singer and Bilyk (2002) compared alum with combined MIEX and alum treatment of nine surface waters, and observed that values of total organic carbon (TOC) concentration, UV absorbance, THM, and haloacetic acid concentrations in treated water were lower when MIEX resin was also used. The ability of MIEX to remove hydrophobic and hydrophilic acids was observed as well. The effect of water hydrophobicity on DOC removal by MIEX was studied, and it was found that with increased water hydrophobicity, DOC removal decreases. Thus, DOC elimination of 56%, 33%, and 25% was observed for waters containing 21%, 50%, and 75% of hydrophobic NOM, respectively (Mergen et al., 2008).

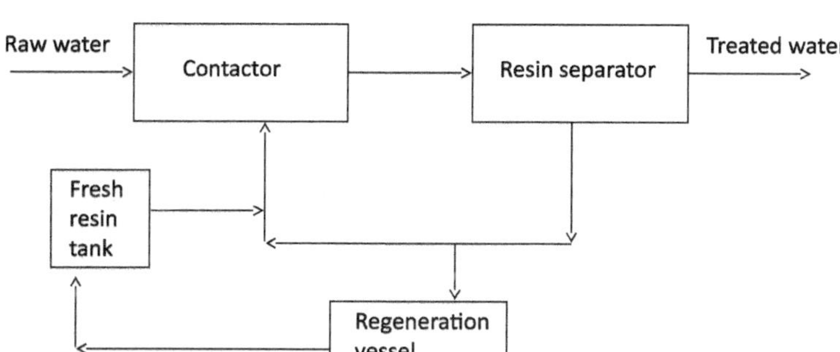

Figure 10.2 A flow chart of the magnetic ion exchange resin (MIEX) process. *(Adapted from Boyer and Singer (2006).)*

Monitoring of different parameters at a water treatment plant in South Australia suggested that changes in water quality and the character of the NOM affect the performance of MIEX significantly (Morran et al., 2004) and confirm the efficiency of a combined MIEX and coagulation process. Improved NOM removal efficiency using MIEX in comparison with coagulation was also reported by Fearing et al. (2004) and Humbert et al. (2005). Apart from improving NOM removal from water, the incorporation of the MIEX process to a conventional water treatment scheme (coagulation—flocculation) leads to a decrease in the required dose of coagulants and other chemicals. For instance, five different raw drinking water samples from the City of Istanbul were treated with a MIEX process prior to coagulation on laboratory scale, to investigate the effect of different doses of MIEX and contact time, as well as coagulant dose on the quality of treated water (Kitis et al., 2007). After treatment with a resin dose of 5—10 mL settled resin L^{-1} and a contact time 10—20 min, the values of DOC and specific UV absorbance ($SUVA_{254}$) were <1.5 mg L^{-1} and <21 mg^{-1} DOC-m, respectively. Moreover, removal of nitrates and sulfates from all raw water samples was detected at 17%—42% and 9%—24%, respectively; the dose of MIEX was 10 mL settled resin L^{-1} at 10 min of contact time. Depending on the quality of raw water, the decrease in the required coagulant (alum) dose by treating water with MIEX prior to coagulation was estimated to be 0—30 mg L^{-1} (Kitis et al., 2007). Another study was conducted with four types of drinking water from California characterized by low turbidity, moderate organic carbon concentration, moderate to high concentration of bromide, and a wide range of alkalinities (Boyer and Singer, 2005), to determine if MIEX treatment is more efficient than coagulation with alum. The findings suggested that MIEX treatment and MIEX treatment followed by coagulation performed in a similar way, while DOC and SUVA removal efficiency by coagulation was significantly lower, especially for raw waters with high SUVA and low anionic strength (Boyer and Singer, 2005). Interestingly, bromide was removed from all tested water samples with MIEX, and it was established that bromide removal efficiency was dependent on the alkalinity of the raw water and initial bromide concentration. No changes in bromide concentration after coagulation were determined (Boyer and Singer, 2005). MIEX was reported to be very efficient in combination with conventional treatment (coagulation, sedimentation, filtration) for treatment of raw water with very high NOM concentration (DOC = 15.5 mg L^{-1}) (Lin et al., 2015).

The effect of bicarbonate, chloride, and bromide presence in water on DOC removal with MIEX and polystyrene resins A-641 and IRA910 was studied by Hsu and Singer (2010). Bromide removal was reported to be more efficient with use of A-641 and IRA910 resins, but the level of DOC elimination using MIEX was higher: DOC removal efficiencies by MIEX, A-641, and IRA910 resins after 30 min were 46%, 13%, and 7%, respectively (Hsu and Singer, 2010). Increasing the contact time to 5 h resulted in higher DOC elimination for all tested resins but MIEX still performed better. The simultaneous removal of NOM, nitrates, sulfates, bromides, and pesticides from high DOC-content surface water was performed using four commercial strong anion exchange resins: MIEX, DOWEX-11, DOWEX-MSA, and IRA938 (Humbert et al., 2005). All the tested resins demonstrated significant DOC removal, but the kinetics of DOC removal from water with MIEX and IRA938 were faster than that of DOWEX-11 and DOWEX-MSA. After only 5 min of contact time with 8 mL L^{-1} of MIEX and IRA938, the residual DOC concentration in water varied from 1 to 2 mg L^{-1}. Both high and partially low MW UV-absorbing NOM were removed with all tested resins. Pesticide removal by MIEX, DOWEX-11, and DOWEX-MSA was estimated to be 10%, while 35% was eliminated by IRA938. The optimal contact time for DOC removal by MIEX with doses 2–8 mL L^{-1} was estimated to be 10–20 min (Humbert et al., 2005).

Recently, the efficiency of NOM removal was tested by MIEX in bicarbonate form and compared with MIEX chloride form, which has been thoroughly studied (Walker and Boyer, 2011). It was observed that the performance of both resins was similar for removal of DOC, UV_{254}, and sulfate, but bicarbonate-form MIEX demonstrated higher performance for bromide removal. The sequence of affinity for dissolved organic matter (DOM) and anion removal was similar for fresh MIEX-HCO_3 and virgin MIEX-Cl, and can be presented in following way: sulfate $\sim UV_{254} > DOC >$ bromide. A decrease in removal efficiency for both resins was noted with an increased number of regeneration cycles. The DOC and bromide elimination with the use of MIEX-HCO_3 resin was reported to be 69% and 34%, correspondingly, from average initial concentrations of 18 mg C L^{-1} and 780 μg L^{-1}, respectively.

The results of NOM removal studies with MIEX for batch and continuous operations conducted before 2008 are excellently summarized by Mergen et al. (2008). The effect of MIEX on membrane fouling reduction was demonstrated in a number of studies (Fabris et al., 2007; Humbert et al., 2007; Huang et al., 2011; Kim and Dempsey, 2010).

The first water plant utilizing a MIEX-DOC process was launched at Mount Pleasant, South Australia, in 2001, with a capacity of 2.5 ML day^{-1} (Drikas et al., 2011) for the treatment of potable water. This plant was divided into two streams with a capacity of 1.25 ML day^{-1} each. The first stream consists of a MIEX-DOC process followed by coagulation, flocculation, sedimentation, and filtration, while the second stream contains a submerged microfiltration step that is implemented after the MIEX-DOC process. Monitoring of different parameters and comparison with similar water treatment schemes without the MIEX-DOC were conducted over a 2-year period at this plant (Drikas et al., 2011). The results showed a significant improvement in DOC removal from treated water when MIEX was used. Also, water quality after treatment was less affected by DOC changes in raw water when MIEX was used.

Experiments in NOM removal from water conducted with MIEX and a continuous-flow pilot plant operated at 7.6 L min^{-1} demonstrated substantial removal of UV-absorbing substances and DOC, and the efficient dose of MIEX was estimated to be 0.16 mL L^{-1} (Boyer and Singer, 2006). Another MIEX pilot with a flow of 1 ML day^{-1} was successfully tested, and a plant scale-up to 225 ML day^{-1} was then conducted (Bolto et al., 2002).

Fluidized ion exchange (FIX) is another type of process using IE resin for NOM removal (Cornelissen et al., 2009). The main principle of this method is that the rate of sedimentation of suspended solids in feed water is lower than that of IE resin. Thus, when water is pumped into an upflow reactor configuration at a certain speed, IE resin is fluidized, while suspended solids present in raw water are removed. A test was conducted with real surface water and FIX followed by ultrafiltration, demonstrating the feasibility of this process (Cornelissen et al., 2009). It was also shown that use of the FIX process prior to ultrafiltration significantly decreases membrane fouling.

A process consisting of a fluidized bed reactor with MIEX followed by hydrophilic (NE70) and hydrophobic (NE90) nanofiltration (NF) membranes was tested for NOM removal from raw water from the Nakdong River, Korea (Kaewsuk and Seo, 2011). The raw water was characterized by low-MW NOM content with hydrophilic compounds, such as carboxylic, amino sugar, and aliphatic hydrocarbon compounds. With MIEX, the removal efficiency of phenolic compounds and compounds with low SUVA values (aliphatic hydrocarbon and amino sugar) was reported to be significant, while the removal of carboxylic compounds was not that efficient (Kaewsuk and Seo, 2011). The best performance for NOM elimination from raw Nakdong River water was achieved with NE90 membrane filtration followed by MIEX.

Several investigations have been conducted into MIEX processes combined with other water treatment techniques, such as coagulation (Singer and Bilyk, 2002; Fearing et al., 2004; Allpike et al., 2005), lime softening (Hsu and Singer, 2009), activated carbon (Drikas et al., 2009; Humbert et al., 2008), and filtration (Fabris et al., 2007; Huang et al., 2012). For instance, the effect of MIEX pretreatment on ozone demand, ozone exposure for disinfection, and bromate formation was studied for three water samples collected from raw water supplies impacted by the San Francisco Bay Delta (Kingsbury and Singer, 2013). After treatment with MIEX, 41%–68% of TOC was removed from raw water, while alum treatment eliminated only 12%–44%. The reduction of bromide concentration in water by MIEX was estimated to be 20%–50%. Ozone demand was significantly decreased after MIEX and alum pretreatment. Chlorination of raw water and of samples after MIEX, alum, and ozonation treatment was conducted to estimate trihalomethane formation potential (THMFP). The results demonstrate that MIEX performed better than alum and ozonation, attaining THMFP removal efficiencies of 39%–85%, 16%–56%, and 35%–45%, respectively (Kingsbury and Singer, 2013). Based on these results, it was concluded that pretreatment with MIEX prior to ozonation is more efficient than coagulation for bromide-rich waters (Kingsbury and Singer, 2013).

An attempt to simultaneously remove DOC and hardness from water was made by Apell and Boyer (2010). For this purpose, MIEX was used together with cation IE resin in a single completely mixed reactor, and removal of DOC and hardness was estimated to be 70% and >55%, respectively. Thus, the feasibility of combined IE water treatment was demonstrated, and it was recommended as a possible pretreatment technique to decrease membrane fouling during the subsequent filtration step. In a recent study of a combined IE process (MIEX and cation exchange resin in a single completely mixed vessel), DOC and hardness removal from groundwater was reported to be 76% and 97%, respectively (Comstock and Boyer, 2014). The successful regeneration of exhausted anion and cation exchange resins was obtained by using NaCl in a single vessel (Table 10.1). Summary of studies conducted in last 10–20 years for NOM removal from water by IE is presented in Table 10.1.

Table 10.1 Summary of studies on NOM removal by ion exchange (IE) conducted in the last 10–20 years.

Treatment	Water quality	NOM removal (%)	Reaction conditions	References
MIEX-C	Wanneroo raw water appeared to be similar to northern European swamp and lake waters, such as water from some lakes in Finland.	After MIEX treatment: DOC = 2.46 mg L^{-1}; color (at 400 nm) = 29; UV254 = 0.175 cm^{-1}; SUVA = 7.1 mg C L^{-1}; pH = 7.46.	Data were obtained from a full-scale (110 mL day^{-1}) potable water treatment plant operating in Perth, Western Australia.	Allpike et al. (2005)
MIEX combined with coagulation	The following parameters of raw water are available: DOC = 5.70 mg L^{-1}; color (at 400 nm) = 45; UV254 = 0.326 cm^{-1}; SUVA = 5.7 mg C L^{-1}; pH = 7.14; Mn (method A DOC) = 2405; Mn (method A UV254) = 4273; Mn (method B UV254) = 3407; Mw (method A DOC) = 5319; Mw (method A UV254) = 6371; Mw (method B UV254) = 5294.	After MIEX treatment combined with coagulation: Alum dose = 55 mg L^{-1}; polyelectrolyte dose = 0.75 mg L^{-1}; DOC = 1.59 mg L^{-1}; color (at 400 nm) = 2; UV254 = 0.025 cm^{-1}; SUVA = 1.6 mg C L^{-1}; pH = 6.61.		
	Cedar key groundwater characteristics: DOC = 5.6 mg L	Water quality after combined IE treatment: DOC = 1.7 mg L^{-1};	A Phipps and Bird PB 700 jar	

Combined IE treatment (MIEX-Cl and MIEX-Na)	$^{-1}$; UV254 = 0.171 cm^{-1}; pH = 7.6; alkalinity = 244 mg L^{-1} $CaCO_3$; hardness = 275 mg L^{-1} $CaCO_3$; Cl− = 11.8 mg L^{-1}; SO_4^{2-} = 20.9 mg L^{-1}; Ca^{2+} = 103 mg L^{-1}; >90% of the hardness was as calcium.	pH = 7.7; Cl− = 48.8 mg L^{-1}; SO_4^{2-} = 3.1 mg L^{-1}. The following trends were observed for DOC and UV254 removal when fresh resins were used: sequence 1 ~ sequence 2 > simultaneous IE. After the third regeneration cycle, no difference between simultaneous and sequential IE treatment was observed.	tester with 2 L square jars was used for experiments. Simultaneous performance of anion and cation exchange resins, sequential anion exchange followed by cation exchange (sequence 1) and sequential cation exchange followed by anion exchange (sequence 2) were conducted.	Apell and Boyer (2010)
20 different resins were tested	Two types of waters were prepared using reverse osmosis to concentrate NOM in surface water samples from the east Moorabool river near Anakie, Victoria, and from the Mount Zero reservoir at Horsham,	NOM removal from Hope Valley water was 98%–100%, whether gel or macroporous (MP) styrene or MP acrylic resins was used.	Normally, dose of a resin equal to 3 g L^{-1} was used in all experiments and contact time was 16 h.	Bolto et al. (2002)

Continued

Table 10.1 Summary of studies on NOM removal by ion exchange (IE) conducted in the last 10–20 years.—cont'd

Treatment	Water quality	NOM removal (%)	Reaction conditions	References
	Victoria. Also, regeneration effluent obtained from MIEX plant at Hope Valley, South Australia, was used.			
	Fractionation and percentage of UV-absorbing contribution for various NOM fractions were performed for tested water samples. Four fractions were distinguished: Very hydrophobic acids (VHA), slightly hydrophobic acids (SHA), charged hydrophilic acids, and neutral hydrophilic compounds. UV-absorbance contribution for each fraction is presented below.	NOM removal efficiencies from Moorabool water (based on UV absorber removal): ResinTech SIR 22P—85%; Amberlite IRA 458—90%; Amberlite IRA 958—90%.		
	Water from Hope Valley: VHA—57%; SHA—17%; charged—26%; neutral—0%.	NOM removal efficiencies from Hope Valley water (based on UV absorber removal): ResinTech SIR 22P—98%; Amberlite IRA 958—99%.		
	Water from Moorabool: VHA—40%; SHA—17%;	NOM removal efficiencies from Horsham water (based on UV		

	charged—38%; neutral—5%. Water from Horsham: VHA—70%; SHA—13%; charged—11%; neutral—5%.	absorber removal): ResinTech SIR 22P—84%; Amberlite IRA 958—73%.		
Lewatit MP 500; Lewatit S6328A; Lewatit SP112;	Water samples from Seine river (France) with following characteristics were used: DOC = 1.9–2.7 mg L^{-1}; BDOC = 0.5–0.8 mg L^{-1}; UV254 = 6.8–8.6 m^{-1}; pH = 7.8–8.3; conductivity = 460 μS cm^{-1}; ammonia ≤0.1 mg L^{-1}.	Percentage of NOM removal, from DOC: MP 500—59%; S6328—73%; SP112—not significant Percentage of NOM removal, from UV254 (m^{-1}): MP 500—90%; S6328—93%; SP112—not significant.		Afcharian et al. (1997)
MIEX	Raw water from Little river Reservoir was used. Characteristics of water (mean value calculated from may to december): Turbidity = 4.3 NTU; pH = 6.8; alkalinity (as $CaCO_3$) = 24.1 mg L^{-1}; hardness (as $CaCO_3$) = 28.5 mg L^{-1}; conductivity = 95 μS cm^{-1}; Cl− = 9.5 mg L^{-1}; UV254 = 0.205 cm^{-1}; DOC = 5.9 mg L^{-1}; SUVA = 3.4 mg C L^{-1}.	Variations in removal efficiency of DOC and UV254 were observed depending on operational conditions. Operational conditions: 20 min HRT; 10% regeneration ratio; steady-state MIEX concentration of 20 mL L^{-1} (ERD = 2 mL L^{-1}); removal efficiency: DOC = 74%; UV254 = 79%.	Experiments were conducted on continuous-flow pilot scale.	Boyer and Singer (2006)

Continued

Table 10.1 Summary of studies on NOM removal by ion exchange (IE) conducted in the last 10–20 years.—cont'd

Treatment	Water quality	NOM removal (%)	Reaction conditions	References
		Operational conditions: 20 min HRT; 1% regeneration ratio; ERD = 0.16 mL L^{-1}; MIEX concentration in contactors = 16 mL L^{-1}; removal efficiency: DOC = 64%; UV254 = 71%		
		Operational conditions: 20 min HRT; 2% regeneration ratio; ERD = 0.2 mL L^{-1}; MIEX concentration in contactors = 10 mL L^{-1}; removal efficiency: DOC = 63%; UV254 = 65%.		
		Operational conditions: 20 min HRT; 2% regeneration ratio; ERD = 0.3 mL L^{-1}; MIEX concentration in contactors = 15 mL L^{-1}; removal efficiency: DOC = 67%; UV254 = 65%.		
		Operational conditions: 20 min HRT; 2% regeneration ratio; ERD = 0.4 mL L^{-1}; MIEX		

		concentration in contactors = 20 mL L^{-1}; removal efficiency: DOC = 69%; UV254 = 72%. Operational conditions: 20 min HRT; 5% regeneration ratio; ERD = 0.75 mL L^{-1}; MIEX concentration in contactors = 15 mL L^{-1}; removal efficiency: DOC = 73%; UV254 = 72%.		
Prechlorination -MIEX® combined process	Raw water from Hongze Lake (China) was used. pH = 8.1; UV254 = 0.11 cm^{-1}; SUVA = 2.37 L mg^{-1} m^{-1}; ρ(DOC) = 4.68 mg L^{-1}; conductivity 48 mS m^{-1}; turbidity 46.71 NTU.	Effect of prechlorination on performance of MIEX® and combination of MIEX® with coagulation was studied. It was reported that prechlorination slightly enhance performance of MIEX®, thus it was suggested that prechlorination is not necessary in tested water. If MIEX is applied prior coagulation, it can decrease amount of needed coagulant and increase elimination of turbidity and organic compounds. Optimal resin dose and contact time were 8 mL L^{-1} and 15 min.	Kinetic jar tests and multiple-loading jar tests.	Han et al. (2010)

Continued

Table 10.1 Summary of studies on NOM removal by ion exchange (IE) conducted in the last 10–20 years.—cont'd

Treatment	Water quality	NOM removal (%)	Reaction conditions	References
MIEX – DOC continuous operation	The study was performed from June 2005 until June 2007. Raw water quality was as follows:DOC = 2.7–5.8 mg L^{-1}; turbidity 8.7–60 NTU; colour 6–23 HU. VHA fraction was dominant in raw water.	Average MIEX resin dose was 12 mL L^{-1}. Average DOC removal was significantly improved when MIEX pre-treatment was applied. Thus, about 54% and 57% of DOC removal was achieved for treatment trains with MIEX in comparison with 33% and 11% of DOC removal for conventional and microfiltration treatment, respectively. The highest VHA removal from raw water was achieved when MIEX was applied. Importantly, final water quality was less affected by variations of concentrations and character of NOM when MIEX was incorporated in treatment train.	Experiments were performed in pilot treatment plant. Following treatment trains were compared: (1) conventional, consisting of coagulation, flocculation, sedimentation and rapid filtration; (2) MIEX pretreatment followed by conventional treatment train (1); (3) microfiltration; (4) MIEX pretreatment followed by microfiltration.	Drikas et al. (2011)

MIEX® DOC process	Raw surface water was used in this study was collected from Duoro River (from an intake point of Lever WTP). Due to seasonal variation of water quality, collected water was divided into three classes (where class I correspond to higher water quality). Initial DOC varied from 4.19 to 8.18 mg L^{-1}. Variation of other parameters are shown below: UV254 = 4.8–10.2 m^{-1}; TOC = 4.44–8.84 mg L^{-1}; SUVA254 = 1.12–1.84 L m^{-1} mg^{-1}; UV272 = 3.9–8.4 m^{-1}; SUVA272 = 0.93–1.46 L m^{-1} mg^{-1}.	Reported optimal resin dose and contact time were 15 mL L^{-1} and 10 min. At optimal conditions DOC removal was higher than 90%. MIEX® led to removal of 60–92% of anionic compounds in DOC.	MIEX® kinetic jar tests (dose-response) were done.	Karpinska et al. (2013)
MIEX, DOWEX 11	Raw water from Camlidere and Kesikkopru were used. Camlidere water: UV254 = 0.091 cm^{-1}, SUVA254 = 1.68 L (mg of C)$^{-1}$ m^{-1}, DOC = 5.4 mg L^{-1}, pH = 8; alkalinity 84 mg of $CaCO_3$ L^{-1}; conductivity 200 μS cm^{-1}; total hardness 95 mg of $CaCO_3$ L^{-1}; turbidity 7 NTU;	Results obtained for Camlidere water:70% of DOC and almost complete elimination of UV254 and sulfate was reached at optimal dose of MIEX resin (10 mL L^{-1}). Sequence of removal and affinity was UV254 ~SO_4^{2-} > DOC > bicarbonate. Removal of DOC, UV254 and	Batch and continuous-flow tests were conducted.	Ates and Incetan (2013)

Continued

Table 10.1 Summary of studies on NOM removal by ion exchange (IE) conducted in the last 10–20 years.—cont'd

Treatment	Water quality	NOM removal (%)	Reaction conditions	References
	total dissolved solids 100 mg L^{-1}; Br^{-} = 20 μg L^{-1}; SO_4^{2-} = 17 mg L^{-1}; NO_3^{-} = 20 mg L^{-1}.Kesikkopru water: UV254 = 0.034 cm^{-1}, SUVA254 = 0.71 L (mg of C)$^{-1}$ m^{-1}, DOC = 4.8 mg L^{-1}, pH = 8.1; alkalinity 115 mg of $CaCO_3$ L^{-1}; conductivity 1637 μS cm^{-1}; total hardness 350 mg of $CaCO_3$ L^{-1}; turbidity 8 NTU; total dissolved solids 818 mg L^{-1}; Br^{-} = 50 μg L^{-1}; SO_4^{2-} = 390 mg L^{-1}; NO_3^{-} < 0.06 mg L^{-1}.	sulfate with DOWEX 11 resin was 60% (dose 300 mg L^{-1}), 85–90% (dose 1000 mg L^{-1}) and > 80%, respectively. Results obtained for Kesikkopru water: 54% of DOC and about 45% of UV254 and sulfate was reached at optimal dose of MIEX resin. Sequence of removal and affinity (MIEX) was DOC > UV254 ~ sulfate > bicarbonate. 40% of DOC removal and about 60% of UV254 and sulfate elimination was achieved with optimal dose of DOWEX 11. Sequence of removal and affinity (DOWEX 11): UV254 > DOC > sulfate > bicarbonate.		
Lewatit VP OC 1071	Schie canal (Delft, The Netherlands) raw water was filtered through 1 μm filter prior NOM removal. Sint Jansklooster	The increase of resin dose (from 2 to 10 mL L^{-1}) led to increase of NOM removal for all tested waters. Further increase of resin	Batch mode.	Hu et al. (2014)

	(The Netherlands) groundwater was used for NOM removal after rapid sand filtration.	dose (up to 12 mL L^{-1}) had negligible effect on NOM elimination. DOC and UV254 removal (dose: 10 mL L^{-1}, 1 hour) was 69% and 86% for Shie water, while 80% and 91%, respectively, was reached for underground water. Preferences of tested resin towards aromatic constituents of NOM was reported.		
Nanoscale anion exchange resin	Water from Myrtle Beach, SC, was collected and diluted to obtain following parameters: DOC = 11.52 mg L^{-1}; UV254 = 0.396 cm^{-1}; SUVA = 4.36 L mg-1 m-1.	Contact time was 5 min, after which about 20% of DOC and about 79% of UV254 removal was observed. Elimination of SUVA was negligible.	Experiments were conducted using film of NanoResin deposited onto polypropylene support membrane.	Johnson et al. (2016)
MIEX	Raw water samples were obtained from New Port, Richey, FL; Charleston, SC; Palmdale, CA; Northern KY; Southern NV.	Removal of NOM was evaluated as DOC removal in percent. Data are shown below.	Experiments were conducted on continuous-flow pilot scale	Budd et al. (2005) and Fonseca et al. (2005)

Continued

Table 10.1 Summary of studies on NOM removal by ion exchange (IE) conducted in the last 10–20 years.—cont'd

Treatment	Water quality	NOM removal (%)	Reaction conditions	References
	Raw water characteristics are presented below. New Port, Richey, FL: DOC = 1.9 mg L^{-1}; SUVA = 5.6 mg C L^{-1}; sulfate = 18 mg L^{-1}; alkalinity (as $CaCO_3$) = 200 mg L^{-1}.	New Port, Richey, FL: DOC = 65%–77%.		
	Charleston, SC: DOC = 5.0 mg L^{-1}; SUVA = 4.3 mg C L^{-1}; sulfate = 15 mg L^{-1}; alkalinity (as $CaCO_3$) = 27 mg L^{-1}.	Charleston, SC: DOC = 58%–70%.		
	Palmdale, CA: DOC = 4.2 mg L^{-1}; SUVA = 2.1 mg C L^{-1}; sulfate = 29–41 mg L^{-1}; alkalinity (as $CaCO_3$) = 126 mg L^{-1}.	Palmdale, CA: DOC = 52%–60%.		
	Northern KY: DOC = 1.8 mg L^{-1}; SUVA = 4.6 mg C L^{-1}; sulfate = 75 mg L^{-1}; alkalinity (as $CaCO_3$) = 113 mg L^{-1}.	Northern KY: DOC = 47%–61%.		
	Southern NV: DOC = 2.3 mg L^{-1}; SUVA = 1.2 mg C L^{-1}; sulfate = 246 mg L^{-1}; alkalinity (as $CaCO_3$) = 288 mg L^{-1}.	Southern NV: DOC = 23%–44%.		

MIEX	Four types of drinking water in California were used: North Bay Aqueduct (NBA), South Bay Aqueduct (SBA), Castaic Lake (CL), and Sweetwater Lake (SL). Characteristics of water are shown below. NBA water: pH = 7.5; turbidity = 20 NTU; alkalinity (as $CaCO_3$) = 149 mg L^{-1}; Br− = 76 μg L^{-1}; UV254 = 0.193 cm^{-1}; TOC = 5.5 mg L^{-1}; DOC = 5.1 mg L^{-1}; SUVA = 3.8 mg C L^{-1}; THM4FP = 294 μg L^{-1}; HAA9FP = 224 μg L^{-1}. SBA water: pH = 7.8; turbidity = 6 NTU; alkalinity (as $CaCO_3$) = 57 mg L^{-1}; Br− = 83 μg L^{-1}; UV254 = 0.064 cm^{-1}; TOC = 1.9 mg L^{-1}; DOC = 1.9 mg L^{-1}; SUVA = 3.4 mg C L^{-1}; THM4FP = 111 μg L^{-1}; HAA9FP = 90.5 μg L^{-1}.	Removal of DOC and UV254 from NBA water: DOC = 70%; UV254 ~80%. Removal of DOC and UV254 from SBA water: DOC = 50%; UV254 ~60%. Removal of DOC and UV254 from CL water: DOC = 30%; UV254 ~50%.		Boyer and Singer (2005)

Continued

Table 10.1 Summary of studies on NOM removal by ion exchange (IE) conducted in the last 10–20 years.—cont'd

Treatment	Water quality	NOM removal (%)	Reaction conditions	References
	CL water: pH = 8.6; turbidity = 7 NTU; alkalinity (as $CaCO_3$) = 92 mg L^{-1}; Br− = 240 μg L^{-1}; UV254 = 0.074 cm^{-1}; TOC = 2.3 mg L^{-1}; DOC = 2.5 mg L^{-1}; SUVA = 3.0 mg C L^{-1}; THM4FP = 150 μg L^{-1}; HAA9FP = 65.3 μg L^{-1}. SL water: pH = 8.1; turbidity = 6 NTU; alkalinity (as $CaCO_3$) = 188 mg L^{-1}; Br− = 540 μg L^{-1}; UV254 = 0.102 cm^{-1}; TOC = 5.2 mg L^{-1}; DOC = 5.1 mg L^{-1}; SUVA = 2.0 mg C L^{-1}; THM4FP = 283 μg L^{-1}; HAA9FP = 127 μg L^{-1}.	Removal of DOC and UV254 from SL water: DOC ~40%; UV254 ~50%.		

Combined IE (MIEX-Cl and amberlite 200C –Na	Groundwater was used. Characteristics of water are shown below. Cedar key water (average): pH = 7.4; UV254 = 0.222 cm^{-1}; DOC = 6.3 mg L^{-1}; SUVA = 3.6 mg C L^{-1}; total hardness (as $CaCO_3$) = 315 mg L^{-1}; calcium hardness (as $CaCO_3$) = 250 mg L^{-1}; alkalinity (as $CaCO_3$) = 245 mg L^{-1}; sulfate = 17 mg L^{-1}. Yankeetown water: pH = 7.2; UV254 = 0.081 cm^{-1}; DOC = 2.9 mg L^{-1}; SUVA = 2.8 mg C L^{-1}; total hardness (as $CaCO_3$) = 440 mg L^{-1}; calcium hardness (as $CaCO_3$) = 420 mg L^{-1}; alkalinity (as $CaCO_3$) = 320 mg L^{-1}; sulfate = 130 mg L^{-1}.	The best performance was achieved at 200 BV, therefore DOC removal at 200 BV only is shown below. DOC removal efficiency was obtained for fresh and regenerated resin, but the results were very close. Cedar key water: DOC ~84%. Yankeetown water: DOC ~83%.	A Phipps and Bird PB 700 jar tester was used for experiments. Multiple loading was conducted. Resin: 5 mL MIEX-Cl/20 mL A200C–Na; regeneration with 2% NaCl and 20% NaCl. Five 1 L water in series; 30 min mixing at 200 rpm; 5 min settling.	Comstock and Boyer (2014)

Continued

Table 10.1 Summary of studies on NOM removal by ion exchange (IE) conducted in the last 10–20 years.—cont'd

Treatment	Water quality	NOM removal (%)	Reaction conditions	References
	Palm Springs water: pH = 7.6; UV254 = 0.370 cm^{-1}; DOC = 9.7 mg L^{-1}; SUVA = 3.8 mg C L^{-1}; total hardness (as $CaCO_3$) = 300 mg L^{-1}; calcium hardness (as $CaCO_3$) = 295 mg L^{-1}; alkalinity (as $CaCO_3$) = 280 mg L^{-1}; sulfate = 10 mg L^{-1}.	Palm springs water: DOC ~95%.		
Combined treatment by IE resin (MIEX; IRA938; IRA958; DOWEX-11; DOWEX-MSA; AMBERSORB) and powdered activated carbon (PAC)	Raw water from the Villejean/Rennes drinking water treatment plant was used. The main quality parameters are: pH = 7.0–7.9; conductivity = 220–250 μS cm^{-1}; alkalinity (mg L^{-1} as $CaCO_3$) = 20–30; NO_3^- = 7–29 mg L^{-1}; SO_4^{2-} = 13–29 mg L^{-1}; DOC = 5.6–6.7 mg L^{-1}; UV254 = 0.14–0.16 cm^{-1}; SUVA = 2.2–2.9 mg C L^{-1}.	All tested resins removed more than 75% of DOC (resin dose 8 mL L^{-1}) after pseudoequilibrium time (time varied from 15 min to more than 1 h depending on resin). Less than 28% of DOC was removed by PAC (20–40 mg L^{-1} doses, pseudoequilibrium time of about 2 h). Implementation of PAC treatment after resins led to a slight improvement in DOC removal and higher uptake of pesticides.	Batch tests	Humbert et al. (2008)

MIEX; DOWEX-11; DOWEX-MSA; IRA- 938	Raw, clarified and postozonated water from the Villejean/Rennes drinking water treatment plant was used. Characteristics of clarified water: pH = 7.7; alkalinity (mg L^{-1} as $CaCO_3$) = 29; NO_3^- = 8 mg L^{-1}; SO_4^{2-} = 22 mg L^{-1}; DOC = 4.8 mg L^{-1}; UV254 = 0.1 cm^{-1}; SUVA = 2.0 mg C L^{-1}. Characteristics of postozonated water: pH = 7.1; alkalinity (mg L^{-1} as $CaCO_3$) = 13; DOC = 2.15 mg L^{-1}; UV254 = 0.021 cm^{-1}; SUVA = 1.0 mg C L^{-1}.	DOC was successfully removed by all studied resins. MIEX and IRA938 demonstrated faster kinetics after 5 min of contact time with dose of these resins of 8 mL L^{-1}; level of DOC was 1–2 mg L^{-1}. The studied resins were able to remove high- and partly low-MW UV-absorbing NOM.	Bench-scale experiments	Humbert et al. (2005)
MIEX–NF combined process	Influent of a water treatment plant (Changwon city, Korea) was used. Characteristics the of water: Turbidity = 10 NTU; pH = 7.0; DOC = 2.5 mg L^{-1}.	Water treated with MIEX was reported to have the following parameters: DOC = 1.28 mg L^{-1}, UVA = 0.010 cm^{-1}, SUVA = 0.39 mg C L^{-1}. Better performance was attributed to NE90 after MIEX treatment.	The fluidized bed MIEX column (150 mL of resin) was operated at flow rate 50 mL min^{-1}.	Kaewsuk and Seo (2011)

Continued

Table 10.1 Summary of studies on NOM removal by ion exchange (IE) conducted in the last 10–20 years.—cont'd

Treatment	Water quality	NOM removal (%)	Reaction conditions	References
DAX-8; IRA958; DEAE (preswollen); MIEX	DAX-8; IRA958; DEAE (preswollen); MIEX	Tests in fixed bed column demonstrated 23%–67% of the initial TOC removal, with removal increasing in the order DAX-8 < IRA958 < DEAE < MIEX. It should be noted that all tested experimental set-ups were equally efficient for TOC removal.	Packed column, in-line addition and addition to complete-mix tank tests were conducted.	Kim and Dempsey (2010) and Kennedy et al. (2008)
MIEX	Raw waters from the North Bay Aqueduct, South Bay Aqueduct, spiked and nonspiked water from Lake Campbell were tested. Water characteristics are shown below. North Bay Aqueduct (NBA): TOC = 3.7 mg L^{-1}; DOC = 3.5 mg L^{-1}; UV254 = 0.113 cm^{-1}; SUVA = 3.2 mg C L^{-1}; fluorescence index = 1.39; pH = 7.9; bromide = 40 μg L^{-1}.	In NBA water after treatment with 2.0 mL L^{-1} of MIEX/Cl, concentration of TOC decreased by 46% and UV absorbance by 58%. In SBA water treated with 2.0 mL L^{-1} of MIEX/Cl and MIEX/HCO_3, TOC was decreased by 36% and 25% and UV absorbance by 44% and 42%, respectively. For lake water, MIEX/Cl and MIEX/HCO_3 at the doses of 6.0 mL L^{-1} reduced TOC by 63% and 54%, respectively, and UV	Bulk tests were performed in a 16-L glass carboy fitted with a variable-speed motor and an approximately 20-cm by 3-cm rectangular paddle.	Kingsbury and Singer (2013)

		absorbance by 80%. It was demonstrated that for treatment of bromide-rich waters, MIEX pretreatment prior to ozonation is more efficient than alum. South Bay Aqueduct (SBA): TOC = 2.4 mg L^{-1}; DOC = 2.4 mg L^{-1}; UV254 = 0.071 cm^{-1}; SUVA = 3.0 mg C L^{-1}; fluorescence index = 1.43; pH = 8.2; bromide = 360 μg L^{-1}. Lake Campbell: TOC = 8.7 mg L^{-1}; DOC = 8.5 mg L^{-1}; UV254 = 0.256 cm^{-1}; SUVA = 3.0 mg C L^{-1}; fluorescence index = 1.43; pH = 7.8; bromide = 33 μg L^{-1}. Lake Campbell, spiked: TOC = 8.7 mg L^{-1}; DOC = 8.5 mg L^{-1}; UV254 = 0.256 cm^{-1}; SUVA = 3.0 mg C L^{-1}; fluorescence index = 1.43; pH = 7.8; bromide = 380 μg L^{-1}.		

Continued

Table 10.1 Summary of studies on NOM removal by ion exchange (IE) conducted in the last 10–20 years.—cont'd

Treatment	Water quality	NOM removal (%)	Reaction conditions	References
MIEX	Raw waters from five drinking water plants in city of Istanbul were tested. The physicochemical characteristics are shown below. Elmali: DOC = 4.3 mg L^{-1}; UV254 = 0.220 cm^{-1}; SUVA = 5.11 mg C L^{-1}; alkalinity = 30 mg $CaCO_3$ L^{-1}; total hardness = 110 mg $CaCO_3$ L^{-1}; TDS = 140 mg L^{-1}; conductivity = 279 μS cm^{-1}; pH = 7.62; turbidity = 21.2 NTU. B.Cekmece: DOC = 3.1 mg L^{-1}; UV254 = 0.084 cm^{-1}; SUVA = 2.71 mg C L^{-1}; alkalinity = 35 mg $CaCO_3$ L^{-1}; total hardness = 150 mg $CaCO_3$ L^{-1}; TDS = 230 mg L^{-1}; conductivity = 460 μS cm^{-1}; pH = 7.92; turbidity = 1.4 NTU. Ömerli: DOC = 2.6 mg L^{-1}; UV254 = 0.102 cm^{-1}; SUVA = 3.92 mg C L^{-1}; alkalinity = 35 mg $CaCO_3$ L^{-1}; total hardness = 90 mg $CaCO_3$ L^{-1}; TDS = 113 mg L^{-1};	Optimal dose and contact time for NOM removal from all waters was 10 min and 10 mL settled resin per liter.	Multiple-loading jar tests were conducted.	Kitis et al. (2007)

	conductivity = 225 μS cm^{-1}; pH = 7.52; turbidity = 4.2 NTU Ikitelli: DOC = 3.1 mg L^{-1}; UV254 = 0.107 cm^{-1}; SUVA = 3.45 mg C L^{-1}; alkalinity = 38 mg $CaCO_3$ L^{-1}; total hardness = 137 mg $CaCO_3$ L^{-1}; TDS = 170 mg L^{-1}; conductivity = 339 μS cm^{-1}; pH = 8.05; turbidity = 1.0 NTU. Kagithane: DOC = 2.8 mg L^{-1}; UV254 = 0.108 cm^{-1}; SUVA = 3.86 mg C L^{-1}; alkalinity = 37 mg $CaCO_3$ L^{-1}; total hardness = 123 mg $CaCO_3$ L^{-1}; TDS = 221 mg L^{-1}; conductivity = 441 μS cm^{-1}; pH = 7.63; turbidity = 21.5 NTU.			
MIEX	Three types of water were used: Barcombe, Draycote, and Albert. Barcombe: DOC = 9.6 mg L^{-1}; UV254 = 16.5 m^{-1}; SUVA = 1.73 mg C L^{-1}; pH = 7.7; turbidity = 13.1 NTU; conductivity = NA; alkalinity = 112 mg $CaCO_3$ L^{-1}; zeta potential = −12.7; charge	Removal of DOC in Draycote, Barcome, and Albert waters was 56%, 33%, and 25% respectively. It was observed that both hydrophilic and hydrophobic NOM can be removed by resin. However, high-MW NOM quickly saturated or blocked the resin.	Bench-scale tests	Mergen et al. (2008)

Continued

Table 10.1 Summary of studies on NOM removal by ion exchange (IE) conducted in the last 10–20 years.—cont'd

Treatment	Water quality	NOM removal (%)	Reaction conditions	References
	density = 1.9 meq g^{-1} DOC. Draycote: DOC = 10.7 mg L^{-1}; UV254 = 13.9 m^{-1}; SUVA = 1.3 mg C L^{-1}; pH = 8.1; turbidity = 1.4 NTU; conductivity = 692 μS m^{-1}; alkalinity = 157 mg $CaCO_3$ L^{-1}; zeta potential = −10.8; charge density = 0.3 meq g^{-1} DOC. Albert: DOC = 9.4 mg L^{-1}; UV254 = 60.1 m^{-1}; SUVA = 6.4 mg C L^{-1}; pH = 5.9; turbidity = 1.8 NTU; conductivity = 70 μS m^{-1}; alkalinity <10 mg $CaCO_3$ L^{-1}; zeta potential = −12.4; charge density = 6.4 meq g^{-1} DOC.			
MIEX for enhanced coagulation	Nine surface waters were tested. Manchester water works (TOC >2–4 mg L^{-1} and alkalinity 0–60 mg L^{-1} as $CaCO_3$); Metropolitan water District of	Enhanced coagulation was reported to be more efficient than coagulation only. Thus, THM and HAA formation potential was reduced by more than 60% in all studied waters.	Batch MIEX treatment, followed by alum coagulation.	Singer and Bilyk (2002)

	Southern California (TOC >2–4 mg L^{-1} and alkalinity >60–120 mg L^{-1} as $CaCO_3$); Davis water treatment plant (TOC >2–4 mg L^{-1} and alkalinity >120 mg L^{-1} as $CaCO_3$); Brown water treatment plant (TOC >4–8 mg L^{-1} and alkalinity 0–60 mg L^{-1} as $CaCO_3$); Haworth water treatment plant (TOC >4–8 mg L^{-1} and alkalinity >60–120 mg L^{-1} as $CaCO_3$); White river filtration plant (TOC >4–8 mg L^{-1} and alkalinity >120 mg L^{-1} as $CaCO_3$); Manatee water treatment plant (TOC >8 mg L^{-1} and alkalinity 0–60 mg L^{-1} as $CaCO_3$); Tampa water department (TOC >8 mg L^{-1} and alkalinity >60–120 mg L^{-1} as $CaCO_3$); Sioux falls water purification plant (TOC >8 mg L^{-1} and alkalinity >120 mg L^{-1} as $CaCO_3$).	TOC and UV absorbance of MIEX pretreated waters were subsequently lower than that of water treated by coagulation.		

BDOC, biodegradable dissolved organic carbon; DOC, dissolved organic carbon; ERD, effective resin dose; HAA, haloacetic acid; HRT, hydraulic residence time; MIEX, magnetic ion exchange resin; MW, molecular weight; NOM, natural organic matter; PAC, powdered activated carbon; SUVA, specific UV absorbance; THM, trihalomethane; TOC, total organic carbon; UVA, UV absorbance.

10.3 Conclusion

When designing a drinking water treatment plant, one of the most crucial issues to be addressed is the presence of NOM. It reduces water quality in terms of color, odor, and taste, and increases the dose of coagulant required for treatment, which in turn leads to high volumes of sludge formation. Another problem resulting from the presence of NOM in water is bacterial proliferation in drinking water distribution systems.

The complex matrix of water, often containing humic and/or fulvic acids as well as different organic compounds coming from urban or industrial activities, is a major reason for difficulties that arise during NOM characterization. Despite this, it is important to know the fractions of NOM present in treated water to increase the efficiency of the treatment process. Yet, usually only 10%–30% of the species of NOM present in water are identified.

The most common technique for NOM removal from water is coagulation and flocculation followed by sedimentation and sand filtration. Yet, the coagulant dose strongly depends on the nature of the NOM present. Filtration, which is sometimes used as an alternative treatment method for NOM removal, often suffers from membrane fouling.

IE is another example of an alternative treatment for water that has a relatively high content of NOM. Depending on the nature of the water, 10%–40% of NOM was reported to remain unremoved after IE treatment; this was speculated to be composed of uncharged species. As various studies demonstrate, IE is more efficient than coagulation with alum for the elimination of charged species contained in NOM. IE is proposed to be an effective choice for their treatment, given its efficiency in treating the transphilic fraction of NOM (Bond et al., 2011).

Taking into account the fact that polar components of NOM can lead to high formations of DBPs, IE can be considered more beneficial than coagulation. Yet, industrialization of the IE process normally requires the use of IE resin in a packed bed, which is only possible at the polishing stages of treatment.

Since the development of magnetic ion exchange resin, known as MIEX, it has become possible to utilize IE resin in slurry mode prior to coagulation. Due to the presence of magnetic IOPs within the core of MIEX resin, the separation step is significantly simplified, so this resin can be utilized with different reactor designs. The high stability of this resin and the fact that it can be used without pretreatment makes MIEX an extremely promising option. It is not surprising that, currently, MIEX is the most studied ion exchanger for NOM removal from water. Already in 2001, the first potable water treatment plant using a MIEX-DOC process was launched in Australia. In this plant, the MIEX-DOC step was introduced prior to conventional treatment, and a significant improvement in water quality was observed. It should be noted that MIEX is efficient for NOM removal and the prevention of chlorinated and brominated DBP formation (Hsu and Singer, 2010). Yet, NOM removal efficiency is affected by the presence of other anions in water. Therefore, further studies should be conducted to optimize the performance of IE processes and decrease the costs for industrial-scale operations.

References

A. Afcharian, Y. Levi, L. Kiene, P. Scribe, Fractionation of dissolved organic matter from surface waters using macroporous resins, Water Res. 31 (1997) 2989–2996.

B.P. Allpike, A. Heitz, C.A. Joll, R.I. Kagi, G. Abbt-Braun, F.H. Frimmel, T. Brinkmann, N. Her, G. Amy, Size exclusion chromatography to characterize DOC removal in drinking water treatment, Environ. Sci. Technol. 39 (2005) 2334–2342.

J.N. Apell, T.H. Boyer, Combined ion exchange treatment for removal of dissolved organic matter and hardness, Water Res. 44 (2010) 2419–2430.

N. Ates, F.B. Incetan, Competition impact of sulfate on nom removal by anion exchange resins in high-sulfate and low-SUVA waters, Ind. Eng. Chem. Res. 52 (39) (2013) 14261–14269.

B. Bolto, D. Dixon, R. Eldridge, S. King, K. Linge, Removal of natural organic matter by ion exchange, Water Res. 36 (2002) 5057–5065.

B. Bolto, D. Dixon, R. Eldridge, Ion exchange for the removal of natural organic matter, React. Funct. Polym. 60 (2004) 171–182.

T. Bond, E.H. Goslan, S.A. Parsons, B. Jefferson, Treatment of disinfection by-product precursors, Environ. Technol. 32 (2011) 1–25.

T.H. Boyer, P.C. Singer, A pilot-scale evaluation of magnetic ion exchange treatment for removal of natural organic material and inorganic anions, Water Res. 40 (2006) 2865–2876.

T.H. Boyer, P.C. Singer, Bench-scale testing of a magnetic ion exchange resin for removal of disinfection by-product precursors, Water Res. 39 (2005) 1265–1276.

H. Brattebø, H. Ødegaard, O. Halle, Ion exchange for the removal of humic acids in water treatment, Water Res. 21 (1987) 1045–1052.

G.C. Budd, B. Long, J.C. Edwards, P.C. Singer, M. Meisch, Evaluation of MIEX Process Impacts on Different Source Waters, American Water Works Association Research Foundation, 2005, p. 91067F.

S.E.H. Comstock, T.H. Boyer, Combined magnetic ion exchange and cation exchange for removal of DOC and hardness, Chem. Eng. J. 241 (2014) 366–375.

E.R. Cornelissen, E.F. Beerendonk, M.N. Nederlof, J.P. van der Hoek, L.P. Wessels, Fluidized ion exchange (FIX) to control NOM fouling in ultrafiltration, Desalination 236 (2009) 334–341.

E.R. Cornelissen, N. Moreau, W.G. Siegers, A.J. Abrahamse, L.C. Rietveld, A. Grefte, M. Dignum, G. Amy, L.P. Wessels, Selection of anionic exchange resins for removal of natural organic matter (NOM) fractions, Water Res. 42 (2008) 413–423.

M. Drikas, C.W.K. Chow, D. Cook, The impact of recalcitrant organic character on disinfection stability, trihalomethane formation and bacterial regrowth: an evaluation of magnetic ion exchange resin (MIEX) and alum coagulation, J. Water Supply Res. Technol. Aqua 52 (2003) 475–487.

M. Drikas, J.Y. Morran, C. Pelekani, C. Hepplewhite, D.B. Bursill, Removal of natural organic matter—a fresh approach, Water Sci. Technol. Water Supply 2 (2002) 71–79.

M. Drikas, M. Dixon, J. Morran, Long term case study of MIEX pre-treatment in drinking water; understanding NOM removal, Water Res. 45 (2011) 1539–1548.

M. Drikas, M. Dixon, J. Morran, Removal of MIB and geosmin using granular activated carbon with and without MIEX pre-treatment, Water Res. 43 (2009) 5151–5159.

W. Eckenfelder, Industrial Water Pollution Control, third ed., McGraw-Hill Series in Water Resources and Environmental Engineering, 2000.

R. Fabris, E.K. Lee, C.W.K. Chow, V. Chen, M. Drikas, Pre-treatments to reduce fouling of low pressure micro-filtration (MF) membranes, J. Membr. Sci. 289 (2007) 231–240.

D.A. Fearing, J. Banks, S. Guyetand, C. Monfort Eroles, B. Jefferson, D. Wilson, P. Hillis, A.T. Campbell, S.A. Parsons, Combination of ferric and MIEX for the treatment of a humic rich water, Water Res. 38 (2004) 2551–2558.

J. Fettig, Removal of humic substances by adsorption/ion exchange, Water Sci. Technol. 40 (1999) 173–182.

A.C. Fonseca, J. Kennon, S. Booth, B. Labisi, J. Meyerhofer, P. White, Testing bromide and DBP precursors removal in California State Project Water using a magnetized ion exchange resin, in: Proceedings American Water Works Association Annual Conference, 2005.

O. Gibert, B. Lefèvre, M. Fernández, X. Bernat, M. Paraira, M. Calderer, X. Martínez-Lladó, Characterizing biofilm development on granular activated carbon used for drinking water production, Water Res. 47 (2013a) 1101–1110.

O. Gibert, B. Lefèvre, M. Fernández, X. Bernat, M. Paraira, M. Pons, Fractionation and removal of dissolved organic carbon in a full-scale granular activated carbon filter used for drinking water production, Water Res. 47 (2013b) 2821–2829.

Z. Han, W. Chen, L. Li, Z. Cao, Combination of chlorine and magnetic ion exchange resin for drinking water treatment of algae, J. Cent. S. Univ. Technol. 17 (2010) 979–984.

C.E. Harland, Ion Exchange. Theory and Practice, second ed., The Royal Society of Chemistry, Bath, 1994.

F. Helfferich, Ion Exchange, Dover Publications, USA, 1995.

D. Hongve, J. Baann, G. Becher, O. Beckmann, Experiences from operation and regeneration of an anionic exchanger for natural organic matter (NOM) removal, Water Sci. Technol. 40 (1999) 215–221.

S. Hsu, P.C. Singer, Application of anion exchange to control NOM interference on lime softening, J. Am. Water Works Assoc. 101 (2009) 85–94.

S. Hsu, P.C. Singer, Removal of bromide and natural organic matter by anion exchange, Water Res. 44 (2010) 2133–2140.

J. Hu, A. Martin, R. Shang, W. Siegers, E. Cornelissen, B. Heijman, L. Rietveld, Anionic exchange for NOM removal and the effects on micropollutant adsorption competition on activated carbon, Separ. Purif. Technol. 129 (2014) 25–31.

H. Huang, H. Cho, K.J. Schwab, J.G. Jacangelo, Effects of magnetic ion exchange pretreatment on low pressure membrane filtration of natural surface water, Water Res. 46 (2012) 5483–5490.

H. Huang, H. Cho, K. Schwab, J.G. Jacangelo, Effects of feedwater pretreatment on the removal of organic microconstituents by a low fouling reverse osmosis membrane, Desalination 281 (2011) 446–454.

H. Humbert, H. Gallard, V. Jacquemet, J. Croué, Combination of coagulation and ion exchange for the reduction of UF fouling properties of a high DOC content surface water, Water Res. 41 (2007) 3803–3811.

H. Humbert, H. Gallard, H. Suty, J. Croué, Natural organic matter (NOM) and pesticides removal using a combination of ion exchange resin and powdered activated carbon (PAC), Water Res. 42 (2008) 1635–1643.

H. Humbert, H. Gallard, H. Suty, J. Croué, Performance of selected anion exchange resins for the treatment of a high DOC content surface water, Water Res. 39 (2005) 1699–1708.

B.R. Johnson, T.B. Eldred, A.T. Nguyen, W.M. Payne, E.E. Schmidt, A.Y. Alansari, J.E. Amburgey, J.C. Poler, High-capacity and rapid removal of refractory NOM using nanoscale Anion exchange resin, ACS Appl. Mater. Interfaces 8 (28) (2016) 18540–18549.

J. Kaewsuk, G.T. Seo, Verification of NOM removal in MIEX-NF system for advanced water treatment, Sep. Purif. Technol. 80 (2011) 11–19.

A.M. Karpinska, R.A. Boaventura, V.J. Vilar, A. Bilyk, M. Molczan, Applicability of MIEX DOC process for organics removal from NOM laden water, Environ. Sci. Pollut. Res. 20 (6) (2013) 3890–3899.

M.D. Kennedy, J. Kamanyi, B.G.J. Heijman, G. Amy, Colloidal organic matter fouling of UF membranes: role of NOM composition and size, Desalination 220 (2008) 200–213.

H. Kim, B.A. Dempsey, Membrane fouling due to alginate, SMP, EfOM, humic acid, and NOM, J. Membr. Sci. 428 (2013) 190–197.

H. Kim, B.A. Dempsey, Removal of organic acids from EfOM using anion exchange resins and consequent reduction of fouling in UF and MF, J. Membr. Sci. 364 (2010) 325–330.

R.S. Kingsbury, P.C. Singer, Effect of magnetic ion exchange and ozonation on disinfection by-product formation, Water Res. 47 (2013) 1060–1072.

M. Kitis, B. İlker Harman, N.O. Yigit, M. Beyhan, H. Nguyen, B. Adams, The removal of natural organic matter from selected Turkish source waters using magnetic ion exchange resin (MIEX), React. Funct. Polym. 67 (2007) 1495–1504.

N. Lee, G. Amy, J. Croué, H. Buisson, Identification and understanding of fouling in low-pressure membrane (MF/UF) filtration by natural organic matter (NOM), Water Res. 38 (2004) 4511–4523.

P. Lin, X. Zhang, J. Wang, Y. Zeng, S. Liu, C. Chen, Comparison of different combined treatment processes to address the source water with high concentration of natural organic matter during snowmelt period, J. Environ. Sci. 27 (2015) 51–58.

R. Malekian, J. Abedi-Koupai, S.S. Eslamian, S.F. Mousavi, K.C. Abbaspour, M. Afyuni, Ion-exchange process for ammonium removal and release using natural Iranian zeolite, Appl. Clay Sci. 51 (2011) 323–329.

M.R.D. Mergen, B. Jefferson, S.A. Parsons, P. Jarvis, Magnetic ion-exchange resin treatment: impact of water type and resin use, Water Res. 42 (2008) 1977–1988.

J.Y. Morran, M. Drikas, D. Cook, D.B. Bursill, Comparison of MIEX treatment and coagulation on NOM character, Water Sci. Technol. Water Supply 4 (2004) 129–137.

T.I. Nkambule, R.W. Krause, B.B. Mamba, J. Haarhoff, Removal of natural organic matter from water using ion-exchange resins and cyclodextrin polyurethanes, Phys. Chem. Earth Parts A/B/C 34 (2009) 812–818.

P.C. Singer, K. Bilyk, Enhanced coagulation using a magnetic ion exchange resin, Water Res. 36 (2002) 4009–4022.

M.R. Teixeira, V.S. Sousa, Fouling of nanofiltration membrane: effects of NOM molecular weight and microcystins, Desalination 315 (2013) 149–155.

C. Tian, R. Liu, H. Liu, J. Qu, Disinfection by-products formation and precursors transformation during chlorination and chloramination of highly-polluted source water: significance of ammonia, Water Res. 47 (2013) 5901–5910.

A.M. Wachinski, J.E. Etzel, Environmental Ion Exchange: Principles and Design, Lewis Publishers, USA, 1997.

K.M. Walker, T.H. Boyer, Long-term performance of bicarbonate-form anion exchange: removal of dissolved organic matter and bromide from the St. Johns River, FL, USA, Water Res. 45 (2011) 2875–2886.

X. Yang, W. Guo, W. Lee, Formation of disinfection byproducts upon chlorine dioxide preoxidation followed by chlorination or chloramination of natural organic matter, Chemosphere 91 (2013) 1477–1485.

A.W. Zularisam, A.F. Ismail, M.R. Salim, M. Sakinah, O. Hiroaki, Fabrication, fouling and foulant analyses of asymmetric polysulfone (PSF) ultrafiltration membrane fouled with natural organic matter (NOM) source waters, J. Membr. Sci. 299 (2007) 97–113.

CHAPTER 11

Integrated methods

Mika Sillanpää[1,2,3,4,5], Sari Metsämuuronen[6], Anu Matilainen[7] and Mika Mänttäri[8]
[1]Department of Chemical Engineering, School of Mining, Metallurgy and Chemical Engineering, University of Johannesburg, Doornfontein, South Africa; [2]Department of Applied Physics, Faculty of Science and Technology, Universiti Kebangsaan Malaysia, Bangi, Selangor, Malaysia; [3]Zhejiang Rongsheng Environmental Protection Paper Co. LTD, Zhejiang, PR China; [4]Department of Civil Engineering, University Centre for Research & Development, Chandigarh University, Gharuan, Mohali, Punjab, India; [5]International Research Centre of Nanotechnology for Himalayan Sustainability (IRCNHS), Shoolini University, Solan, Himachal Pradesh, India; [6]Department of Chemical Technology, LUT University, Lappeenranta, Finland; [7]Finnish Safety and Chemicals Agency, Tampere, Finland; [8]Department of Separation Science, LUT University, Lappeenranta, Finland

Abstract

Integrating conventional coagulation with novel treatment methods is expected to enhance natural organic matter (NOM) removal in water treatment. The great variability in NOM composition makes it difficult to completely remove from drinking water by any single technique. This chapter discusses NOM removal by integrating coagulation with other treatment processes (ion exchange, oxidation, adsorption, and membrane technology) and hybrid processes combining membrane techniques with other unit processes (coagulation, adsorption, oxidation, and membrane bioreactors).

Keywords: Adsorption; Coagulation; Coupling; Filtration; Ion exchange; Membrane technology; Natural organic matter (NOM); Oxidation; Process integration; Water treatment.

11.1 Introduction

The great variability in the composition of natural organic matter (NOM) makes it difficult to remove NOM from drinking water completely by conventional coagulation or by any single technique. This chapter discusses the integration of various processes for NOM removal (partially reviewed earlier in Matilainen et al. (2010) and Metsämuuronen et al. (2014)).

11.2 Coupling coagulation with other processes

11.2.1 Magnetic ion exchange resin (MIEX)

The magnetic ion exchange resin (MIEX) technique is based on the traditional ion exchange (IE) process, but it uses microsized resin particles dispersed as slurry on a stirred contactor, thus allowing maximum surface

Natural Organic Matter in Water
ISBN 978-0-12-824274-2
https://doi.org/10.1016/B978-0-12-824274-2.00011-9

area for adsorption. It removes both organic and inorganic contaminants such as sulfate, nitrate, or phosphate. The negatively charged NOM compounds are removed efficiently by this strong-base anion resin. A high density and the magnetic properties of the resin provide rapid clarification following the contact stage (Morran et al., 2004). MIEX can be easily regenerated, and even after several regenerations, its organic matter removal efficiency is almost the same. Moreover, it is suitable for both small and large plants, and complements other water treatment technologies such as coagulation or membrane filtration. MIEX has been used in pretreatment prior to coagulation to enhance the efficiency of coagulation and reduce the coagulant dose (up to 60%), with reduced sludge formation and turbidity load (Shorrock and Drage, 2006; Humbert et al., 2007; Jarvis et al., 2008; Sani et al., 2008). In the MIEX process, the NOM compounds with increasing charge density are removed, the most and the presence of functional groups has been noted to be more important than aromatic structures (Mergen et al., 2009). Moreover, MIEX can also remove hydrophilic, low molecular mass (LMM) NOM, and thus can remove a fraction of NOM recalcitrant to coagulation (Mergen et al., 2008, 2009). The flocs formed after pretreatment with MIEX have been larger and more resistant to shearing effects than those formed by conventional processes (Jarvis et al., 2008; Sani et al., 2008). Hence, MIEX in combination with coagulation has increased the dissolved organic carbon (DOC) and disinfection by-product (DBP) precursor reduction (Jarvis et al., 2008; Sani et al., 2008; Mergen et al., 2009), although this has not been evident in all cases (Bond et al., 2010). According to the literature, an addition of 8%–63% and about 10%–30% of DOC removal has been observed in jar (Mergen et al., 2008) and pilot tests (Shorrock and Drage, 2006), respectively.

11.2.2 Oxidation

When advanced oxidation processes (AOPs) are used as a pretreatment method, oxidation changes the structural properties of NOM constituents, affecting both the adsorption and coagulation characteristics of NOM (Uyguner et al., 2007b). Photocatalysis as a pre-oxidation method before coagulation has been found to result in approximately <15% decrease in removal efficiency, measured as $color_{436}$ and UV_{254}. Coagulation prior to oxidation has enhanced the removal of NOM about 5%–32% and 8%–33% as DOC and UV_{254}, respectively, compared to coagulation alone. Also, decreases in specific UV absorbance (SUVA) values were noted (Uyguner et al., 2007a).

Pre-oxidation has been traditionally used to improve the coagulation process. Prechlorination was widely utilized until the discovery of DBP formation during chlorination. Some other oxidants have since been in use and, for example, chloramine pre-oxidation has been noticed to enhance the coagulation of surface water (Yang et al., 2007). Ozonation has become a preferential pre-oxidation method prior the coagulation, and is used for elimination of taste, odor, and color, and inactivation of microorganisms (Yan et al., 2007). The influence of pre-ozonation in coagulation has been closely related to NOM characteristics and basic properties of water to be treated (Liu et al., 2007, 2009). Moreover, the ozone dosage in pre-ozonation has been observed to be highly important. At lower ozone dosage, ozone has produced some hydrophobic neutral and intermedial MM NOM compounds, which are removed efficiently in the following coagulation process, but at higher ozone doses, the NOM has been oxidized further into more hydrophilic and LMM compounds, thus being more difficult to coagulate (Bose and Reckhow, 2007; Yan et al., 2007; Liu et al., 2007, 2009; Chinag et al., 2009). Accordingly, the ozonation process placement should be carefully determined, since post-ozonation might get better results than pre-ozonation in combination with biological activated carbon (BAC) in some cases (Liu et al., 2009). The beneficial effect of ozonation toward better coagulation of nonhumic fraction has also been observed. Hence, to maximize DOC and SUVA removals, two-staged coagulation with intermediate ozonation has been proposed for waters containing both humic and nonhumic NOM (Bose and Reckhow, 2007). The composite polyaluminum chloride (HPAC) coagulant efficiency has not declined even after high ozone doses (Yan et al., 2007), and PACl has been found to be suitable in coagulation with pre-ozonation (Liu et al., 2007).

Advanced oxidation processes are powerful techniques in disinfection and organic compounds (e.g., contaminants) degradation (Rämö and Sillanpää, 2001; Pirkanniemi and Sillanpää, 2002; Pirkanniemi et al., 2007). The photo-Fenton process, in which iron species can play a role as enhanced catalyst as well as coagulant, has been investigated for NOM removal with rather good results (Murray and Parson, 2004; Goslan et al., 2006). NOM removal improved compared to conventional coagulation with iron salts; in particular, LMM compounds were removed more efficiently (Murray and Parson, 2004). Photocatalytic oxidation prior to coagulation reduced the coagulation efficiency by 15%, most likely because the oxidation changes the characteristics of NOM and degrades NOM molecules toward LMM compounds (Uyguner et al., 2007b). Oxidation after coagulation, however, enhanced DOC and UV_{254} removal by about 5%–32% and 8%–33%, respectively (Uyguner et al., 2007a). Photocatalytic coagulation with nanoparticles of copper-doped titania has been observed to provide both better NOM removal and reduction of membrane fouling (Choi et al., 2007).

11.2.3 Activated carbon filtration

Activated carbon (AC) is an effective adsorbent used widely in drinking water treatment. Micropollutants such as pesticides, industrial chemicals, tastes, odors, and algal toxins are removed efficiently in the filtration process. NOM may also be removed in filtration, although it often competes for adsorption sites, decreasing the removal of other pollutants. If enhanced coagulation is insufficient for NOM reduction, additional treatment with granular or powdered AC (GAC/PAC) can maintain the required water quality in terms of, e.g., DBP formation (Szlachta and Adamski, 2009; Uyak et al., 2007; Garcia and Moreno, 2009). AC can also be colonized by a heterotropic biomass, creating BAC, thus reducing the proportion of easily assimilated and biodegradable elements. In addition, efficient removal of LMM and uncharged fractions of NOM is gained (Uyak et al., 2007). Yet, in most cases, a part of LMM NOM is not removed, even after AC filtration (Szlachta and Adamski, 2009). The water can be filtered also only when necessary, e.g., during elevated NOM or alkalinity periods (Smith and Kamal, 2009). PAC-enhanced coagulation has also been noted to reduce the required coagulant dose, while increasing the process efficiency (Szlachta and Adamski, 2009).

11.2.4 Membrane filtration

Membranes have become increasingly common in water treatment. Pressure-driven membrane processes include reverse osmosis (RO), nanofiltration (NF), ultrafiltration (UF), and microfiltration (MF), which all have different NOM and particle removal potentials (Jacangelo et al., 1995; Zularisam et al., 2006). The fouling of membranes and decline of flux are the major problems associated with membrane techniques (Zularisam et al., 2006). Several studies have indicated that NOM—especially the hydrophobic fraction and high molar mass (HMM) compounds—has a great impact on fouling and decline of flux (Zularisam et al., 2006; Chen et al., 2007; Dong et al., 2007; Howe et al., 2006; Jung et al., 2006). Pretreatment with coagulation is an effective method to prevent these adverse effects (Chen et al., 2007; Dong et al., 2007; Howe et al., 2006; Jung et al., 2006;

Qiao et al., 2008), and a combination of coagulation and membrane filtration also reduces the amount of coagulant needed while improving turbidity and DOC reduction (Blankert et al., 2007). Moreover, coagulation has been observed to provide an effective hygienic barrier in combination with membranes (Leiknes, 2009), although contradictory results have been obtained (Barbot et al., 2008), and the removal of NOM has been significantly affected by the type of coagulant, coagulation conditions, type of membrane, filtration conditions, and characteristics of the water to be treated (Bond et al., 2009; Meyn et al., 2008; Park and Yoon, 2009; Barbot et al., 2008; Howe et al., 2006; Cai et al., 2008; Cho et al., 2006; Choo et al., 2008; Kim et al., 2006; Tran et al., 2006; Zularisam et al., 2009).

A combination of the coagulation with UF (Barbot et al., 2008; Blankert et al., 2007; Zularisam et al., 2009; Konieczny et al., 2009) or MF (Meyn et al., 2008; Cho et al., 2006; Loi-Brugger et al., 2006) is an in-line hybrid process. The coagulant dose and the duration of water treatment are reduced by batching the coagulant to the feed immediately before membrane (Konieczny et al., 2009). Moreover, a novel submerged UF membrane coagulation bioreactor (MCBR) process has been studied for treating slightly contaminated surface water to produce potable water (Tian et al., 2008a,b). A proposed new way of reducing membrane fouling is the immobilization of TiO_2 photocatalyst on the membrane surface, thus creating a composite membrane (Ma et al., 2009). The use of nanoparticles (including alumina, silica, silver, and iron oxide) in membrane structures has gained a lot of interest, but their use in water treatment requires further research due to the possibility of nanoparticles being released into the water (Kim and Van der Bruggen, 2010).

11.3 Coupling membrane technology with other processes

Coupling membrane techniques such as RO, NF, UF, and MF, with other unit processes has become a common way to comply with drinking water legislation. It is estimated that MF and UF plant capacity is around 20 million m^3 per day, 60% of which is for drinking water production (Schrotter and Bozkaya-Schrotter, 2010), and UF applications represent about 74% of the total capacity (Fane et al., 2008). MF and UF plants are primarily developed for turbidity, bacteria, and protozoa removal as standalone processes, as retrofits of existing conventional treatment plants, or as pretreatments to NF and RO (Fane et al., 2008).

An essential advantage of membrane filtration is its compact and modular design that allows for a high backing density and small footprint. The fouling of membranes and decline of flux are the major challenges associated with membrane techniques (Zularisam et al., 2006). These problems can be reduced by coupling membranes with other techniques.

11.3.1 Coupling with filters

Coarse materials can be detrimental to membrane filtration performance. Hence, membranes are normally protected at least by an upstream screen, cartridge, or backed bed filters. An upstream screen with a mesh from 500 to 50 μm may be high enough protection for the MF membranes (Bonnélye et al., 2008), but the smaller pore size membranes need at least a 50-μm cartridge filter to ensure stable performance in NF of groundwater (Nederlof et al., 2000) or surface waters (Ødegaard et al., 2010).

A typical flow diagram of a small NF plant that has operated successfully for soft water characterized by high color and low turbidity is shown in Fig. 11.1. The pretreatment consists only of a self-cleaning cartridge filter with 50-μm mesh (Ødegaard et al., 2010). A sand filter can be installed upstream of the cartridge when high turbidity raw water is treated. Approximately 150 plants of this type have been opened worldwide by 2005, about 70% of them in Norway. Most of them are small, with capacities in the range of 100–15,000 m^3 per day (Thorsen and Fløgstad, 2006).

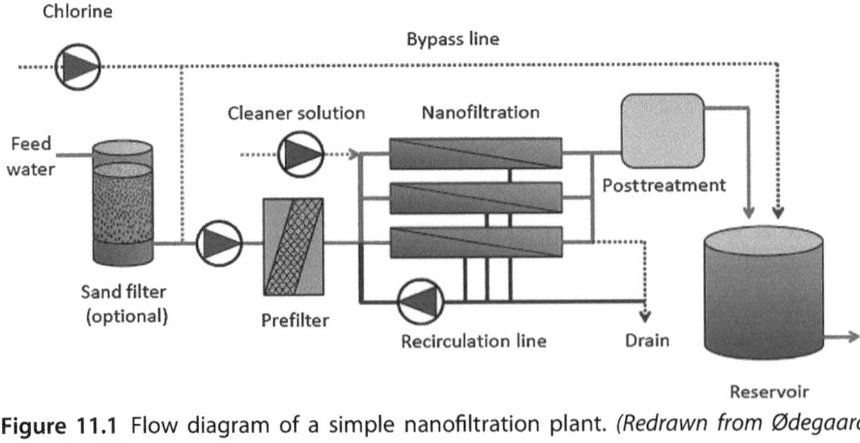

Figure 11.1 Flow diagram of a simple nanofiltration plant. *(Redrawn from Ødegaard et al. (2010)).*

Packed bed filters are effective in capturing particles of large size distribution and, depending on the properties of the downstream membrane, both the membrane surface and pore fouling may be reduced. Conventional packed bed filters can be effectively restored by backwashing, whereas a disadvantage of granular filters is that they may deteriorate and be difficult to recover. In principle, the removal mechanism for packed bed filters is either physical sieving or chemical adsorption/deposition (Huang et al., 2009).

Sand filtration-UF plants have been successfully operated for the production of drinking water (Botes et al., 1998; Pryor et al., 1998; Alborzfar et al., 1998; Mierzwa et al., 2008, 2012; Ødegaard et al., 2010). A periodic relaxation and flushing procedure with a chlorine dosage has been observed as necessary to maintain stable operational conditions in a sand filtration-UF process (Mierzwa et al., 2012). Dual-media filters (sand and anthracite), followed by a 5-μm cartridge prior to membrane filtration with spiral-wound and tubular membranes, are applied in a number of small water treatment plants, producing high-quality drinking water for remote communities (Grose et al., 1998).

Ahn et al. (2004) investigated a coagulation/sedimentation-sand filtration-NF process (Fig. 11.2), and found that the installation of MF between the sand filter and the NF process can significantly increase the permeability of the NF membrane. Several researchers have compared the coagulation/sedimentation-sand filtration pretreatment with the UF pretreatment for NF or RO (Glucina et al., 2000; Kim and Yu, 2005; Chang et al., 2009). It has been observed that the former reduced the NOM effectively with lower energy consumption than the latter, but in general,

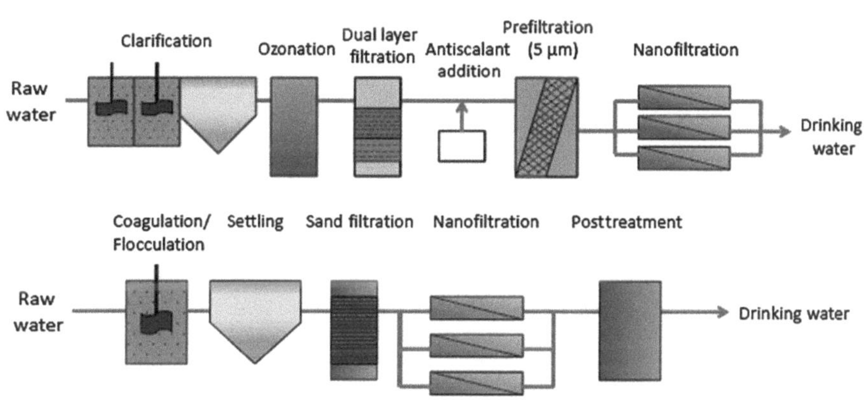

Figure 11.2 Flow chart of typical pretreatments for nanofiltration.

the UF-NF process exhibited better removal efficiency for NOM and lower fouling (Chang et al., 2009). Also, Kim and Yu (2005) obtained the best pretreatment results with the UF-NF process for trihalomethane (THM) and haloacetic acid (HAA) formation potential removal (85% and 89%, respectively).

In RO for seawater and brackish water desalination plants, MF, UF, and NF pretreatments have shown great advantages over conventional pretreatment processes (Hilal et al., 2006). Yet, Glucina et al. (2000) found that both pretreatments were effective to control RO fouling and that both combinations produced water that exceeded the most stringent quality level proposed in the EU Directives. Altogether, MF and UF membranes have increasingly been applied as prefilters for NF and RO membranes. They remove suspended particles, colloids, *Giardia lamblia* and *Cryptosporidium* cysts, algae, some bacterial species, and a certain amount of HMM NOM (Kajitvichyanukul et al., 2008). Lamsal et al. (2012) reported that UF pretreatment removed 66% of DOC and 54% of THM formation potential from river water, and that NF was able to remove most of the remaining DOC with excellent performance.

Low-pressure membranes have been observed to be more effective than a cartridge MF filter for the pretreatment of NF membrane processes (Siddiqui et al., 2000). This is because these membranes are able to remove colloids, which is crucial for preventing NF fouling (Lee and Lee, 2007).

Lee and Lee (2006) found that UF can efficiently remove NF foulants, and that the efficiency of the NF membrane improves with decreasing prefilter pore size. Yet, pretreatment cost also increases with decreasing prefilter pore size.

11.3.2 Coupling with coagulation

Coagulation has remained the most common process for water treatment, normally for the removal of turbidity and NOM (Liu et al., 2011). The conventional process incorporates several physicochemical processes including rapid mixing, slow mixing (flocculation), sedimentation, filtration, and disinfection. Coagulants are used to increase the size of aquatic substances, making them large enough to settle or to be removed by sand, dual layer filtration, or the membrane. Coagulation is based on charge effects: Particles suspended in water usually carry a negative charge due to the ionization of surface groups, adsorption of ions, or ion substitution; hence, they can be coagulated by adding positively charged coagulants.

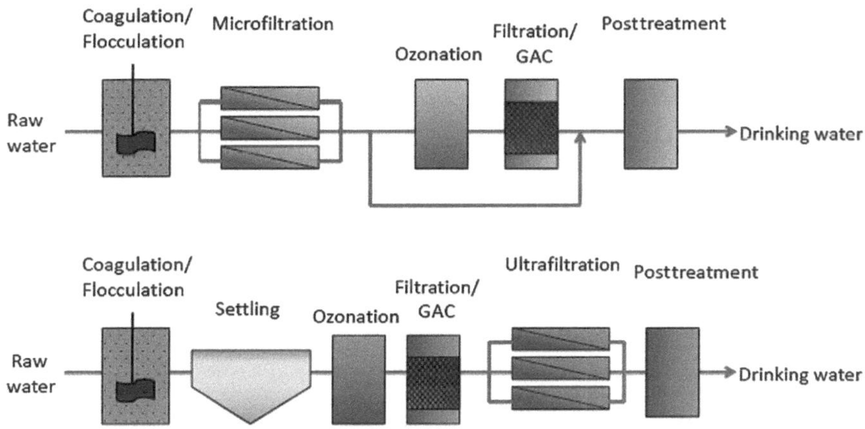

Figure 11.3 Flow chart of coagulation without settling before microfiltration and with settling before ultrafiltration.

Integration of coagulation into membrane filtration has two main advantages: Enhanced removal of NOM molecules and reduction of membrane fouling. These two unit operations can be combined with or without settling (Fig. 11.3). The most conventional way is to use a coagulation tank, which may be divided to rapid mixing and slow mixing/flocculation parts, and settling to remove flocs from the feed water. In this case, the supernatant is conveyed to membrane filtration. The advantage of the sedimentation step is that it smoothens out the sudden increase in solid loading before the membrane system (Moon et al., 2009). Another approach is to convey water directly from flocculation to the membrane unit without removal of flocs (direct filtration).

The most recent method of combining coagulant with MF (Meyn et al., 2008; Cho et al., 2006; Loi-Brugger et al., 2006) or UF (Barbot et al., 2008; Blankert et al., 2007; Zularisam et al., 2009; Konieczny et al., 2009) is to patch coagulant into the feed stream immediately prior to the membrane process without removing coagulated solids (in-line coagulation). These integrated processes composed of MF or UF and in-line coagulation have become particularly attractive because they are expected to enhance operational efficiency (Bergamasco et al., 2011). The advantages of in-line coagulation are reduced treatment plant footprint, reduced water treatment time, and the possibility of lower coagulant dosing, since flocs do not need to settle and instead will be removed by the membrane (Jung and Kang, 2003; Konieczny et al., 2009). Furthermore, coagulant selection and dosing can be optimized specifically for NOM removal as particle removal is assured by the membrane (Vickers et al., 1995). Charge neutralization and sweep coagulation conditions have both been found to be effective for the removal of NOM and turbidity by in-line coagulation-UF (Choi and Dempsey, 2004).

In water treatment plants, MF membranes are generally positioned toward the front of a treatment chain after a light pretreatment such as coagulation/flocculation (Fig. 11.3). Flocculated water is clarified typically with outside-in hollow-fiber MF membranes, which are able to treat a high solid loading. This process is adapted to treat karstic water and is mainly developed to remove turbidity and bacteria (Schrotter and Bozkaya-Schrotter, 2010). Tight inside-out hollow-fiber UF membranes are commonly situated at the end of the treatment chain because they require a more advanced pretreatment.

An example of a full-scale integrated water treatment plant is in Heemskerk, The Netherlands (Kamp et al., 2000). It is a direct surface water treatment plant based on the combined application of UF and RO from the Lake Ijssel (River Rhine) water. Its major quality control objectives are the removal of pathogenic microorganisms, organic micropollutants, and inorganics, as well as maintenance of biological stability. The treatment chain consists in coagulation, sedimentation, filtration, GAC (optional), UF, and RO. The permanent fouling of UF membrane by coagulant residuals was avoided by a temporary application of GAC filtration.

The removal of NOM has been significantly affected by the type and dosage of coagulant, coagulation conditions, type of membrane, filtration conditions, and characteristics of the water to be treated (Kabsch-Korbutowicz, 2005; Howe et al., 2006; Kim et al., 2006; Tran et al., 2006; Cho et al., 2006; Chooi et al., 2008; Cai et al., 2008; Meyn et al., 2008; Barbot et al., 2008; Bond et al., 2009; Park and Yoon, 2009; Zularisam et al., 2009). A sufficient dose of coagulant is required to produce large enough flocs to avoid pore blocking, but increased coagulation dose has been observed to lead to increased fouling (Tran et al., 2006). Small flocs result in rapid blocking of the membrane (Rojas et al., 2010). Therefore, coagulant dosing might be difficult if raw water quality varies significantly. It is also reported that floc strength is more important than floc size for membrane performance (Kim et al., 2013).

The most commonly used coagulants in practice are the hydrolyzing metal ions Al^{3+} and Fe^{3+} (Capodaglio et al., 2011). Polyaluminum chloride (PACl), aluminum sulfate (alum, $Al_2(SO_4)_3 \cdot 14H_2O$), and ferric chloride

($FeCl_3 \cdot 6H_2O$) have been widely adopted in drinking water plants. NOM removal with these coagulants is achieved through several mechanisms, which include double layer compression, charge neutralization, sweep coagulation, and interparticle bridging (Capodaglio et al., 2011).

Aluminum-based and iron-based coagulants are ideal for charge neutralization and sweep flocculation, and are known to preferentially remove hydrophobic, charged, and larger sized (>10 kDa) substances (Choi and Dempsey, 2004; Jung and Kang, 2003; Tran et al., 2006; Maartens et al., 1999; Carroll et al., 2000; Wang and Wang, 2006; Chen et al., 2007; Zhang et al., 2011a). Several studies have indicated that these compounds of NOM have great impact on fouling and decline of flux (Zularisam et al., 2006; Chen et al., 2007; Dong et al., 2007; Howe et al., 2006; Jung et al., 2006). Thus, pretreatment with coagulation is an effective method to prevent these adverse effects (Chen et al., 2007; Dong et al., 2007; Howe et al., 2006; Jung et al., 2006; Qiao et al., 2008).

Chen et al. (2007) reported that the flux was higher for coagulated water than for supernatant water when the coagulant doses were low, and vice versa at higher coagulant doses. The higher flux that was observed with coagulated water was supposed to be due to the formation of a permeable cake layer that prevented membrane fouling, leading to the deposition of hydrophilic neutral substances on the cake layer instead of on the membrane surface. This layer is then effectively removed by backwashing to maintain the flux at desired level (Tran et al., 2006; Chen et al., 2007; Liu et al., 2006). Other researchers have reported that filtration performance was enhanced when flocs are removed from membrane feed water (Carroll et al., 2000; Chen et al., 2007).

Two-stage coagulant addition is believed to be an effective strategy to reduce membrane fouling in the coagulation-UF process (Liu et al., 2011). This is reported to be due to larger flocs with lower density compared to one-stage addition. Microflocs formed in the first stage might fill the cake layer and/or block the membrane pores, whereas floc aggregates formed in the second stage formed a permeable cake layer.

Coagulation has shown poor removal efficiency in the neutral hydrophilic and small MM fractions (Li et al., 2011; Chen et al., 2007). A coagulation-MF-NF chain has been reported to exhibit DOC removal of 75%–87%, HAA formation potential removal of >72%, and final DOC and HAA formation potential concentrations of $<1\ mg\ L^{-1}$ and $<30\ mg\ L^{-1}$, respectively (Ahn et al., 2004). Another study of a coagulation-sedimentation-MF-NF process observed DOC, biodegradable dissolved organic carbon (BDOC), and assimilable organic carbon (AOC) removals of 68%–80%, 33%–80%, and 45%–67%, respectively (Park et al., 2005).

Härmä (1999) performed filtration experiments using tight NF membranes in six water treatment plants where raw water was coagulated with alum, PACl, or ferric coagulants. All permeates contained total organic carbon (TOC) of less than 0.3 mg L^{-1} and NOM removal efficiency (>90%) was independent of water recovery (50%–80%), *trans*-membrane pressure (500–1000 kPa), and temperature (0–20°C). The removal of AOC was lower, varying between 25% and 94% depending on the raw water. Tran et al. (2006) have compared different coagulants in a coagulation-UF process. They noticed that polysilicato-iron (PSI) coagulant was more effective in removing NOM and improving fouling conditions than aluminum-based coagulants: A DOC removal of 82%–89% was obtained with PSI coagulant in comparison to the 67%–86% removal with aluminum-based coagulants.

Novel electrocoagulation (Ben-Sasson et al., 2013; Dubrawski et al., 2013) and submerged coagulation membrane bioreactor (Tian et al., 2008a,b; Li et al., 2011) processes for the production of drinking water have also been tested. A new proposal for reducing membrane fouling is to coat membranes by coagulants (Lin et al., 2012) and to immobilize TiO_2 photocatalyst on the membrane surface, thus creating a composite membrane (Ma et al., 2009). The use of nanoparticles (e.g., nano-alumina, silica, silver, or iron oxide) in membrane structures has raised a great deal of interest, but the risk is that nanoparticles may be released into the water; thus, further research is needed (Kim and van der Bruggen, 2010).

11.3.3 Coupling with adsorption

Another extensively used drinking water treatment method is the hybrid adsorption-membrane process. Suspended powders have been widely used to remove organic compounds that are scarcely removed by UF and MF membrane processes. Similar to the hybrid coagulation process, membrane filtration can be performed with clarification as a pretreatment or without clarification as a combined adsorbent/membrane process. Adsorbents generally require long contact times or high concentrations to make the process effective. Therefore, regeneration of the adsorbent is required in many cases to make the process feasible. The membrane provides a physical barrier retaining the adsorbent and the organic compounds adsorbed on it. Two configurations exist: external loop and submerged membrane tank. In the latter, the membrane module is immersed into the adsorption tank, which significantly reduces the size and space of the equipment compared to the traditional configuration.

The performance of the hybrid adsorption-membrane process has been shown to depend on adsorbent dosing, membrane properties, backwashing frequency, reactor size and configuration, filtration mode, and NOM concentration and characteristics (Campos et al., 1998; Campinas and Rosa, 2010a; Mavrov et al., 1998; Lebeau et al., 1998; Oh et al., 2006).

The most widely studied adsorbent for MF and UF is PAC. It has been reported to remove small (>500 Da) and nonpolar NOM molecules (Zhang et al., 2011a). PAC/UF systems demonstrated about 90% humic acid and 100% phenol removal (Tomaszewska and Mozia, 2002), and very effective removal of fine contaminants such as trichloroethene, tetrachloroethane, and atrazine (Pianta et al., 1998). Yet, NOM removal from drinking water has not been as successful as removal of the above-mentioned small surrogate compounds (Xia et al., 2007). Enhanced NOM removal from micromolecular sizes up to 17 kDa has been observed when UF is combined with PAC (Kang and Choo, 2010; Lin et al., 1999; Lohwacharin et al., 2010).

PAC was also observed to have a lesser impact on the removal of algal organic matter. PAC enhanced the TOC rejection of algogenic organic matter in UF (35% vs. 55%), but had no significant effect on rejections of the highly hydrophilic compounds of extracellular organic matter (Campinas and Rosa, 2010b). It was suggested that PAC was not able to adsorb the highly hydrophilic compounds of polysaccharide-like substances and, as such, did not lower the membrane reversible fouling. This is in accordance with the findings that the hydrophilic fraction exhibits the worst flux decline during UF, regardless of whether the PAC treatment was applied (Lin et al., 2000). In the latter study, both the PAC pretreated UF and combined PAC/UF system rejected 35% of DOC of the hydrophilic fraction. The hydrophobic fraction exhibited ~55% and ~40% DOC rejection with the PAC pretreatment and combined PAC/UF system, respectively.

In most of the above-mentioned studies, 100 kDa cut-off membranes were used. Yet, a comparison of different cut-off membranes in the range of 1−100 kDa has shown that the highest cut-off membrane exhibited the highest flux decline (60%) and the lowest UV_{254} removal (21%) of humic acid in the PAC/UF process (Li and Chen, 2004). The highest removal (89%) and the lowest flux decline (<5%) were observed when a 1 kDa cut-off membrane was used. These findings favor the use of finer UF membranes in integrated membrane systems based on adsorbents.

GAC has been reported to be effective in the removal of the lower MM and hydrophilic fractions of NOM (Yee et al., 2009). Other researchers have reported that the main removed compounds have been mostly hydrophobic (Kim et al., 2009a,b; Gur-Reznik et al., 2008; Tsujimoto et al., 1998).

The GAC filters that are not regenerated contain microbiological communities that make them effective in removing the biodegradable compounds, hence reduce BDOC and AOC. Biological GAC filtering is a high-performance, cost-effective and chemical-free method, which has been widely used for drinking water pretreatment (Hallé et al., 2009). It removes large fractions of polysaccharides and proteins (Hallé et al., 2009; Huang et al., 2011) and colloidal NOM (Kwon et al., 2005). Huang et al. (2011) found that the biodegradation of NOM not only depends on its chemical composition, but is also affected by molecule size and hydrophobicity. GAC adsorption is more suitable after pretreatment to remove large molecules or reduce them in size sufficiently to enter the pores of the adsorbent (e.g., coagulation or ozonation). As a biological process, it is influenced by ambient conditions and contact time, such that a higher temperature can enhance biopolymer removal (Hallé et al., 2009).

Other adsorbent particles that have been tested in hybrid systems include freshly precipitated iron and aluminum oxide particles, heated aluminum and iron oxide particles, ferrihydrite, and carbon black nanoparticles. Heated aluminum and iron oxide particles have consistently been reported to reduce fouling caused by NOM (Zhang et al., 2003; Shi and Benjamin, 2008; Kim et al., 2008). They adsorb foulants preferentially over nonfouling particles, whereas the opposite is true for PAC. As a result, higher foulant removal has been achieved with these oxide particles than with PAC, even though the highest NOM removal was achieved by PAC (Kim et al., 2008). The NOM rejection of 79%–92%, 64%–72%, and 46%–65% were observed with PAC, heated aluminum, and iron oxide particles, respectively (Kim et al., 2008).

Higher NOM removal efficiencies have been observed when oxide particles were predeposited on the membrane surface than when adsorbent particles were suspended in raw water (Kim et al., 2008; Ha et al., 2004; Yao et al., 2009). It has been hypothesized that the formation of adsorbent cake layers at the membrane would contribute to further NOM rejection due to the sieving effect during UF. Heated aluminum oxide particles deposited on the membrane surface have been observed to effectively remove NOM and foulants such as polysaccharides and proteins (Cai and Benjamin, 2011).

Ferrihydrite (Kang and Choo, 2010) and carbon black nanoparticles (Lohwacharin et al., 2010) have been reported to mitigate fouling more effectively than PAC. They remove more macromolecular NOM than PAC and form more porous cake layers that are more easily detachable by backwashing than the PAC cake layer.

11.3.4 Coupling with ion exchange

Combinations of fluidized IE and MIEX with membrane filtration have been studied in the past few years. NOM can be removed by macroporous anion exchangers because of the negative charge of the humic substances. The MIEX particles contain a high proportion of magnetic iron oxides that agglomerate into larger, fast-settling particles.

Cornelissen et al. (2009, 2010) applied fluidized IE treatment before UF and NF, and observed removals of 60% for NOM and >90% for humic substances, respectively. The fluidized IE treatment reduced fouling of a 100-kDa UF membrane, but not fouling of an NF membrane. Humbert et al. (2012) used a combined anion exchange resin and UF with a 100-kDa membrane. They observed that small resin particles fouled the membrane, but that irreversible fouling was significantly lowered.

Kaewsuk and Seo (2011) treated river water with a MIEX-NF process. They observed that MIEX could significantly remove NOM compounds with a low SUVA value (phenolic compounds, aliphatic hydrocarbon, and amino sugars), but it showed low efficiency for removing carboxylic compounds. Protein, polysaccharides, and amino sugars were completely retained by the NF membranes. They also noted that membrane roughness influences rejection.

Also, Humbert et al. (2007) and Kaewsuk and Seo (2011) observed that the MIEX selectively adsorbs the low and medium MM fractions of NOM, but is not effective in removing the high MM compounds. The combination of MIEX-UF with a 100-kDa membrane was observed to remove up to 80% of DOC (Humbert et al., 2007, 2012). The same removal rate was observed when MIEX was combined with much tighter (5, 10, and 30 kDa) membranes (Kabsch-Korbutowicz et al., 2008).

Kaewsuk and Seo (2011) used a fluidized bed type MIEX column as a pretreatment process for submerged NF membranes. The MIEX process could remove the NOM MM above 500 Da and remove ~60% of polysaccharide compound (>43 kDa). A hydrophilic 350-Da membrane removed 80% and a hydrophobic 550-Da membrane 88% of DOC.

Bond et al. (2010) compared ferric sulfate coagulation, MIEX anion exchange resin, and two NF membranes in the removal of DBP formation potential from NOM surrogates. MIEX experiments were performed with a resin dose of 10 mL L^{-1} and a bed volume of 1500. Coagulation and MIEX preferentially removed hydrophobic organic acids, which are typically highly negatively charged, whereas neutral compounds were not significantly removed. Hydrophobic anionic tannic acid was well removed with all these methods: coagulation removed 89% and MIEX and NF both 92%. With NF (150−430 Da) membranes, removal generally increased with molecular weight from glycine (75 g mol^{-1}, 24% removal) to tannic acid (1701 g mol^{-1}). The researchers concluded that full-scale MIEX would not offer improved performance over coagulation.

Mergen et al. (2008) noted that MIEX treatment is capable of removing both hydrophilic and hydrophobic NOM. Water dominated by hydrophilic acids showed consistently high levels (60%) of removal onto the resin. Yet, HMM NOM quickly saturated or blocked the resin and algal less charged NOM was poorly removed. To mitigate resin blockage, raw water was pretreated by a 50-μm microsieve or by a rapid sand filter when the turbidity was higher than 0.5−1.0 FTU (Ødegaard et al., 2010).

11.3.5 Coupling with oxidation

Ozonation is the preferred oxidation method for humic water where the main goals are color removal and disinfection (Ødegaard et al., 2010). Ozone can act directly or via an indirect reaction where it is decomposed to hydroxyl radicals. Direct oxidation targets mainly activated aromatics and double bonds, resulting in high color removal, but little TOC removal. Hydroxyl radicals are more powerful but less selective oxidants than ozone; they remove less color and more TOC.

The reaction of ozone and humic material produces a number of LMM polar compounds that are more biodegradable than the original compounds (Karnik et al., 2005; Malek et al., 2006; Song et al., 2010; Yan et al., 2010; Yang et al., 2010). Therefore, the UV_{254} removal of NOM can be much higher than the corresponding DOC removal. In an ozone-MF treatment, 71% UV_{254} removal, but only 10% DOC removal were observed (Song et al., 2010). Ozonation does not lead to full mineralization of most compounds, thus the initial DOC concentration declines only slightly. Karnik et al. (2005) noticed that an ozonation-UF process resulted in 50%

reduction in humic substances and an increase of 20% in nonhumic substances. Furthermore, partially oxidized compounds that were less reactive with chlorine were formed, hence the total THMs and HAAs distributed in the water were reduced by 80% and 65%, respectively.

Byun et al. (2011) studied the effect of ozone dosage on DOC removal and fouling of a 200—400-Da NF membrane and a 30-kDa UF membrane. It was found that DOC removal by the NF membrane was >92% regardless of the ozone dosage, whereas DOC removal by the UF membrane increased slightly from 11% to 26% with an increase in the ozone dosage. The researchers reported that increased fouling due to ozonation was caused by the changes in the chemistry, not the size of the foulants.

Ribau Teixeira et al. (2011) compared conventional sequence, ozonation-NF and ozonation-coagulation/flocculation-sedimentation-NF for the removal of NOM and DBPs. They treated hard water with a low organic matter content (DOC 1.31 mg L^{-1}) consisting of mainly hydrophilic, less aromatic material of a lower molecular weight. All treatment sequences removed large molecules and reduced DOC content to below 1 mg L^{-1}. When NF was applied, smaller MM compounds were removed more than in the conventional sequence.

One disadvantage of pre-oxidation is the formation of small assimilable compounds that may block membrane pores and increase biofouling in the downstream membrane processes and distribution systems. The increased degree of fouling has been reported to be due to the increase in AOC and extracellular biopolymers released by algae upon ozonation (Her et al., 2007). Another disadvantage is the formation of harmful by-products, such as formaldehyde, glyoxal, and glyoxylic acid, which have been reported to be mutagenic (Koltuniewich and Drioli, 2008).

The above-mentioned problems can be avoided by combining oxidation with downstream biologically active filters, BAC or GAC (Fig. 11.3) (Ødegaard et al., 2010; Yan et al., 2010; Yang et al., 2010). Biofiltration removes biodegradable organic matter produced by the oxidation step, improving the biostability of the water. Yang et al. (2010) reported that the conventional treatment followed by the O_3/BAC process exhibited DOC removal of 63%, which was further increased to 89% when UF-NF was added at the end of the treatment chain. Moreover, very high removals were observed for THM formation potential (84%) and HAA formation potential (98%).

A treatment facility combining UF with ozonation to treat surface water is one of the world's largest UF plants, located in Mississauga, Ontario, and

Canada (Byun et al., 2011). The UF membrane filtration unit at the Mississauga plant is combined with ozonation and biologically activated carbon contactors to meet water quality regulations and improve the water's aesthetic quality.

The combination of an AOP and membrane separation has been shown to have significant potential in mitigating both NOM fouling of membranes and the formation of DBPs (Le-Clech et al., 2006; Choo et al., 2008; Huang et al., 2008; Bai et al., 2009; Yao et al., 2009).

Among AOPs, TiO_2/UV oxidation has been observed to effectively remove HMM and hydrophobic organic compounds, and to reduce membrane fouling not only through NOM mineralization, but also by altering NOM characteristics and therefore their fouling potential. Thus, photocatalytic oxidation prior to membrane filtration has been considered quite an efficient method regarding membrane performance and controlling of NOM fouling (Choo et al., 2008; Huang et al., 2008; Bai et al., 2009), although contradictory results have also been observed (Syafei et al., 2008). When membranes are coated using iron oxide particles, the NOM removal performance of the photocatalysis/membrane system is further improved (Yao et al., 2009).

11.3.6 Membrane bioreactor

The hybrid membrane bioreactor (MBR) process is quite a new technique in drinking water production. As indicated previously, different pretreatment technologies often preferentially remove certain components of NOM. In the MBR process, several removal mechanisms can be organized in a single step. Biomass contains microorganisms that can degrade biodegradable organic matter, adsorbents can be used to remove small-medium MM NOM, coagulants can be used to flocculate hydrophobic large MM NOM and, finally, the membrane provides a physical barrier to separate the entire sludge containing these substances and chemicals. Furthermore, an oxidation step can be added to a submerged MBR process to remove the residual nonadsorbable, nonbiodegradable fraction of DOC (Treguer et al., 2010).

MBR systems are implemented based on two main configurations (Fig. 11.4). The submerged configuration consists of outer skin membranes that are situated inside the bioreactor. The driving force across the membrane is achieved by suction pumps or by pressurizing the bioreactor. It operates at a much lower *trans*-membrane pressure and at a lower liquid cross-flow velocity than the external MBR. Plate and frame modules or hollow-fiber membranes in bundles or curtains have been used in submerged membrane systems (Fane et al., 2008).

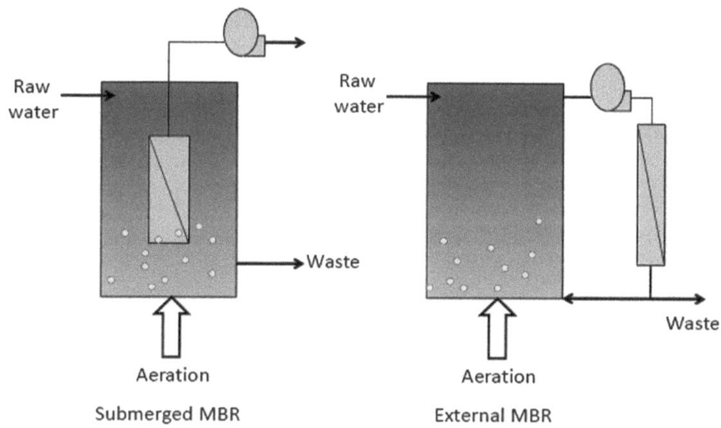

Figure 11.4 Flow charts for submerged and external membrane bioreactors.

In the external configuration, mixed liquid is recirculated through membrane modules that are situated outside the bioreactor. MBRs operate at high concentration of suspended solids, so back transport of solids is necessary to sustain membrane permeability. MBRs can produce high-quality water from polluted raw water, but marginally higher capital and operating costs are considered disadvantages of MBRs compared to other biological processes (Fane et al., 2008).

Li and Chu (2003) used a laboratory MBR equipped with a submerged polyethylene hollow-fiber membrane module with a pore size of 0.4 μm and a total surface area of 0.2 m^2 for the treatment of raw water slightly polluted by domestic sewage (3–5 mg TOC L^{-1}). The MBR ran continuously for over 500 days with a ≤1-h hydraulic retention time. Reductions of 60% and 75% were achieved for TOC and THM formation potential, respectively. Furthermore, the biostability of the effluent improved considerably, as the AOC removal was 81%. Visvanathan et al. (2003) studied a membrane bioreactor as a pretreatment for seawater RO. They noted a 78% DOC removal and quasitotal biodegradable DOC removal when treating sodium acetate-spiked seawater (6–8 mg DOC L^{-1}).

MBR removal rates improved with the addition of coagulant (Tian et al., 2008a) or adsorbent (Tian et al., 2008b). Adding biomass did not improve removal as much as other aids and BDOC removal was even reduced compared to plain MBR (Tian et al., 2009). In this system, DOC was removed through a combination of adsorption and biodegradation mechanisms. The best results were achieved with the combination of coagulant-adsorbent-MBR (Tian et al., 2010). The retentions observed were: DOC 63%, UV254 76%, THM formation potential 55%, HAA formation potential 56%, BDOC 67% and AOC 76%.

Sagbo et al. (2008) compared MBR and MBR-PAC hybrid processes for drinking water production. They achieved a TOC removal of 25% for MBR and 50% for MBR-PAC (2 g L^{-1}) system when compared to the MBR. Ravindran et al. (2009) reported TOC removals of 63% when an MBR was operated without biomass and PAC, 70% with biomass alone, and 84% with a combination of PAC and biomass. In the latter case, the THM formation potential removal was 98%.

Zhang et al. (2011b) filtered algal-rich water with a submerged hollow-fiber PVC UF membrane. UF could achieve absolute removal of *Microcystis aeruginosa* cells but poor removal of algogenic organic matter that was released into the water. The addition of 4 g PAC L^{-1} to the reactor increased DOC removal from 54% to 65%. Yet, PAC had little effect on the rejection of hydrophilic high molecular weight assimilable organic matter such as carbohydrates and proteins. Williams and Pirbazari (2007) found that MBR combined with PAC and biomass was effective in the removal of AOC and aldehydes from pre-ozonated water. Yet, Huang et al. (2011) noted that protein-like substances are more readily eliminated by microorganisms than humic-like substances, indicating that the effect of adsorbent is less important.

It was suggested that an MBR with a short hydraulic retention time could be developed as an effective biological water treatment process for seriously contaminated surface water resources in developing countries (Wang and Menon, 2008).

11.4 Conclusions

Different treatment combinations with coagulation processes have been investigated. Pre-oxidation has been used to enhance the coagulation efficiency, with rather contradictory results (Liu et al., 2009). Also, pre-treatment with MIEX has been studied widely with promising results (Jarvis et al., 2008). After coagulation, the adsorption process with AC filtration has proved rather efficient in enhancing NOM reduction, particularly the better removal of LMM and NOM recalcitrant to coagulation (Uyak et al., 2007). Membrane filtrations have also become more common, and hybrid coagulation/membrane systems have been tested by several researchers. Postfiltration (NF) has reduced recalcitrant NOM and its corrosion aggressive constituents significantly (Bond et al., 2009; Shi and Taylor, 2007). Accordingly, for high DOC waters, the combination of MIEX, coagulation, and membrane filtration (UF) is highly efficient for NOM reduction (Humbert et al., 2007). Optimized coagulation may sometimes be sufficient for DBP control alone.

References

W. Ahn, S. Yim, B. Son, G.T. Kim, H. Ahn, D. Lee, S. Lee, B. Kwon, J. Cho, Experimental evaluations of a pilot nanofiltration system with respect to NOM and BOM removals and stable permeability with appropriate pre-treatments: a case study in Korea, J. Water Supply Res. Technol. Aqua 53 (2004) 169–181.

M. Alborzfar, G. Jonsson, C. Grøn, Removal of natural organic matter from two types of humic ground waters by nanofiltration, Water Res. 32 (1998) 2983–2994.

H. Bai, X. Zhang, J. Pan, D. Sun, J. Shao, Combination of nano TiO_2 photocatalytic oxidation with microfiltration (MF) for natural organic matter removal, Water Sci. Technol. Water Supply 9 (2009) 31–37.

E. Barbot, S. Moustier, J. Bottero, P. Moulin, Coagulation and ultrafiltration: understanding of the key parameters of the hybrid process, J. Membr. Sci. 325 (2008) 520–527.

M. Ben-Sasson, Y. Zidon, R. Calvo, A. Adin, Enhanced removal of natural organic matter by hybrid process of electrocoagulation and dead-end microfiltration, Chem. Eng. J. 232 (2013) 338–345.

R. Bergamasco, L.C. Konradt-Moraes, M.F. Vieira, M.R. Fagundes-Klen, A.M.S. Vieira, Performance of a coagulation–ultrafiltration hybrid process for water supply treatment, Chem. Eng. J. 166 (2011) 483–489.

B. Blankert, B. Betlem, B. Roffel, Development of a control system for in-line coagulation in an ultrafiltration process, J. Membr. Sci. 301 (2007) 39–45.

T. Bond, E.H. Goslan, B. Jefferson, F. Roddick, L. Fan, S.A. Parsons, Chemical and biological oxidation of NOM surrogates and effect on HAA formation, Water Research 43 (10) (2009) 2615–2622.

T. Bond, E. Goslan, S. Parsons, B. Jefferson, Disinfection by-product formation of natural organic matter surrogates and treatment by coagulation, MIEX and nanofiltration, Water Res. 44 (2010) 1645–1653.

V. Bonnélye, L. Guey, J. Del Castillo, UF/MF as RO pre-treatment: the real benefit, Desalination 222 (2008) 59–65.

P. Bose, D. Reckhow, The effect of ozonation on natural organic matter removal by alum coagulation, Water Res. 41 (2007) 1516–1524.

J.P. Botes, E.P. Jacobs, S.M. Bradshaw, Long-term evaluation of a UF pilot plant for potable water production, Desalination 115 (1998) 229–238.

S. Byun, J.S. Taurozzi, A.L. Alpatova, F. Wang, V.V. Tarabara, Performance of polymeric membranes treating ozonated surface water: effect of ozone dosage, Separ. Purif. Technol. 81 (2011) 270–278.

Z. Cai, M.M. Benjamin, NOM fractionation and fouling of low-pressure membranes in microgranular adsorptive filtration, Environ. Sci. Technol. 45 (2011) 8935–8940.

Z. Cai, J. Kim, M. Benjamin, NOM removal by adsorption and membrane filtration using heated aluminum oxide particles, Environ. Sci. Technol. 42 (2008) 619–623.

M. Campinas, M.J. Rosa, Removal of microcystins by PAC/UF, Separ. Purif. Technol. 71 (2010a) 114–120.

M. Campinas, M.J. Rosa, Assessing PAC contribution to the NOM fouling control in PAC/UF systems, Water Res. 44 (2010b) 1636–1644.

C. Campos, B.J. Mariñas, V.L. Snoeyink, I. Baudin, J.M. Laîné, Adsorption of trace organic compounds in CRISTAL(R) process, Desalination 117 (1998) 265–271.

A.G. Capodaglio, A. Callegari, P. Sauvignet, Advanced separation technology application for NOM removal from a freshwater supply, in: IWA Specialty Conference on Natural Organic Matter, July 27–29, 2011, Costa Mesa, California, 2011.

T. Carroll, S. King, S.R. Gray, B.A. Bolto, N.A. Booker, The fouling of microfiltration membranes by NOM after coagulation treatment, Water Res. 34 (2000) 2861–2868.

E.E. Chang, Y. Chen, Y. Lin, P. Chiang, Reduction of natural organic matter by nanofiltration process, Chemosphere 76 (2009) 1265–1272.

Y. Chen, B. Dong, N. Gao, J. Fan, Effect of coagulation pretreatment on fouling of an ultrafiltration membrane, Desalination 204 (2007) 181–188.

P.-C. Chiang, E.-E. Chang, P.-C. Chang, C.-P. Huang, Effects of pre-ozonation on the removal of THM precursors by coagulation, Sci. Total Environ. 407 (2009) 5735–5742.
M.-H. Cho, C.-H. Lee, S. Lee, Effect of flocculation conditions on membrane permeability in coagulation–microfiltration, Desalination 191 (2006) 386–396.
I.-H. Choi, S. Lee, I.-C. Kim, B.-R. Min, K.-H. Lee, Effect of new photocatalytic coagulant on NF membrane fouling, Ind. Eng. Chem. Res. 46 (2007) 2280–2285.
K.Y. Choi, B.A. Dempsey, In-line coagulation with low-pressure membrane filtration, Water Res. 38 (2004) 4271–4281.
K.-W. Choo, R. Tao, M.-J. Kim, Use of photocatalytic membrane reactor for the removal of natural organic matter in water: effect of photoinduced desorption and ferrihydrite adsorption, J. Membr. Sci. 322 (2008) 368–374.
E.R. Cornelissen, E.F. Beerendonk, M.N. Nederlof, J.P. van der Hoek, L.P. Wessels, Fluidized ion exchange (FIX) to control NOM fouling in ultrafiltration, Desalination 236 (2009) 334–341.
E.R. Cornelissen, D. Chasseriaud, W.G. Siegers, E.F. Beerendonk, D. van der Kooij, Effect of anionic fluidized ion exchange (FIX) pre-treatment on nanofiltration (NF) membrane fouling, Water Res. 44 (2010) 3283–3293.
B. Dong, Y. Chen, N. Gao, J. Fan, Effect of coagulation pretreatment on the fouling of ultrafiltration membrane, J. Environ. Sci. 19 (2007) 278–283.
K.L. Dubrawski, M. Fauvel, M. Mohseni, Metal type and natural organic matter source for direct filtration electrocoagulation of drinking water, J. Hazard Mater. (2013) 244–245, 135–141.
A.G. Fane, R. Wang, Y. Jia, Membrane technology: past, present and future, in: L.K. Wang, J.P. Chen, Y. Hung, N.K. Shammas (Eds.), Handbook of Environmental Engineering. Membrane and Desalination Technologies, vol. 13, Springer Science & Business Media, New York, 2008, pp. 1–45.
I. Garcia, L. Moreno, Use of GAC after enhanced coagulation for the removal of natural organic matter from water for purification, Water Sci. Technol. Water Supply 9 (2009) 173–180.
K. Glucina, A. Alvarez, J.M. Laîné, Assessment of an integrated membrane system for surface water treatment, Desalination 132 (2000) 73–82.
E. Goslan, F. Gurses, J. Banks, S. Parsons, An investigation into reservoir NOM reduction by UV photolysis and advanced oxidation processes, Chemosphere 65 (2006) 1113–1119.
A.B.F. Grose, A.J. Smith, A. Donn, J. O'Donnell, D. Welch, Supplying high quality drinking water to remote communities in Scotland, Desalination 117 (1998) 107–117.
S. Gur-Reznik, I. Katz, C.G. Dosoretz, Removal of dissolved organic matter by granular-activated carbon adsorption as a pretreatment to reverse osmosis of membrane bioreactor effluents, Water Res. 42 (2008) 1595–1605.
T. Ha, K. Choo, S. Choi, Effect of chlorine on adsorption/ultrafiltration treatment for removing natural organic matter in drinking water, J. Colloid Interface Sci. 274 (2004) 587–593.
C. Hallé, P.M. Huck, S. Peldszus, J. Haberkamp, M. Jekel, Assessing the performance of biological filtration as pretreatment to low pressure membranes for drinking water, Environ. Sci. Technol. 43 (2009) 3878–3884.
V. Härmä, Tuloksia suomalaisilla pintavesilaitoksilla tehdyistä nanosuodatuskokeista, Vesitalous 6 (1999) 5–10.
N. Her, G. Amy, A. Plottu-Pecheux, Y. Yoon, Identification of nanofiltration membrane foulants, Water Res. 41 (2007) 3936–3947.

N. Hilal, M. Al-Abri, H. Al-Hinai, Enhanced membrane pre-treatment processes using macromolecular adsorption and coagulation in desalination plants: a Review, Separ. Sci. Technol. 41 (2006) 403–453.
K. Howe, A. Marwah, K.-P. Chiu, S. Adham, Effect of coagulation on the size of MF and UF membrane foulants, Environ. Sci. Technol. 40 (2006) 7908–7913.
G. Huang, F. Meng, X. Zheng, Y. Wang, Z. Wang, H. Liu, M. Jekel, Biodegradation behavior of natural organic matter (NOM) in a biological aerated filter (BAF) as a pretreatment for ultrafiltration (UF) of river water, Appl. Microbiol. Biotechnol. 90 (2011) 1795–1803.
H. Huang, K. Schwab, J.G. Jacangelo, Pretreatment for low pressure membranes in water treatment: a review, Environ. Sci. Technol. 43 (2009) 3011–3019.
X. Huang, M. Leal, Q. Li, Degradation of natural organic matter by TiO_2 photocatalytic oxidation and its effect on fouling of low-pressure membranes, Water Res. 42 (2008) 1142–1150.
H. Humbert, H. Gallard, J. Croué, A polishing hybrid AER/UF membrane process for the treatment of a high DOC content surface water, Water Res. 46 (2012) 1093–1100.
H. Humbert, H. Gallard, V. Jacquemet, J. Croué, Combination of coagulation and ion exchange for the reduction of UF fouling properties of a high DOC content surface water, Water Res. 41 (2007) 3803–3811.
J. Jacangelo, J. DeMarco, D. Owen, S. Randtke, Biofiltration performance: part 2, effect of backwashing, J. Am. Water Works Assoc. 87 (1995) 64–70.
P. Jarvis, M. Mergen, J. Banks, B. McIntosh, S. Parsons, B. Jefferson, Pilot scale comparison of enhanced coagulation with magnetic resin plus coagulation systems, Environ. Sci. Technol. 42 (2008) 1276–1282.
C. Jung, L. Kang, Application of combined coagulation-ultrafiltration membrane process for water treatment, Kor. J. Chem. Eng. 20 (2003) 855–861.
C.-W. Jung, H.-J. Son, L.-S. Kang, Effects of membrane material and pretreatment coagulation on membrane fouling: fouling mechanism and NOM removal, Desalination 197 (2006) 154–164.
M. Kabsch-Korbutowicz, Effect of Al coagulant type on natural organic matter removal efficiency in coagulation/ultrafiltration process, Desalination 185 (2005) 327–333.
M. Kabsch-Korbutowicz, K. Majewska-Nowak, T. Winnicki, Water treatment using MIEX-DOC/ultrafiltration process, Desalination 221 (2008) 338–344.
J. Kaewsuk, G.T. Seo, Verification of NOM removal in MIEX-NF system for advanced water treatment, Separ. Purif. Technol. 80 (2011) 11–19.
P. Kajitvichyanukul, N.K. Shammas, Y. Hung, L.K. Wang, J. Ananpattarachai, Portable water biotechnology, membrane filtration and biofiltration, in: L.K. Wang, J.P. Chen, Y. Hung, N.K. Shammas (Eds.), Handbook of Environmental Engineering. Membrane and Desalination Technologies, vol. 13, Springer Science & Business Media, New York, 2008, pp. 477–523.
P.C. Kamp, J.C. Kruithof, H.C. Folmer, UF/RO treatment plant Heemskerk: from challenge to full scale application, Desalination 131 (2000) 27–35.
S. Kang, K. Choo, Why does a mineral oxide adsorbent control fouling better than powdered activated carbon in hybrid ultrafiltration water treatment? J. Membr. Sci. 355 (2010) 69–77.
B.S. Karnik, S.H. Davies, M.J. Baumann, S.J. Masten, The effects of combined ozonation and filtration on disinfection by-product formation, Water Res. 39 (2005) 2839–2850.
H. Kim, J.-H. Hong, S. Lee, Fouling of microfiltration membranes by natural organic matter after coagulation treatment: a comparison of different initial mixing conditions, J. Membr. Sci. 283 (2006) 266–272.

J. Kim, B. Van der Bruggen, The use of nanoparticles in polymeric and ceramic membrane structures: review of manufacturing procedures and performance improvement for water treatment, Environ. Pollut. 158 (2010) 2335–2349.

J. Kim, Z. Cai, M.M. Benjamin, Effects of adsorbents on membrane fouling by natural organic matter, J. Membr. Sci. 310 (2008) 356–364.

J. Kim, W. Shan, S. Davies, M. Baumann, S. Masten, V. Tarabara, Interactions of aqueuos NOM with nanoscale TiO_2: implications for ceramic membrane filtration-ozonation hybrid process, Environ. Sci. Technol. 43 (2009) 5488–5494.

J.H. Kim, C.H. Lee, E.J. Lee, K.H. Lee, S.B. Kwon, H. Park, H.S. Kim, A. Jang, The effect of re-aggregated floc by additional coagulant on membrane, J. Taiwan Inst. Chem. Eng. 44 (2013) 802–807.

K. Kim, H. Kim, J. Kim, J. Nam, J. Kim, S. Son, A hybrid microfiltration–granular activated carbon system for water purification and wastewater reclamation/reuse, Desalination 243 (2009) 132–144.

M.H. Kim, M.J. Yu, Characterization of NOM in the Han River and evaluation of treatability using UF–NF membrane, Environ. Res. 97 (2005) 116–123.

A.B. Koltuniewich, E. Drioli, Membranes in Clean Technologies, Theory and Practice, Wiley-VCH Verlag, Weinheim, 2008.

K. Konieczny, D. Sakol, J. Plonka, M. Rajca, M. Bodzek, Coagulation—ultrafiltration system for river water treatment, Desalination 240 (2009) 151–159.

B. Kwon, S. Lee, J. Cho, H. Ahn, D. Lee, H.S. Shin, Biodegradability, DBP formation, and membrane fouling potential of natural organic matter: characterization and controllability, Environ. Sci. Technol. 39 (2005) 732–739.

R. Lamsal, K.R. Montreuil, F.C. Kent, M.E. Walsh, G.A. Gagnon, Characterization and removal of natural organic matter by an integrated membrane system, Desalination 303 (2012) 12–16.

T. Lebeau, C. Lelièvre, H. Buisson, D. Cléret, L.W. Van de Venter, P. Côté, Immersed membrane filtration for the production of drinking water: combination with PAC for NOM and SOCs removal, Desalination 117 (1998) 219–231.

P. Le-Clech, E.-K. Lee, V. Chen, Hybrid photocatalysis/membrane treatment for surface waters containing low concentrations of natural organic matters, Water Res. 40 (2006) 323–330.

S. Lee, C. Lee, Microfiltration and ultrafiltration as a pretreatment for nanofiltration of surface water, Separ. Sci. Technol. 41 (2006) 1–23.

S. Lee, C. Lee, Effect of membrane properties and pretreatment on flux and NOM rejection in surface water nanofiltration, Separ. Purif. Technol. 56 (2007) 1–8.

T. Leiknes, The effect of coupling coagulation and flocculation with membrane filtration in water treatment: a review, J. Environ. Sci. 21 (2009) 8–12.

C. Li, Y. Chen, Fouling of UF membrane by humic substance: effects of molecular weight and powder-activated carbon (PAC) pre-treatment, Desalination 170 (2004) 59–67.

X. Li, H.P. Chu, Membrane bioreactor for the drinking water treatment of polluted surface water supplies, Water Res. 37 (2003) 4781–4791.

Y. Li, J. Wang, W. Zhang, X. Zhang, C. Chen, Effects of coagulation on submerged ultrafiltration membrane fouling caused by particles and natural organic matter (NOM), Chin. Sci. Bull. 56 (2011) 584–590.

C. Lin, Y. Huang, O.J. Hao, Ultrafiltration processes for removing humic substances: effect of molecular weight fractions and PAC treatment, Water Res. 33 (1999) 1252–1264.

C. Lin, T. Lin, O.J. Hao, Effects of humic substance characteristics on UF performance, Water Res. 34 (2000) 1097–1106.

J.C. Lin, A. Cheng, Y. Lin, D. Lee, W. Guo, Dense membranes precoated with aluminium tridecamer (Al_{13}) for water treatment, Desalination 284 (2012) 349–352.

H. Liu, F. Cheng, D. Wang, Interaction of ozone and organic matter in coagulation with inorganic polymer flocculant-PACl: role of organic components, Desalination 249 (2009) 596–601.

H. Liu, D. Wang, M. Wang, H. Tang, M. Yang, Effect of pre-ozonation on coagulation with IPF–PACls: role of coagulant speciation, Colloids Surf. A Physicochem. Eng. Asp. 294 (2007) 111–116.

Q. Liu, K. Seung-Hyun, Y. Jong-Sup, M. Seong-Yong, Impacts of coagulation pretreatment on MF filtration and a comparative study of different membrane module types, J. Ocean. Univ. China 5 (2006) 295–299.

T. Liu, Z. Chen, W. Yu, J. Shen, J. Gregory, Effect of two-stage coagulant addition on coagulation-ultrafiltration process for treatment of humic-rich water, Water Res. 45 (2011) 4260–4268.

J. Lohwacharin, K. Oguma, S. Takizawa, Use of carbon black nanoparticles to mitigate membrane fouling in ultrafiltration of river water, Separ. Purif. Technol. 72 (2010) 61–69.

A. Loi-Brugger, S. Panglisch, P. Buchta, K. Hattori, H. Yonekawa, Y. Tomita, R. Gimbel, Ceramic membranes for direct river water treatment applying coagulation and microfiltration, Water Sci. Technol. Water Supply 6 (2006) 89–98.

N. Ma, X. Quan, Y. Zhang, S. Chen, H. Zhao, Integration of separation and photocatalysis using an inorganic membrane modified with Si-doped TiO_2 for water purification, J. Membr. Sci. 335 (2009) 58–67.

A. Maartens, P. Swart, E.P. Jacobs, Feed water pretreatment: methods to reduce membrane fouling by natural organic matter, J. Membr. Sci. 163 (1999) 51–62.

F. Malek, J.L. Harris, F.A. Roddick, Interrelationship of photooxidation and microfiltration in drinking water treatment, J. Membr. Sci. 281 (2006) 541–547.

A. Matilainen, M. Vepsäläinen, M. Sillanpää, Natural organic matter removal by coagulation during drinking water treatment: a review, Adv. Colloid Interface Sci. 159 (2010) 189–197.

V. Mavrov, H. Chmiel, J. Kluth, J. Meier, F. Heinrich, P. Ames, K. Backes, P. Usner, Comparative study of different MF and UF membranes for drinking water production, Desalination 117 (1998) 189–196.

M. Mergen, B. Adams, G. Vero, T. Price, S. Parsons, B. Jefferson, P. Jarvis, Characterization of natural organic matter (NOM) removed by magnetic ion exchange resin (MIEX Resin), Water Sci. Technol. Water Supply 9 (2009) 199–205.

M.R.D. Mergen, B. Jefferson, S.A. Parsons, P. Jarvis, Magnetic ion-exchange resin treatment: impact of water type and resin use, Water Res. 42 (2008) 1977–1988.

S. Metsämuuronen, M. Sillanpää, A. Bhatnagar, M. Mänttäri, Natural organic matter removal from drinking water by membrane technology, Separ. Purif. Rev. 43 (2014) 1–61.

T. Meyn, A. Bahn, T. Leiknes, Significance of flocculation for NOM removal by coagulation–ceramic membrane microfiltration, Water Sci. Technol. Water Supply 8 (2008) 691–700.

J.C. Mierzwa, M.C.C. da Silva, L.R.V. Veras, E.L. Subtil, R. Rodrigues, T. Li, K.R. Landenberger, Enhancing spiral-wound ultrafiltration performance for direct drinking water treatment through operational procedures improvement: a feasible option for the Sao Paulo metropolitan region, Desalination 307 (2012) 68–75.

J.C. Mierzwa, I. Hespanhol, M.C.C. da Silva, L.D.B. Rodrigues, C.F. Giorgi, Direct drinking water treatment by spiral-wound ultrafiltration membranes, Desalination 230 (2008) 41–50.

J. Moon, M. Kang, J. Lim, C. Kim, H. Park, Evaluation of a low-pressure membrane filtration for drinking water treatment: pretreatment by coagulation/sedimentation for the MF membrane, Desalination 247 (2009) 271–284.

J. Morran, M. Drikas, D. Cook, D. Bursill, Comparison of MIEX treatment and coagulation on NOM character, Water Sci. Technol. Water Supply 4 (2004) 129–137.
C. Murray, S. Parsons, Removal of NOM from drinking water: fenton's and photo-Fenton's processes, Chemosphere 54 (2004) 1017–1023.
M.M. Nederlof, J.C. Kruithof, J.S. Taylor, D. van der Kooij, J.C. Schippers, Comparison of NF/RO membrane performance in integrated membrane systems, Desalination 131 (2000) 257–269.
H. Ødegaard, S. Østerhus, E. Melin, B. Eikebrokk, NOM removal technologies – Norwegian experiences, Drink. Water Eng. Sci. 3 (2010) 1–9.
H. Oh, M. Yu, S. Takizawa, S. Ohgaki, Evaluation of PAC behavior and fouling formation in an integrated PAC–UF membrane for surface water treatment, Desalination 192 (2006) 54–62.
N. Park, B. Kwon, M. Sun, H. Ahn, C. Kim, C. Kwoak, D. Lee, S. Chae, H. Hyung, J. Cho, Application of various membranes to remove NOM typically occurring in Korea with respect to DBP, AOC and transport parameters, Desalination 178 (2005) 161–169.
S. Park, T. Yoon, Effects of iron species and inert minerals on coagulation and direct filtration for humic acid removal, Desalination 239 (2009) 146–158.
R. Pianta, M. Boller, M. Janex, A. Chappaz, B. Birou, R. Ponce, J. Walther, Micro- and ultrafiltration of karstic spring water, Desalination 117 (1998) 61–71.
K. Pirkanniemi, M. Sillanpää, Heterogeneous water phase catalysis as an environmental application: a review, Chemosphere 48 (2002) 1047–1060.
K. Pirkanniemi, S. Metsärinne, M. Sillanpää, Degradation of EDTA and novel complexing agents in pulp and paper mill process and waste waters by Fenton's reagent, J. Hazard Mater. 147 (2007) 556–561.
M.J. Pryor, E.P. Jacobs, J.P. Botes, V.L. Pillay, A low pressure ultrafiltration membrane system for potable water supply to developing communities in South Africa, Desalination 119 (1998) 103–111.
X. Qiao, Z. Zhang, N. Wang, V. Wee, M. Low, C. Loh, T. Ng, Coagulation pretreatment for a large-scale ultrafiltration process treating water from the Taihu River, Desalination 230 (2008) 305–313.
J. Rämö, M. Sillanpää, Degradation of EDTA by hydrogen peroxide in alkaline conditions, J. Clean. Prod. 9 (2001) 191–195.
V. Ravindran, H. Tsai, M.D. Williams, M. Pirbazari, Hybrid membrane bioreactor technology for small water treatment utilities: process evaluation and primordial considerations, J. Membr. Sci. 344 (2009) 39–54.
M. Ribau Teixeira, S.M. Rosa, V. Sousa, Natural organic matter and disinfection by-products formation potential in water treatment, Water Resour. Manag. 25 (2011) 3005–3015.
J.C. Rojas, B. Moreno, G. Garralon, F. Plaza, J. Perez, M.A. Gomez, Influence of velocity gradient in a hydraulic flocculator on NOM removal by aerated spiral-wound ultrafiltration membranes (ASWUF), J. Hazard Mater. 178 (2010) 535–540.
O. Sagbo, Y. Sun, A. Hao, P. Gu, Effect of PAC addition on MBR process for drinking water treatment, Separ. Purif. Technol. 58 (2008) 320–327.
B. Sani, E. Basile, L. Rossi, C. Lubello, Effects of pre-treatment with magnetic ion exchange resins on coagulation/flocculation process, Water Sci. Technol. 57 (2008) 57–64.
J. Schrotter, B. Bozkaya-Schrotter, Current and emerging membrane processes for water treatment, in: K. Peinemann, S. Pereira Nunes (Eds.), Membranes for Water Treatment, Wiley-VCH Verlag, Weinheim, 2010, pp. 53–91.
W. Shi, M.M. Benjamin, Membrane interactions with NOM and an adsorbent in a vibratory shear enhanced filtration process (VSEP) system, J. Membr. Sci. 312 (2008) 23–33.

B. Shi, J.S. Taylor, Potential impact of enhanced coagulation on corrosion by-product release in a distribution system, Desalination 208 (1–3) (2007) 260–268.
K. Shorrock, B. Drage, A pilot plant evaluation of the magnetic ion exchange process for the removal of dissolved organic carbon at Draycote water treatment works, Water Environ. J. 20 (2006) 65–70.
M. Siddiqui, G. Amy, J. Ryan, W. Odem, Membranes for the control of natural organic matter from surface waters, Water Res. 34 (2000) 3355–3370.
E. Smith, Y. Kamal, Optimizing treatment for reduction of disinfection by-product (DBP) formation, Water Sci. Technol. Water Supply 9 (2009) 191–198.
Y. Song, B. Dong, N. Gao, S. Xia, Huangpu River water treatment by microfiltration with ozone pretreatment, Desalination 250 (2010) 71–75.
A. Syafei, C.-F. Lin, C.-H. Wu, Removal of natural organic matter by ultrafiltration with TiO_2-coated membrane under UV irradiation, J. Colloid Interface Sci. 323 (2008) 112–119.
M. Szlachta, W. Adamski, Effects of natural organic matter removal by integrated processes: alum coagulation and PAC-adsorption, Water Sci. Technol. 59 (2009) 1951–1957.
T. Thorsen, H. Fløgstad, Nanofiltration in Drinking Water Treatment, Literature Review, Research Report for Techneau Project, 2006. December11, 2006.
J. Tian, Z. Chen, J. Nan, H. Liang, G. Li, Integrative membrane coagulation adsorption bioreactor (MCABR) for enhanced organic matter removal in drinking water treatment, J. Membr. Sci. 352 (2010) 205–212.
J. Tian, H. Liang, X. Li, S. You, S. Tian, G. Li, Membrane coagulation bioreactor (MCBR) for drinking water treatment, Water Res. 42 (2008a) 3910–3920.
J. Tian, H. Liang, J. Nan, Y. Yang, S. You, G. Li, Submerged membrane bioreactor (sMBR) for the treatment of contaminated raw water, Chem. Eng. J. 148 (2009) 296–305.
J. Tian, H. Liang, Y. Yang, S. Tian, G. Li, Membrane adsorption bioreactor (MABR) for treating slightly polluted surface water supplies: as compared to membrane bioreactor (MBR), J. Membr. Sci. 325 (2008b) 262–270.
M. Tomaszewska, S. Mozia, Removal of organic matter from water by PAC/UF system, Water Res. 36 (2002) 4137–4143.
T. Tran, S. Gray, R. Naughton, B. Bolto, Polysilicato-iron for improved NOM removal and membrane performance, J. Membr. Sci. 280 (2006) 560–571.
R. Treguer, R. Tatin, A. Couvert, D. Wolbert, A. Tazi-Pain, Ozonation effect on natural organic matter adsorption and biodegradation – application to a membrane bioreactor containing activated carbon for drinking water production, Water Res. 44 (2010) 781–788.
W. Tsujimoto, H. Kimura, T. Izu, T. Irie, Membrane filtration and pre-treatment by GAC, Desalination 119 (1998) 323–326.
V. Uyak, S. Yavuz, I. Toroz, S. Ozaydin, E. Genceli, Disinfection by-products precursors removal by enhanced coagulation and PAC adsorption, Desalination 216 (2007) 334–344.
C. Uyguner, M. Bekbolet, H. Selcuk, A comparative approach to the application of a physico-chemical and advanced oxidation combined system to natural water samples, Separ. Sci. Technol. 42 (2007a) 1405–1419.
C. Uyguner, S. Suphandag, A. Kerc, M. Bekbolet, Evaluation of adsorption and coagulation characteristics of humic acids preceded by alternative advanced oxidation techniques, Desalination 210 (2007b) 183–193.
J.C. Vickers, M.A. Thompson, U.G. Kelkar, The use of membrane filtration in conjunction with coagulation processes for improved NOM removal, Desalination 102 (1995) 57–61.

C. Visvanathan, N. Boonthanon, A. Sathasivan, V. Jegatheesan, Pretreatment of seawater for biodegradable organic content removal using membrane bioreactor, Desalination 153 (2003) 133–140.

J. Wang, X. Wang, Ultrafiltration with in-line coagulation for the removal of natural humic acid and membrane fouling mechanism, J. Environ. Sci. 18 (2006) 880–884.

L.K. Wang, R. Menon, Treatment of industrial effluents, municipal wastes, and portable water by membrane bioreactors, in: L.K. Wang, J.P. Chen, Y. Hung, N.K. Shammas (Eds.), Handbook of Environmental Engineering. Membrane and Desalination Technologies, vol. 13, Springer Science & Business Media, New York, 2008, pp. 201–236.

M.D. Williams, M. Pirbazari, Membrane bioreactor process for removing biodegradable organic matter from water, Water Res. 41 (2007) 3880–3893.

S. Xia, Y. Liu, X. Li, J. Yao, Drinking water production by ultrafiltration of Songhuajiang River with PAC adsorption, J. Environ. Sci. 19 (2007) 536–539.

M. Yan, D. Wang, X. Ma, J. Ni, H. Zhang, THMs precursor removal by an integrated process of ozonation and biological granular activated carbon for typical Northern China water, Separ. Purif. Technol. 72 (2010) 263–268.

M. Yan, D. Wang, B. Shi, M. Wang, Y. Yan, Effect of pre-ozonation on optimized coagulation of a typical North-China source water, Chemosphere 69 (2007) 1695–1702.

J. Yang, D. Yuan, T. Weng, Pilot study of drinking water treatment with GAC, O_3/BAC and membrane processes in Kinmen Island, Taiwan, Desalination 263 (2010) 271–278.

W. Yang, J. Chen, X. Li, H. Liang, W. He, G. Li, Using chloramine as a coagulant aid in enhancing coagulation of Yellow River water in China, J. Zhejiang Univ. - Sci. 8 (2007) 1475–1481.

P. Yao, K. Choo, M. Kim, A hybridized photocatalysis–microfiltration system with iron oxide-coated membranes for the removal of natural organic matter in water treatment: effects of iron oxide layers and colloids, Water Res. 43 (2009) 4238–4248.

L.F. Yee, P.M. Abdullah, A. Abdullah, B. Ishak, K.N.Z. Abidin, Hydrophobicity characteristics of natural organic matter and the formation of THM, Malay. J. Anal. Sci. 13 (2009) 94–99.

M. Zhang, C. Li, M.M. Benjamin, Y. Chang, Fouling and natural organic matter removal in adsorbent/membrane systems for drinking water treatment, Environ. Sci. Technol. 37 (2003) 1663–1669.

W. Zhang, X. Zhang, Y. Li, J. Wang, C. Chen, Membrane flux dynamics in the submerged ultrafiltration hybrid treatment process during particle and natural organic matter removal, J. Environ. Sci. 23 (2011) 1970–1976.

Y. Zhang, J. Tian, J. Nan, S. Gao, H. Liang, M. Wang, G. Li, Effect of PAC addition on immersed ultrafiltration for the treatment of algal-rich water, J. Hazard Mater. 186 (2011) 1415–1424.

A. Zularisam, A. Ismail, R. Salim, Behaviours of natural organic matter in membrane filtration for surface water treatment – a review, Desalination 194 (2006) 211–231.

A. Zularisam, A. Ismail, M. Salim, M. Sakinah, T. Matsuura, Application of coagulation–ultrafiltration hybrid process for drinking water treatment: optimization of operating conditions using experimental design, Separ. Purif. Technol. 65 (2009) 193–210.

Index

Note: 'Page numbers followed by "f" indicate figures and "t" indicate tables.'

E

F

G

H

N

O

P

R

S

T

U

V

W

Kolibri 360 GmbH
Ferdinand-Jühlke-Straße 7
99095 Erfurt, Deutschland
produktsicherheit@kolibri360.de